Arsenic in Groundwater

About the Authors

Prof. (Dr.) Sankar Kr Acharya, presently, Dean, Post Graduate Studies; former Head, Dept. of Agril Extension and Director, Extension Education, Bidhan Chandra Krishi Viswavidyalaya, Mohanpur, WB, born on 6th October, 1960, started his career as Assistant Professor at BCKV in 1988 and has been in teaching, research and extension over 35 years. An erudite teacher as well as an elegant speaker, is internationally acclaimed for his unique research domain of Social Entropy and Energy Metabolism, Social Ecology and Environmental Sociology, Enterprise Ecology Framework, Conservation Stewardships: Research Publication: 248 in National and International Journals. Book Publication: 113 books authored. Expert Member, Agriculture Commission; Marketing and Extension Sub Committee, Govt. of West Bengal; Expert member WWF projects in BTR (Buxa Tiger Reserve Project); Expert member; DFID project on Primary Education (DPEP); Rapid Environment Impact Analysis, Visited Italy, France, Germany, China , Sri Lanka and Bangladesh. He has also been honoured to be selected as the Convener of Panel (PE-32) entitled, The hunger, poverty and silence, of IUAES, University of Manchester, UK, 2013 .Co-PI of ICAR, NAHEP and World Bank funded project on Conservation Agriculture. So, far he has successfully guided 21 Ph D scholars,. He is the fellow of ISEE, IARI, New Delhi, Fellow, BICVED; Fellow, Society of community mobilization, IARI, New Delhi He is now acting as editor, reviewer of a good number of national and international journals. Fellow Award, ISEE(ICAR-IARI, New Delhi), 2019 Awards and Distinctions: Distinguish Scholar Award, Krishisanskriti, New Delhi; Certificate of Merit for standing First class first at M Sc(Ag) in Agricultural Extension, 1986; ;Honorary Appointment to the Research Board of Advisors, The American Biographic Institute (2003) ;He has been honored to be selected as the Convener of Panel (PE-32) entitled the hunger, poverty and silence, of World Congress, IUAES, University of Manchester, UK, 2013 ;He has been honored to be selected as the Convener of Panel (P-102) entitled ,' Uncertainties, Unpredictability and Marginalization: The impact and mitigation for survival of agriculture and humanity, of the 19th World Congress, IUAES-WAU, 14-20 October, 2023, New Delhi, India, 2023 ;Honored to be nominated as Expert Member, Working group on Agricultural Extension, West Bengal State Agriculture Commission (2007) ;Honoured to be nominated by the Hon'ble Vice-Chancellor, BCKV to Act as an Expert Member, DPEP, A DFID, UK, project for planning and revamping primary education through POA of GOVT of India.(1995) Expert Member, Buxa Tiger Reserve Project funded by WWF(World Wide Wild Life Funding), 1998 ; Dr Daulat Singh Memorial Award, Society of Extension Education, Distinguished Professor and Academician Award, Research Education Solution (RES), M S Swaminathan School of Agriculture, Centurion University(2022); Achievers of German patent (Utility model no 20202310273) :Novel LOT based inventory management system with radio controlled pallet racking for storage solution. He is the Fellow, West Bengal Academy of Science and Technology (WAST), 2024.

Parban Baidya was born in North 24 Parganas district of West Bengal. He completed my secondary and higher secondary education from Prafullanagar Vidyamandir(H.S.). He obtained my B. Sc Honours in Agriculture from Uttar Banga Krishi Viswavidyalaya in 2020 and completed his M. Sc (Ag) in Agricultural Extension in 2022 from Bidhan Chandra Krishi Viswavidyalaya.

Arsenic in Groundwater
Poison, Man and Ecology

Sankar Kr Acharya
Parban Baidya

1168, Sector 13, Urban Estate
Karnal-132 001, Haryana
Tel: 91-84470 75807, 18440 41168
Email: contentvibesppa@gmail.com
www.contentvibes.in

Print ISBN: 978-81-97682-54-4

ebook ISBN: 978-81-97682-50-6

Preface

Arsenic contamination in ground water has become a global concern , more prominent in south-east Asian countries. This has thrown a challenge to our agriculture depending heavily on ground water depletion. With an unplanned and mindless extraction of ground water for supporting our irrigated agriculture, we have invited arsenic to go for poisoning our groundwater. Arsenic from ground water enters crop, livestock and human bodies and triggers up biological magnifications to do the silent carnage of health and ecosystem. In West Bengal, the eastern part of river Bhagirathi, is extremely vulnerable to arsenic contamination contributed by some geogenic factors. Heavy extraction of ground water during summer rice cultivation for this part of Gangetic alluvial tract keeps on creating highly favourable and conducive conditions for arsenic contamination. The threat also encompasses our adjoin Bangladesh where summer rice has become a main stay for ensuring food security for millions. The farmers and rural population at large suffer from severe health hazards, sometimes leading to untimely deaths, poor market demand for contaminated grains, fruits, vegetables grown on polluted water. The book is created based on ground level empirical researches to elicit the facts and factors impacting on arsenic contamination. The multivariate analytical tools help extract the direct, indirect and residual effect of different factors on arsenic contamination. This has offered huge policy research on safe ground water consumption. It is expected that this unique book will be of extreme interest and utility for the global audience.

Authors

Contents

1

Introduction

Arsenic contamination is one of the serious concerns for public health throughout the world. During the weathering of rocks and minerals, which are followed by subsequent leaching and runoff, arsenic (As) is released into the soil and groundwater. Additionally, human sources can transfer it into the soil and groundwater. Depending on the geological settings, geochemistry, and geo-environmental conditions of an aquifer, different aquifers may have different adsorption and desorption processes, arsenic species, Eh, pH, solid-phase dissolutions, and precipitations. Therefore, rigorous geochemical investigation for an adequate understanding of arsenic geochemistry under different hydrogeological and geo-environmental conditions of aquifers is essentially required for evolving sustainable solutions. All of the arsenic-affected river plains have river routes originated from the Himalayan region. Arsenic in groundwater contamination has far-reaching consequences including its ingestion through the food chain, which are in the form of social disorders, health hazards and socioeconomic dissolution besides its sprawling movement, and exploitation of groundwater. Food crops produced with arsenic-tainted water are marketed to other locations, including areas where people may eat the tainted food and be exposed to arsenic. The arsenic contamination of groundwater is a form of groundwater pollution which is due to naturally occurring high concentrations of arsenic in deeper levels of groundwater.

Groundwater contains arsenic, which is present all over the planet. The Ganga-Brahmaputra area is one of the most significant arsenic contaminated zones in the world. It is a concern since the farmer of Ganges Delta uses tube wells to deliver water, which leads to severe arsenic poisoning. Arsenic poisoning can cause major health complications if not properly treated. Arsenic in the water is extremely dangerous and hard to detect as arsenic has no flavour or odour.

Sometimes arsenic affected people may not be severely impaired but society may point them as dangerous or poisonous ones. According to reports, the majority of arsenic poisoning risks are brought on by ignorance of the metal's source. Numerous families have been split apart by fear of being close, leading to social isolation in schools and a general avoidance of those who reside in severely polluted regions. Arsenic poisoning causes a lot of health issues as well as social and psychological sufferings such as community refusal, social discrimination, unhappy married life, child development issues, mental disorder etc. Drinking arsenic-contaminated tube well water has become a serious health threat in Bangladesh.

A classic example of arsenic (As) contamination can be found in the eastern portion of the Hooghly River in West Bengal as a result of some geogenic causes. Rice, vegetables, meat, eggs, and even breast milk used to nourish newborns have all been negatively impacted by arsenic's multi-directional effects. The depletion of groundwater in the irrigation-based agro system has worst the situation further. This material finds a happy go-round in the operating food chain and subsequently biological magnification results. The current study has focused its investigation on the victims' perceptions of and behaviours in reaction to the threat already posed to the public's health and nutrition.

Agro-ecological dynamics are characterised by the support of pre-existing ecological processes, but the introduction of heavy and toxic elements into the food chain severely jeopardised system operation.

The effect of arsenic contamination has not been designable in the quality of agricultural produce but also has been harshly reflected in the general health status of the poor, social stigma, cultural isolation, withdrawal symptoms to ill-fated victims good enough to consider the gravity of the situation and its protractile effect on the future generation. The purpose of this study is to produce categorised information on how arsenic pollution is perceived, to calculate the impact on rural residents in terms of socio-ecological factors, and to produce micro-level policy implications based on the empirical study.

Objectives of the Study

- To know the health impact of arsenic contaminated water
- To estimate the sociological factor-based impact of arsenic poisoning on the health of rural residents
- To elucidate the effect of a set of socio-ecological variables on the perceived impact of arsenic contamination on health, agriculture and allied sectors
- To extract policy implications based on empirical research for micro-sociological perspectives
- To elucidate the efforts made by the Government to combat the problem.

2

Review of Literature

Source	Year	Author	Key Notions
Contamination of drinking water by arsenic in Bangladesh: a public health emergency. Bulletin of the World Health Organization, 78(9), 1093-1103.	2000	Smith et al.	The contamination of groundwater by arsenic in Bangladesh is the largest poisoning of a population in history, with millions of people exposed. The discovery of arsenic in drinking water in Bangladesh recommends intervention strategies. Tube wells were installed to provide "pure water" to prevent morbidity and mortality from gastrointestinal disease. The water from the millions of tube wells that were installed was not tested for arsenic contamination. Taken together with the discovery of arsenic in groundwater in other countries, the experience in Bangladesh shows that groundwater sources throughout the world that are used for drinking water should be tested for arsenic.
Arsenic contamination of groundwater and drinking water in Vietnam: a human health threat. Environmental science & technology, 35(13), 2621-2626.	2001	Berg et al.	Due to naturally occurring organic matter in the sediments, the ground watersare anoxic and rich in iron. With an average arsenic concentration of 159 µg/L, the contamination levels varied from 1 to 3050 µg/L in rural groundwater samples from private small-scale tube wells. The arsenic in the sediments may be associated with iron oxyhydroxides and released into the groundwater by reductive dissolution of iron. Oxidation of sulphide phases could also release arsenic into the groundwater, but sulphur concentrations in sediments were below 1 mg/g.
Arsenic in groundwater in the Bengal Delta Plain: slow poisoning in Bangladesh. Environmental Reviews, 9(3), 189-220.	2001	Mukherjee & Bhattacharya	The problems concerning the widespread occurrences of arsenic in groundwater in Bangladesh, a land with enormous resources of precipitation, surface water, and groundwater.

Source	Year	Author	Key Notions
			Because of the potential risk of microbiological contamination in the surface water, groundwater was relied on as an alternate source of drinking water. The exploitation of groundwater has increased dramatically in Bangladesh since the 1960s to provide safe water for drinking and to sustain wetland agriculture. The presence of arsenic in the groundwater at elevated concentrations has raised a serious threat to public health in the region. Nearly 60–75 million people inhabiting a large geographical area are at potential risk of arsenic exposure, and several thousand have already been affected by chronic arsenicosis.
Arsenic contamination in Bangladesh groundwater: a major environmental and social disaster. International journal of environmental health research, 12(3), 235-253.	2002	Alam et al.	In attempting to eliminate disease caused by drinking polluted surface water, millions of shallow surface wells were drilled into the Ganges delta alluvium in Bangladesh. The latest statistics indicate that 80% of Bangladesh and an estimated 40 million people are at risk of arsenic poisoning-related diseases because the groundwater in these wells is contaminated with arsenic. The clinical manifestations of arsenic poisoning are myriad, and the correct diagnosis depends largely on awareness of the problem. Patients with melanosis, leuco-melanosis, keratosis, hyperkeratosis, dorsum, non-petting edema, gangrene and skin cancer have been identified.
Analysis of arsenic-contaminated groundwater domain in the Nadia district of West Bengal (India). Hydrological sciences journal, 47(S1), S55-S66.	2002	Majumdar et al.	To conceptualize the groundwater flow domain, to determine groundwater flow paths and groundwater velocities, to study arsenic transport in the flow domain, and to study the well captures zones. The first three objectives and issues are addressed by simulation of steady and transient groundwater flow and contaminant transport using the US Geological Survey three-dimensional finite difference code, MODFLOW, and the three-dimensional advective-dispersive transport code, MT3D.

Source	Year	Author	Key Notions
			Simulated results of the calibrated model replicate the observed monthly water table conditions perfectly. Using the particle tracking algorithm, MODPATH, the possibility of arsenic removal from a sample key location, and the design of wells for withdrawing arsenic-free groundwater.
Worldwide occurrences of arsenic in ground water. Science, 296(5576), 2143-2145.	2002	Nordstrom, D. K.	Revised estimates from the World Health Organization for 1990 indicate that 43% of the world's population do not have adequate sanitation and 22% do not have clean drinking water (1, 2). This has led to increased dependence on groundwater resources in many parts of the world. The consequences of ground-water development often include over drafting, land subsidence, and the use of groundwater unfit for human consumption. The recently increased utilization of groundwater in India and Bangladesh has caused new health issues (3–5). An estimated 36 million people in the Bengal Delta are at risk from drinking arsenic-contaminated water. Numerous other occurrences worldwide have been reported (see table, right) (6), and some of these, such as those in Taiwan, has been recognized for several decades.
Arsenic contamination of groundwater in Bangladesh and its remedial measures. In Arsenic Contamination in Groundwater-Technical and Policy Dimensions. Proceedings of the UNU-NIES International Workshop, United Nations University, Tokyo, Japan (pp. 9-21)	2003	Rahman et al.	Arsenic was first detected in groundwater in the west of Bangladesh in 1993 following reports of extensive contamination of water supplies in the adjoining areas of India and of many people suffering from arsenic related diseases that had been treated medically in India. Several methods for treating water for arsenic reduction are available. The most commonly used methods mostly utilized principles of oxidation, precipitation/co-precipitation, adsorption onto sorptive media, ion exchange and physical separation by synthetic membranes.

Source	Year	Author	Key Notions
Arsenic contamination of groundwater: Mitigation strategies and policies. Hydrogeology Journal, 12(1), 103-114.	2004	Alaerts & Khouri	Contamination of groundwater by arsenic from natural geochemical sources is at present a most serious challenge in the planning of large-scale use of groundwater for drinking and other purposes. Recent improvements in detection limits of analytical instruments are allowing the correlation of health impacts such as cancer with large concentrations of arsenic in groundwater.
Remediation technologies for arsenic contaminated drinking waters (9 pp). Journal of Soils and Sediments, 5(3), 182-190.	2005	Garelick et al.	Arsenic is a toxic metalloid element that is now recognised to be an important contaminant of drinking water – particularly but not exclusively, in poor regions of southern Asia. In affected regions, many millions are at risk of arsenic-induced disease and strategies are required to provide safe water for consumption. The main strategies available are mitigation (the provision of alternative arsenic-free water) and remediation (arsenic removal from extracted water). Consideration of local conditions suggests that for many areas remediation, at an affordable cost, is the only practical option.
Arsenic contamination in drinking water and skin manifestations in lowland Nepal: the first community-based survey. The American journal of tropical medicine and hygiene, 73(2), 477-479.	2005	Maharjan et al.	A community-based, dose-response study on arsenic contamination was conducted in three communities in Terai in lowland Nepal. The arsenic concentration of all the tube wells in use (n 146) and the prevalence of arsenic-induced skin manifestation among 1,343 (approximately 80% of the inhabitants) subjects indicated the existence of a highly contaminated area in Terai.
Groundwater socio-ecology and governance: a review of institutions and policies in selected countries. Hydrogeology Journal, 13(1), 328-345.	2005	Mukherji & Shah	Groundwater is crucial for the livelihoods and food security of millions of people, and yet, knowledge formation in the field of groundwater has remained asymmetrical. While, scientific knowledge in the discipline (hydrology and hydrogeology) has advanced remarkably, relatively little is known about the socio-economic impacts and institutions that govern groundwater use.

Source	Year	Author	Key Notions
			The first is to provide a balanced view of the plus and the downside of groundwater use, especially in agriculture.
Kapaj, S., Peterson, H., Liber, K., & Bhattacharya, P. (2006). Human health effects from chronic arsenic poisoning–a review. Journal of Environmental Science and Health, Part A, 41(10), 2399-2428.	2006	Kapaj et al.	The ill effects of human exposure to arsenic (As) have recently been re-evaluated by government agencies around the world. This has led to a lowering of As guidelines in drinking water, with Canada decreasing the maximum allowable level from 50 to 25 µg/L and the U.S. from 50 to 10 µg/L. There is a strong relationship between chronic ingestion of As and deleterious human health effects.
Toward finding a sustainable solution for arsenic contamination of ground water: A SWOT analysis. Epidemiology, 17(6), S218-S219.	2006	Sarkar et al.	Around 9 million people living in eastern India are at risk of consuming arsenic contaminated water. Even 2 decades after finding the first arsenic case, no comprehensive strategy has been developed yet. While several technologic solutions failed to address the problem, the magnitude of suffering of the people has increased over the period. To find out the sustainable solution by means of SWOT (Strength, Weakness, Opportunity, and Threat) analysis of existing mitigation strategies and alternative opportunities, which are feasible, accepted in local culture, and ecologically appropriate.
Suffering for water, suffering from water: Political ecologies of arsenic, water and* development in Bangladesh. University of Minnesota.	2007	Sultana, F.	The social-ecological implications of a drinking water crisis in Bangladesh, within the context of broader development processes, with particular attention to gendered and classed differences in how people cope with, respond to and adapt to changing water conditions, uses, and rights. Arsenic contamination of groundwater and drinking water sources in Bangladesh is presently posing a significant development and water management challenge, yet the social and gender outcomes are inadequately given attention to by scholars and policymakers.

Source	Year	Author	Key Notions
Arsenic contamination in food chain due to arsenic contaminated groundwater irrigation. International Journal of Sustainable Agricultural Technology, 5(9), 40-45.	2009	Islam et al.	Soil and plant samples from Arsenic contaminated irrigated and Arsenic uncontaminated irrigated soils were collected from upazilla of Bagmara and Mohadebpur under the district of Rajshahi and Noagoan, Bangladesh in 2001. Chemical analysis of soil and plant samples of two Barind areas was accomplished to observe the impact of arsenic-contaminated water irrigation on food chain contamination. Soil pH and EC were increased due to Arsenic contaminated irrigation.
Impacts of arsenic contamination in groundwater: case study of some villages in Bangladesh. Environment, development and sustainability, 11(3), 571-588.	2009	Nahar, N	The contamination of groundwater in Bangladesh by arsenic is a widespread and serious environmental problem, affecting mainly the rural population who rely extensively on groundwater for drinking and cooking. The household survey gathered information on the respondents (affected by arsenic) water usage and sources, knowledge of the arsenic problem, changes in the source of water for drinking and cooking, arsenic mitigation technologies and socio-economic information on the households.
Spatial and temporal variations of groundwater arsenic in South and Southeast Asia. Science, 328(5982), 1123-1127.	2010	Fendorf et al.	Groundwater wells installed in rural areas throughout the major river basins draining the Himalayas have become the main source of drinking water for tens of millions of people. Groundwater in this region is much less likely to contain microbial pathogens than surface water but often contains hazardous amounts of arsenic—a known carcinogen. Arsenic enters groundwater naturally from rocks and sediment by coupled biogeochemical and hydrologic processes, some of which are presently affected by human activity.
Arsenic contamination of ground water and its health impact on population of district of Nadia, West Bengal, India. Indian Journal of Community Medicine:	2010	Mazumder et al.	Arsenic pollution in groundwater, used for drinking purposes, has been envisaged as a problem of global concern. The common symptoms of chronic arsenic toxicity due to prolonged drinking of arsenic-contaminated water are pigmentation and keratosis and

Source	Year	Author	Key Notions
Official Publication of Indian Association of Preventive & Social Medicine, 35(2), 331.			cancer of the skin. In India, significant arsenic contamination in groundwater was detected in the year 1983 in West Bengal, when some villagers were diagnosed to be suffering from arsenicosis due to drinking arsenic-contaminated water. Today, in West Bengal, arsenic contamination in groundwater has been detected in 79 blocks in 8 districts of the state. The major affected districts are Malda, Murshidabad, Nadia, Burdwan and North and South 24 Parganas.
Assessment of ground water quality of Ghazipur district, Eastern Uttar Pradesh, India, special reference to arsenic contamination. Recent Research in Science and Technology, 2(3), 38-41.	2010	Rajesh et al.	Now a days water quality has solicited a major global concern due to ever increasing human developmental activities that over exploit and pollute the water resources on the surface and underground. The situation is very worst in developing as well as developed nations. Water quality of the Gazhipur district of Eastern Uttar Pradesh of India was carried out for the present study to identify the arsenic contamination of groundwater. Arsenic contamination in adjacent district Ballia, the present investigation, therefore, has been undertaken on the possible arsenic contamination in and around Ghazipur district of eastern Uttar Pradesh.
Ecosystem perspective of groundwater arsenic contamination in India and relevance in policy. EcoHealth, 7(1), 114-126.	2010	Sarkar, A	Arsenic-related health problems from an ecosystem perspective through a primary survey conducted in five arsenic-affected villages in the state of West Bengal and a review of existing research and policy documents. Although modern agricultural practices and drinking water policies have resulted in arsenic contamination of groundwater, current mitigation policy is essentially confined to biomedical approaches, which include potable water supply and medical care. Existing disparity, difficulty in coping, inaccessibility to health services and potable water supply and lack of

Source	Year	Author	Key Notions
			participation in decision-making have resulted in more suffering among the poor. On the other hand, the spreading of arsenic contamination in the ecosystem remains unabated.
Human exposure to arsenic through foodstuffs cultivated using arsenic contaminated groundwater in areas of West Bengal, India. Journal of Environmental Science and Health, Part A, 46(11), 1259-1265.	2011	Samal et al.	The widespread incidence of chronic arsenicosis in the Bengal Delta has led to intensive research on arsenic (As) enrichment in groundwater as well as accumulation in foodstuffs, as there are potential health risks associated with exposure to As from both sources. Human As exposure through the drinking of groundwater, consumption of locally grown foodstuffs (e.g., crops and vegetables) and cooked food in Nadia district, West Bengal. Groundwater and foodstuffs were collected and analyzed with FI-HG-AAS to estimate the total As content.
Assessment of surface water quality in an arsenic contaminated village. American Journal of Environmental Sciences, 8(5), 523.	2012	Saikia & Gupta	Arsenic contamination of groundwater has occurred in various parts of the world, becoming a menace in the Ganga-Meghna-Brahmaputra basin (West Bengal and Assam in India and Bangladesh). Recently arsenic has been detected in the Cachar and Karimganj districts of Barak valley, Assam, bordering Bangladesh. In this area, coli form contamination comprises the major constraint towards utilization of its otherwise ample surface water resources. Around 60% of people of the village preferred groundwater for drinking and only 6% were aware of arsenic-related problems.
Political ecology of the arsenic crisis in Bangladesh. Diverting Flow Gender Equity Water South Asia.	2012	Sultana, F. A. R. H. A. N. A.	Bangladesh is often associated with monsoonal floods and an overabundance of water. However, it is also currently facing a severe drinking water crisis. This is due to the naturally-occurring arsenic contamination of groundwater sources that provide drinking water up to 70 million rural people. Arsenic contamination of drinking water emerged as an issue due to the widespread use of groundwater sources in recent decades.

Source	Year	Author	Key Notions
Arsenic mobility and toxicity in South and South-east Asia–a review on biogeochemistry, health and socio-economic effects, remediation and risk predictions. Environmental Chemistry, 11(5), 483-495.	2014	Muehe & Kappler	The dramatic situation caused by high arsenic concentrations in ground and drinking water in South and Southeast Asia has been investigated and discussed by the scientific community in the past twenty years. This introductory review of the research front 'Arsenic Biogeochemistry and Health' gives a broad overview of the current knowledge of As biogeochemistry, exposure, health, toxicity and As-caused socioeconomic effects.
Arsenic contamination in shallow groundwater and agricultural soil of Chakdaha block, West Bengal, India. Frontiers in Environmental Science, 2, 50.	2014	Shrivastava et al.	Each day groundwater is being withdrawn by the village people for the fulfilment of their basic needs and for agricultural purposes. With the groundwater along with a high concentration of arsenic (As), many other heavy metals are also getting introduced into the environment. In the areas with a long history of use of such groundwater, the agricultural lands have been affected severely. The extent of contamination has increased to a level where the crops grown in those lands are becoming a major source for arsenic and other heavy metals poisoning and subsequently transfer to different trophic levels.
Both phosphorus fertilizers and indigenous bacteria enhance arsenic release into groundwater in arsenic-contaminated aquifers. Journal of Agricultural and Food Chemistry, 64(11), 2214-2222.	2016	Lin et al.	Arsenic (As) is a human carcinogen, and arsenic contamination in groundwater is a worldwide public health concern. Arsenic-affected areas are found in many places but are reported mostly in agricultural farmlands, yet the interaction of fertilizers, microorganisms, and arsenic mobilization in arsenic-contaminated aquifers remains uncharacterized. The effects of fertilizers and bacteria on the mobilization of arsenic in two arsenic-contaminated aquifers.

Source	Year	Author	Key Notions
Current status of ground water arsenic contamination in India and recent advancements in removal techniques from drinking water. International Journal of Plant and Environment, 2(1 and 2), 01-15.	2016	Mishra et al.	India is consisting of 29 states and 7 union territories, including a national capital, Delhi. Elevated concentrations (>10 g l-1) of arsenic (As) in groundwater (GW) of many states of India have become a major concern in recent years. Up to now, about 0.2 million GW samples have been analyzed for As contamination from all over India by various researchers and Government agencies. About 90% of these cover only the Eastern part of India while several states and UTs are still unexplored. The current status of GWAs contamination in different states of India and the new developments of mitigation options.
Abundant environmental arsenic contamination: some statistical perspectives. Sankhya B, 78(2), 341-361.	2016	Sen, P. K.	There is a severe impact of the current abundant environmental contamination and toxic pollution on ecology as well as our life sustaining resources: drinkable and safely usable water, respirable air, untainted and edible food, and renewable and affordable energy. A statistical appraisal along with a quantifying contamination severity index constitute the primary objective of this study.
Arsenic Contamination in Bengal Basin: Reinventing Mitigation through Participatory Social Innovations. Washington State University.	2017	Chakrabarti, K. B	Arsenic contamination of groundwater and soil in West Bengal, India is a serious environmental health issue. Approximately 40 million people are exposed to contamination, 20 million being seriously affected. Risk mitigation faces multiple challenges, which are primarily historical, political, socio-cultural, and economic in nature. While some of those are due to the material reality of the water infrastructure, related to issues of access, availability, quantity and quality of water; there are myriad other issues caused by social, cultural, political, economic and structural realities of the region, both of which significantly hinder and is hindered by communicative issues.

Source	Year	Author	Key Notions
Distributive Family Expenditure of Farmers due to Arsenic Contamination: The Interpretation and Inferences. Bangladesh Journal of Extension Education ISSN, 1011, 3916.	2018	Chatterjee et al.	Arsenic contamination in the groundwater of West Bengal takes a serious shape and at least 120 blocks in West Bengal are under the clutch of moderate to the high level of Arsenic contamination. The ill effects of human exposure to arsenic (As) have recently been re-evaluated by government agencies around the world. This has led to a lowering of guidelines in drinking water. It was carried out at Nonaghata Uttar para village of the Haringhata block of Nadia district in West Bengal. The random sampling method with the objective to generate classified information on the distributive family expenditure of the arsenic contaminated farmers, and estimate the level of impact of the expenditure pattern on the health of the rural people in terms of socio-ecological factors
Estimation of impact of arsenic contamination: The severity and contextuality.	2018	Chatterjee et al.	The eastern part of river Hooghly in west Bengal due to some geogenic factor has been a classical example for arsenic (As) contamination. The multi-directional effect of Arsenic Has been harshly reflected through the contaminating effect of rice, vegetables, meat, eggs and even breast milk used to feed newborn babies. The depletion of groundwater in irrigation-based agro systems has worst the situation further. It was carried out at Nonaghata- Uttar para village of the Haringhata block of Nadia district in West Bengal taking seventy respondents randomly. Variables like age, cropping intensity, source of irrigation, and communication variables are taken for the collection of reliable data.
Physio-chemical characteristics and hydrogeological mechanisms in groundwater with special reference to arsenic contamination in	2018	Jain et al.	Deterioration in the groundwater quality is prevalent in many parts of eastern and north-eastern India. In order to evaluate the groundwater suitability in the north-eastern region of India. Hydrogeological studies revealed that regional geological factors might

Source	Year	Author	Key Notions
Barpeta District, Assam (India). Environmental monitoring and assessment, 190(7), 1-16.			be responsible for excess arsenic concentration in the region.
Arsenic Contamination of Groundwater in West Bengal: A Human Health Threat. Page, 1(1).	2019	Roy, M	Arsenic contamination in groundwater in the Ganga-Brahmaputra fluvial plains in India and its consequences on human health have been reported as one of the world's biggest natural calamities. People in these affected states have chronically been exposed to drinking arsenic-contaminated hand tube-wells water. Arsenic toxicity is a great threat to human civilization, especially in West Bengal, India.
Arsenic contaminated groundwater in China and its treatment options, a review. Applied Ecology and Environmental Research, 17(2), 1655-1683.	2019	Sanjrani et al.	A high concentration of arsenic in groundwater is a worldwide problem, high concentration of arsenic in groundwater has been documented as a major health issue around the globe. According to WHO standard, the maximum contamination level for total arsenic in water is 10 µg/L. China is one of the most affected countries facing health issues because of arsenic contamination in groundwater that is often greater than established limits for human health.
Arsenic contamination of India's groundwater: a review and critical analysis. Arsenic Water Resources Contamination, 177-205.	2020	Chiklon, A. K.	The scope of arsenic contamination in India and the efforts are taken by the Central Government of India to address this issue. Arsenic examines its transportation and distribution in the environment and the human population and discusses the impact of arsenic exposure on humans. Arsenic contamination in India and how it is screened and identified. The challenges in mitigating and combating arsenic contamination of groundwater in India. Providing recommendations for risk mitigation and local management of groundwater arsenic contamination.

Source	Year	Author	Key Notions
Geomorphic controls on shallow groundwater arsenic contamination in Bengal basin, India. *Environmental Science and Pollution Research*, *28*(31), 42177-42195.	2021	Das & Mondal	To explore the influence of geomorphic features scattered throughout the area on the occurrence and distribution of arsenic in shallow groundwater. GIS techniques were frequently used to identify the geomorphic features and to correlate with arsenic distribution patterns. The occurrence of geomorphic features and their distribution have a vital role in the heterogeneous distribution pattern of arsenic in shallow groundwater. Arsenic contamination levels are varying in different fluvial plains of the study area following the trend of Older Deltaic Plain (ODP) > Older Flood Plain (OFP) > Active Flood Plain (AFP).

3

Theoretical Orientation

Arsenic contamination of groundwater is a form of groundwater pollution that is often due to naturally occurring high concentrations of arsenic in deeper levels of groundwater. The Ganges Delta's usage of deep tube wells for water delivery has raised awareness of the issue because it has led to widespread arsenic poisoning in significant numbers of people. During the weathering of rocks and minerals, which is followed by subsequent leaching and runoff, arsenic (As) is released into the soil and groundwater. Additionally, anthropogenic sources can transfer it into the soil and groundwater. Six West Bengal districts with a combined area of 34000 km^2 and a population of 30 million have been determined to have groundwater levels of arsenic over the WHO maximum allowable limit of 0.05 mg l1. (Alam et al.,2002)

At present, 37 administrative blocks by the side of the river Ganga and adjoining areas are affected. Areas affected by arsenic contamination in groundwater are all located in the upper delta plain and are mostly in the abandoned meander belt. More than 800 000 people from 312 villages/wards are drinking arsenic-contaminated water and amongst them, at least 175 000 people show arsenical skin lesions. Thousands of tube well water in these six districts have been analysed for arsenic species. Hair, nails, scales, urine, and liver tissue analyses show elevated concentrations of arsenic in people drinking arsenic-contaminated water for a longer period. Arsenic comes from a geological source. Only a few soil strata have high arsenic contents, which are determined to be linked to iron-pyrites according to investigations of bore-hole silt. Arsenical skin lesions in various areas cause a variety of societal issues. (Das et al.,1996)

Arsenic toxicity has been made worse by malnutrition, low socioeconomic conditions, illiteracy, eating habits, and long-term consumption of water contaminated with arsenic. All of these districts rely heavily on groundwater to meet their water needs, and the geochemical response brought on by this heavy water consumption may be what causes arsenic to leak from the source. If alternative water resources are not utilised, a good percentage of the 30 million people in these six districts may suffer from arsenic toxicity in the near future. In India, since the groundwater arsenic contamination first surfaced in West Bengal in 1983, several other States, namely; Jharkhand, Bihar, and Uttar Pradesh in the flood plain of the Ganga River; Assam and Manipur in the flood plain of the Brahmaputra and Imphal rivers, and Rajnandgaon village in Chhattisgarh state have chronically been exposed to drinking arsenic contaminated hand tube-wells water above the permissible limit of 50 µg/L. Many more North-Eastern Hill States in the flood plains are also suspected to have the possibility of arsenic in groundwater. Through a primary survey carried out in five arsenic-affected villages in the state of West Bengal and a study of existing research and policy documents, an analysis of arsenic-related health issues from an ecosystem viewpoint was made. Although modern agricultural practices and drinking water policies have resulted in arsenic contamination of groundwater, current mitigation policy is essentially confined to biomedical approaches, which include potable water supply and medical care.

Arsenic contamination of shallow groundwater aquifers has been documented in more than 20 nations, with West Bengal and Bangladesh's Bengal Delta seeing the worst effects. The widespread incidence of chronic arsenicosis in the Bengal Delta has led to intensive research on arsenic (As) enrichment in groundwater as well as accumulation in foodstuffs, as there are potential health risks associated with exposure to as from both sources. This study deals with human as exposure through the drinking of groundwater, consumption of locally grown foodstuffs (e.g., crops and vegetables) and cooked food in Nadia district, West Bengal. Groundwater and foodstuffs were collected and analyzed with FI-HG-AAS to estimate the total arsenic content. (Ravenscroft et al.,2005)

The Earth's crust contains a lot of the naturally occurring element arsenic. Chemically speaking, arsenic is a metalloid with characteristics of both a metal and a non-metal, but it is commonly referred to as a metal. The solid form of elemental arsenic, often known as metallic arsenic, is steel grey in colour. Arsenic is typically present in the environment in combination with other substances like oxygen, chlorine, and sulphur. Inorganic arsenic is arsenic that has been mixed with these substances. Arsenic combined with carbon and hydrogen is referred to as organic arsenic.Planning the large-scale use of groundwater for drinking and other uses currently faces a very serious difficulty due to the contamination of groundwater by arsenic from natural geochemical sources. Large quantities of arsenic in groundwater can now be linked to health effects including cancer because of recent advancements in analytical equipment detection limits.

3.1. Need of the Study

To calculate how arsenic contamination affects the economy and ecology and is reflected in the health status of the poor, social stain, cultural separation, and what would be the protractile effect on the future generation.

3.2. Sources of Arsenic Contamination in Groundwater

The Ganga-Brahmaputra-Meghna (GBM) river system forms a substantial portion of West Bengal. This sedimentary basin was created by the deposition of significant amounts of sediments containing arsenic, primarily from the Himalayas, which were transported by the powerful GBM rivers during the Pleistocene and Holocene periods. From these sediments, arsenic is leaching into the groundwater aquifers located in the fan deposit areas and Holocene alluvium. Though the exact mechanism of the arsenic leaching into groundwater is not clear, three mechanisms have been suggested in explaining the process of leaching of arsenic in the groundwater in the GBM basin: 1) Arsenical pyrites in the alluvial sediments are oxidized, and as a result, arsenic is released into the groundwater. The oxidation might have occurred due to the entry of atmospheric oxygen into the aquifers after the heavy withdrawal of groundwater through shallow and deep tube wells. 2) Microbial metabolism of organic matter in the underground leads to anoxic conditions, thus leading to the reduction of iron oxyhydroxides ($FeHO_2$), and subsequently resulting in the release of sorbed arsenic into the groundwater. 3) Arsenic anions sorbed to aquifer minerals are being displaced into solution by competitive exchange of phosphate anions resulting in arsenic contamination in groundwater. The overuse of phosphate fertilizer in agriculture, the fermentation or decay of buried peat deposits and other naturally occurring organic compounds, etc. are cited as the sources of phosphate. However, these suggested mechanisms yet remain to be substantiated. (Saha et al.,2016)

Groundwater in West Bengal contains a higher concentration of arsenic than surface water sources and the groundwater contains both forms of inorganic arsenic (As III and As V), and As III is the predominant species. Both inorganic and organic forms of arsenic are common in surface and dug well water, with the oxidized form of arsenic being the dominating species. Although As III and As V are both poisonous, As III is thought to be the most dangerous. Arsenic pollution has been found more commonly in tube wells that were built between 15 and 50 meters underground. However, in some locations, arsenic pollution has been found in the tube wells that were built at deeper depths. It is uncommon to find arsenic contamination in the water of deep tube wells (DTWs; more than 150 m deep). Initially, the arsenic contamination in tube well water of West Bengal was thought to be in the Gangetic delta plain; later, the contamination was detected in almost all the sedimentary areas of West Bengal, except in Hilly and Pleistocene Uplands areas (Terrace Land). (Ahmad et al., 2018)

3.3. Factors Affecting Arsenic Contamination in Groundwater

3.3.1. Biochemical Factors

- **Oxidation:** It has been suggested that one way to release as from geologic materials is by the oxidation of As-bearing sulphides. The existence of sulphide minerals in these aquifers is uncommon and appears to be restricted to biogenic framboidal pyrite, pyrite in woody peat, and pyrite on magnetite, even though they were initially hypothesized as a mechanism for as release from the alluvial deposits of West Bengal. Oxidation of sulphides in mined areas throughout the world is a well-known phenomenon that has led to high concentrations of as in soils, surface water and groundwater. Sulphide oxidation has also resulted in high as concentrations (up to 215 μg/L) in groundwater. A broad diversity of microorganisms oxidizes dissolved arsenic for different reasons including dissimilatory respiration, detoxification, and energy needs.(Santini and Ward, 2012)

- **Reduction:** West Bengal has drawn a lot of interest from the research community due to the severity of the contamination there. The observed mobilization of arsenic from sediments to groundwater is generally explained by reductive dissolution of Fe hydroxides and release of sorbed As. Organic matter in the alluvial aquifers is likely an important component of the reduction process. Field and experimental studies have shown that metal-reducing microbes can enhance mobilization of As and that oxidation of the organic matter drives the redox reactions whereby Fe hydroxides are reductively dissolved and sorbed arsenic released. During the reductive dissolution of Mn oxides, arsenic is also released; however, it may not stay in the groundwater and may instead reabsorb to form Fe hydroxides. (McArthur et al., 2004)

- **Microbially mediated reactions** Bacteria in Cambodian sediments indicated arsenic-respiring bacteria that reduce As (V) (arsenate) to As (III) (arsenite) were fuelled by inputs of organic carbon, similar to findings for bacteria in West Bengal sediments. Additionally, the results of experiments showed a microbially mediated reduction of As (V), and Fe, and release from minerals, demonstrating the involvement of microbes in the reduction-oxidation (redox) reactions. There are various groups of bacteria participating in the reactions. The Geobacter genus of iron-reducing bacteria can reduce Fe in minerals like hydroxides, causing the hydroxides to dissolve and release sorbed As. As reducers have been studied for Geobacter bacteria. As(V) acts as the terminal electron acceptor in the dissimilatory reduction of arsenate and is a characteristic of bacteria known as dissimilatory arsenate respiring prokaryotes (DARPs). All microbial reduction of arsenate is the result of bacterial respiration. The presence of dimethyl arsenate and monomethylarsonate in groundwater, as was found in shallow groundwater in glauconitic sediments was indicative of such microbial activity. (Lear et al., 2007)

3.3.2. Hydrological Factors

- **Residence time:** Several biogeochemical processes that release As from geologic materials have been identified that whether released As remains at problematic levels in groundwater depends not only on whether there are biogeochemical reactions that retard the transport of As but also on the hydrologic and hydrogeologic properties of the aquifer, such as flow velocity and dispersion. As concentrations might not rise to the point where groundwater would be deemed contaminated if the kinetics of as release is slow and the groundwater residence period is brief. Conversely, if reactions that mobilize as are rapid and residence time is long, then as can accumulate in groundwater such that concentrations become hazardous-as seen in Nadia, for example. The source might eventually run out if the biogeochemical factors that cause as to release and mobilize are still in place (within a geologic era). (Smedley and Kinniburgh,2002)

- **Seasonal changes in recharge:** The disposition and transportation of as within groundwater systems can be impacted by natural oscillations. Seasonal variations in recharge may contaminate shallow groundwater during times of heavy precipitation, but they may also transfer minerals generated from the surface to the aquifer. The transport of dissolved organic matter from agricultural land in the Bengal Delta has been suggested to fuel bacterial activity that releases As from the aquifer materials, and in that region, seasonal (monsoon) rainfall has an important effect on recharge rates and the transmission of land-derived substances to depth in the aquifers. Increased recharge during springtime results in more diluted shallow groundwater and low as concentrations, whereas hot, dry weather results in decreased recharge and higher as concentrations. In areas where clay lenses support the stream channel, shallow groundwater levels above the clay decrease during warm, dry spells, and some stream segments may eventually lose water to groundwater. Thus, seasonal hydrologic factors regulate the release of As-rich groundwater to some stream segments. In other segments, As-rich groundwater may discharge on a relatively constant basis with higher concentrations of as being present during warm dry weather. (Barringer et al., 2010).

- **Effects of pumping:** Pumping-induced changes to hydraulic gradients can alter flow paths at regional and local scales and can lead to the introduction of contaminants to otherwise potable water. The shallower of two aquifers is where West Bengal's contaminated well water is located. Although the deeper aquifer's water is often free of as contamination, it was anticipated that replacing shallow tube wells with deep aquifer wells might cause contamination to travel downhill to low-As groundwater. Anoxic water can be drawn upward to oxic zones near the wellhead, resulting in disequilibrium between as species. The converse is also possible-anoxic waters could be introduced by pumping to oxic zones such that Fe (III) in ferric hydroxides is reduced, hydroxides are dissolved and absorbed As is released. In a field experiment, longer periods between pumping episodes allowed longer periods of anoxia, which resulted in higher concentrations of as in groundwater. Pumping could potentially transfer higher pH water into a region of lower pH water, creating a setting for materials that have aquifer-solubilized substances to desorb. Additionally, flow rates at the size of a single borehole can be so high that the amount of time groundwater has in contact with aquifer material is kept to a minimum, and reactions that release may be relatively slow.(Ravenscroft et al, 2001; Harvey et al., 2002; van Geen et al., 2003).

3.4. What Happens to Arsenic when Arsenic Enters the Environment?

Arsenic is present in soils, and as a result, it can enter the water, air, and land through dust carried by the wind. It can also enter water by runoff and leaching. Arsenic also comes from volcanic explosions. During the mining and smelting of these ores, arsenic is released into the environment. Small amounts of arsenic also may be released into the atmosphere from coal-fired power plants and incinerators because coal and waste products often contain some arsenic. Arsenic cannot be destroyed in the environment. It can only change its form or become attached to or separated from particles. It may change its form by reacting with oxygen or other molecules present in the air, water, soil, or by the action of bacteria that live in soil or sediment. Arsenic is typically associated with very minute particles when it is emitted from power plants and other combustion operations. Arsenic contained in wind-borne soil is generally found in larger particles. When linked to tiny particles, arsenic can float through the air for days and cover great distances. In water, several typical arsenic compounds can dissolve. As a result, arsenic can enter lakes, rivers, or underground water through the melting of snow or rain, as well as the release of industrial waste. Some of the arsenic will stick to particles in the water or sediment on the bottom of lakes or rivers, and some will be carried along by the water. Ultimately, most arsenic ends up in the soil or sediment. Despite certain fish and shellfish ingesting arsenic, which may accumulate in tissues, the majority of this arsenic is in the much less dangerous organic form arsenobetaine, sometimes known as "fish arsenic."(Chatterjee et al., 2018)

3.5. Arsenic Levels in the Environment and Food

The U.S. Geological Survey has done multiple studies over the years to map the soil's arsenic content, and this data is easily accessible. Because arsenic is global in the environment, it is not able to be avoided finding traces in our food supply given the obvious need for soil, air and water to grow crops needed for direct human consumption and for feeding livestock.

There has also been a lot of focus on how human activity affects the amount of arsenic in the soil. The levels of arsenic in the soil, and subsequently in food, are thought to be affected by mining, the use of arsenical pesticides, and the usage of groundwater (rather than surface water).

The majority of inorganic arsenic exposure in the American diet comes through food, and that exposure is distributed among a number of food sources, including vegetables, fruit and fruit juices, rice, beer and wine, other grains, meat, and eggs, and a variety of other food types. It would be impractical to remove all of these things from one's diet to completely prevent inorganic arsenic exposure from food. Specific levels of inorganic arsenic in specific food types have been the topic of much media coverage, with most of the focus over the past year on fruit juice, rice and rice products. The FDA keeps an eye on and controls the amount of arsenic in meals, nutritional supplements, and cosmetics. Due to its presence in the environment as a naturally occurring element as well as via consumer and industrial products and processes, arsenic can appear in food. Arsenic levels in the environment are generally low but can vary depending on the natural geological makeup of local areas. For example, volcanic eruptions can bring arsenic from the earth's interior to the surface. Arsenic contamination in some areas is also a result of mining, fracking, coal-fired power plants, arsenic-treated lumber and arsenic-containing insecticides. Arsenic does not disappear from the environment over time, and it is not possible to remove arsenic entirely from the food supply. The FDA, therefore, seeks to limit consumer exposure to arsenic to the greatest extent feasible by monitoring the food supply, setting action levels for arsenic in certain foods, and taking regulatory action when levels of arsenic are too high. The FDA's Toxic Elements Working Group identifies and prioritizes FDA activities to reduce exposure to arsenic from specific foods. (Chatterjee et al.,2018)

3.6. Social Impacts of Arsenic Contamination in Groundwater

Arsenic affected people are also facing serious social problems. The affected villagers are living in very poor conditions. A few people are aware of arsenic pollution and its impacts on human health. A large number of people are ignorant of arsenic pollution. They suffer from arsenic diseases and become the victim of arsenic contamination of water but they do not think of it. Only until the effects of arsenic are severe and a person develops black foot illness do they recognize they are poisoned with arsenic. People who have various skin conditions have very rough-looking bodies and black blotches on their hands and feet. There are, nevertheless, obstacles to social access. While the presence of arsenic in the urine first indicates arsenicosis, physical evidence later appears in the form of melanosis or keratosis, with elevated levels of arsenic apparent in the nail, hair, and skin scales. The results include disfiguring and painful dermatological abnormalities, including hypo and hyperpigmentation, and the emergence of corn-like protrusions on the hands and feet, commonly marked by a darkening of the skin. As a disease arsenicosis thus provides highly visible markers of difference that commonly shape social judgments and inform lay diagnoses.

While it is intuitive that as a highly visible and misunderstood disorder, arsenicosis will lead to social impacts, complicating the question of causation is evidence that prolonged exposure to arsenic can result in depressed verbal intelligence and long-term memory as well as other cognitive effects, proportional to dose. The social effects of arsenicosis may also be mediated by wealth to the extent that it results in a decline in the ability to work. The stigma of arsenicosis is based on widespread misconceptions about the causes and modes of transmission, including the idea that the sickness is contagious. In cultures with low levels of education, the illness may take on the moral and health risks of divine providence. Arsenic patients showed a significant minority reported experiencing a reduction in interaction with neighbours, and worried about the impact of isolation on their children, with some parents withholding children from school to avoid exacerbating social disapproval. The social effects of arsenicosis may change depending on gender, with women possibly being more affected than males. This difference in impact is likely to be at least partially mitigated by the lower literacy rates of rural women than those of men, who are typically responsible for collecting water.

The causal route from disease to impact in this case may be complicated by evidence that malnutrition—associated with both genders is directly associated with the severity of arsenicosis. the social impacts of arsenicosis suggest a broad range of impacts, they all cluster around a central notion of isolation: affected individuals experience difficulty getting married, maintaining friendships, going to school, and advancing themselves through work. Few quantitative research has been done on the social and economic repercussions of arsenic pollution; instead, the focus has been on the health effects. (Nahar,2009)

3.7. Effects of Arsenic on Crops

Heavy metals and toxic substances are released into the environment, contributing to a variety of toxic effects on living organisms in the food chain by accumulation and biomagnifications. Certain pollutants such as arsenic (As) remain in the environment for an extensive period. They gradually build up to concentrations that could damage the physiochemical characteristics of soils, diminish soil fertility, and reduce crop output. If arsenic is not detoxified, it may set off a chain of events that inhibits growth, interferes with photosynthetic and respiratory processes, and stimulates secondary metabolism. A range of mechanisms, including hyper accumulation, the antioxidant defence system, and phytochelatin, are used by plants to respond to as toxicity. There is arbuscular mycorrhizae symbiosis practically everywhere, even in disturbed soils. There is growing evidence that arbuscular mycorrhizae fungi may alleviate metalloid toxicity to the host plant. Irrigating agricultural fields with arsenic-contaminated water produces an accumulation of arsenic in soil and subsequently an increase of arsenic concentration in crops. The concentration of arsenic in crops depends on many factors, for example, type of crop, arsenic concentration of soil and water, soil type, among others. arsenic in irrigation water tends to accumulate in agricultural soil and through several mechanisms is absorbed by crops. (Robles et al.,2018)

3.8. Effects of Arsenic on Livestock

Arsenic is an important heavy metal intoxicant to livestock. Arsenical pesticides present significant hazards to animal health. The toxicity of arsenic varies with several factors--its chemical form, oxidation states, and solubility. The least hazardous phenyl arsenic compounds are included as feed additives to swine and poultry rations. The mechanism of action is related to its reaction with sulfhydryl group values to enzyme function and to its ability to uncouple oxidative phosphorylation. Most animals excrete arsenic quite readily. Toxicoses caused by inorganic and aliphatic organic arsenicals result in a different clinical syndrome than that from the phenyl arsenic compounds.

Arsenic poisoning may be confused with other types of intoxication. Phenylarsonic structures have a carbon ring structure (a phenyl group) with arsenic or a form of arsenic attached to it, whereas an aliphatic organic arsenical has a straight chain of carbon atoms attached to the arsenic. Cattle exposed to monosodium methanearsonate (MSMA) may merely exhibit rapid mortality or they may exhibit hemorrhagic or necrotic rumens as well as a very brilliant red serosal surface of the omasum or abomasum. Cattle and pigs may experience agonizing deaths, watery diarrhoea, extreme dehydration, excruciating stomach agony, and pale, enlarged kidneys. Pigs may show lower weight gain as the early clinical symptoms, which may be followed by unsteadiness, hind limb paralysis, and ultimately quadriplegia. Blindness could develop in swine. A crucial component of treatment is locating the source and removing any animals from it. Unfortunately, there is no specific treatment for organic arsenical exposure. Supportive care may be helpful but is unlikely to be curative. Affective pigs may be prone to consistently high-pitched squealing. If pigs are only blind, they will grow to a normal weight. The specific antidote for inorganic arsenical poisoning is dimercaprol (BAL). (Mandal et al.,2017)

3.9. Effect of Arsenic on Health

- The immediate symptoms of acute arsenic poisoning include vomiting, abdominal pain, and diarrhoea. These are followed by numbness and tingling of the extremities, muscle cramping and death, in extreme cases.
- Poor people are more affected than others by arsenic-related diseases. Arsenic has been linked to “Blackfoot disease”, which is a severe disease of blood vessels leading to gangrene.
- When people are exposed to amounts of arsenic above 50 gg/l in their drinking water, the prevalence of skin lesions and lung disease is higher in males than in women. Women had higher prevalence odds ratios for lung disease.
- Women are more socially damaged than men by arsenic-related illnesses, no doubt because of their generally lower social status. If unmarried, they find it difficult to find a husband; and if married they may be abandoned or divorced. Women are less likely to talk about arsenic-related health problems and are more likely to attend to their health needs.

- Thickening of the skin (hyperkeratosis). Keratin is the protein that makes skin tough and hyper sort of means super-duper, so hyperkeratosis is when someone produces an abnormally high amount of keratin.
- The financial effects of the disease require consideration. Many who are ill are either too weak to work or lose employment opportunities because of widespread fears of contagion.
- It has also been associated with cardiovascular disease and diabetes. In utero and early childhood, exposure has been linked to negative impacts on cognitive development and increased deaths in young adults.

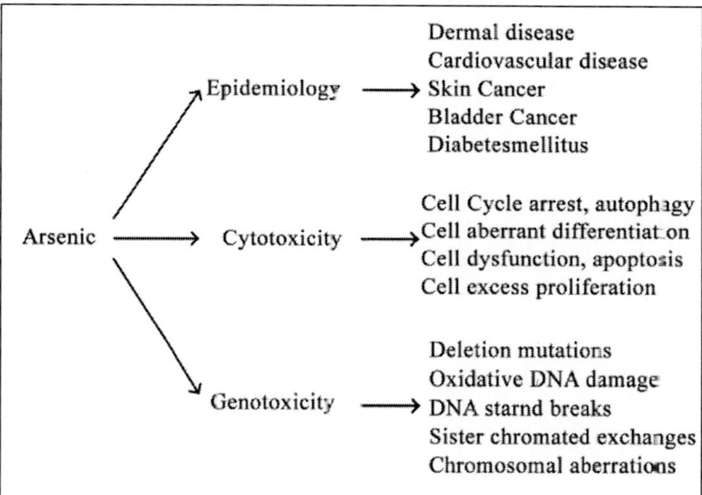

Fig. 3.1: Effects of arsenic on health
Source: *Lena. et al. ("Arsenic and selenium toxicity and their interactive effects in humans," Environment International)*

3.10. Key Findings Related to Social Factors

- People have multiple urgent and competing concerns. Many poor households might not place high importance on arsenic.The perception of arsenicosis as a disease is still a major problem since, according to research, more than half of rural inhabitants believe the sickness may spread from person to person. This has major implications for behaviour change potential about arsenic and suggests that more work needs to be done to improve understanding.
- Red tube wells are shameful, which could have an impact on families who have tube wells that are tainted with arsenic. This is in contrast to pride built up over the years in having and using tube well water, which had been considered to be safe, and which became a status symbol.

4

Research Setting and Social Ecology

The research setting is an important component of the research methodology. Research setting refers to the detailed information of an area where the study was conducted. The research setting can be seen as the physical, social, and cultural site in which the researcher conducts the study. In qualitative research, the participants are observed in their natural environments with a primary focus on meaning-making. The study area generally consists of a particular geographical area viz. a state, district, block, or gram panchayat area selected according to the benefit of the investigator who possesses adequate knowledge regarding the location, communication facility, the period of study, the availability of resources, etc. of the locality so that he can easily approach every corner of the area for data collection. Additionally, to comprehend people's knowledge, attitudes, and behaviour more easily, investigators must have a fundamental understanding of their socio-demographic backgrounds. It is the environment within which studies are run that has important consequences for experimental design, the type of data that can be collected and the interpretation of results.

The study was taken up at Mitrapur village of Mollabelia gram panchayat in Haringhata block in West Bengal. A brief description of the state and district in general and the block and village in particular is given below.

4.1. Area of Study

The area of investigation of this study is situated in the state of West Bengal located in the eastern part of India. The arseniferous tract located mainly within the upper deltaic plain and south eastern part of the delta in the mouth at shallow depth is built of sediments deposited by meandering streams and levees. The majority of West Bengal's arsenic-affected areas are located in alluvial plains that originated during the quaternary period. The area of investigation belongs to the Haringhata block in the Nadia district. The area of the study is village Mitrapur under Mollabelia gram panchayat.

4.2. Profile of the State of West Bengal

West Bengal was created as one of the constituent states or the Indian union among the 29 states in India on 15 August 1947 as the result of the partition of the undivided. British Indian province of Bengal into West Bengal. West Bengal is situated in the north eastern part of India and lies between 21^0 37'-27^0 10' north latitude and 85^0 51 '-89^0 53' east longitude. West Bengal, is a state in the eastern region of India and is the nation's fourth most populous. It is also the seventh most populous sub-national entity in the world. West Bengal is the sixth-largest contributor to India's GDP. It is bordered by the countries of Nepal, Bhutan, and Bangladesh and the states of Orissa, Jharkhand, Bihar, Sikkim, and Assam and has an area of 34,267 sqm (88,752 km2); the capital is Kolkata (Calcutta).

The Gangetic Plain in the south and the sub- and Himalayan region in the north make up the two major natural regions that make up West Bengal. From the 3rd century BCE, the broader region of Bengal formed part of Ashoka's empire. In the 4th century CE, it was absorbed into the Gupta empire. At Indian independence in 1947, Bengal was partitioned, the eastern sector becoming East Pakistan (later Bangladesh) and the western sector becoming India's West Bengal. Although in the area West Bengal ranks as one of the smaller states of India, it is one of the largest in population with over 91 million people. The state's primary economic activity is agriculture. The majority of people in the state are farmers and agricultural labourers. Rice is considered to be the principal food crop of West Bengal. Other major food crops include maize, pulses, oil seeds, wheat, barley, potatoes and vegetables. The state supplies nearly 66% of the jute requirements of India. Another significant cash crop is tea. Tea plantations are a major draw to Darjeeling. It is well known for its creative endeavours, which include filmmaking.

4.3. Climate of West Bengal

West Bengal's climate varies greatly, mostly because of the geography and geographic location of the area in question. For instance, the north of Bengal gets humid subtropical weather, whereas the south of Bengal has a tropical savannah environment. The seasons here can broadly be divided into five main categories: spring, summer, rainy season, autumn, and winter. In summer the maximum temperature ranges between 38°C and 45°C, while the minimum is around 20°C. Winter is moderate in the plains, with average minimum temperatures being somewhere around 15°C. During this period, the humidity is quite low, even on the plains. In different areas of the state, varying amounts of rainfall fall on a yearly average. While the northern region experiences considerable rainfall, averaging between 200 and 400 cm, the coastal regions, Gangetic plains, and the state's central regions only see 150 to 200 cm of precipitation. The western plateau of West Bengal receives low rainfall, around 100 to 125 cm.

4.4. General Information of West Bengal

West Bengal ranks as one of the smaller states in India, it is one of the largest in population. The capital is Kolkata (Calcutta). Area 34,267 square miles (88,752 square km). Pop. (2011) 91,347,738. Population density is 1029/km. In this state male population is higher than the female population. The literacy rate of the female is higher than that of the male. The total no of blocks in West Bengal is 341.Nadia district has 947 females as against 1000 males.

Table 4.1: General information of West Bengal

Total population	91,347,738
Male	46,927,391
Female	44,420347
Population Growth	13.93%
Sex ratio	947
Population density	1029/ km
Literacy rate	77.08%
Male	82.67%
Female	71.16%
No of Sub divisions	66
No of blocks	341
No of Gram Panchayats	3354
Inhabited Villages	37945
Area (sq. Km)	88752

Source: *Office of the Register General and Census Commissioner 2011 Ministry of Home Affairs, GOI.*

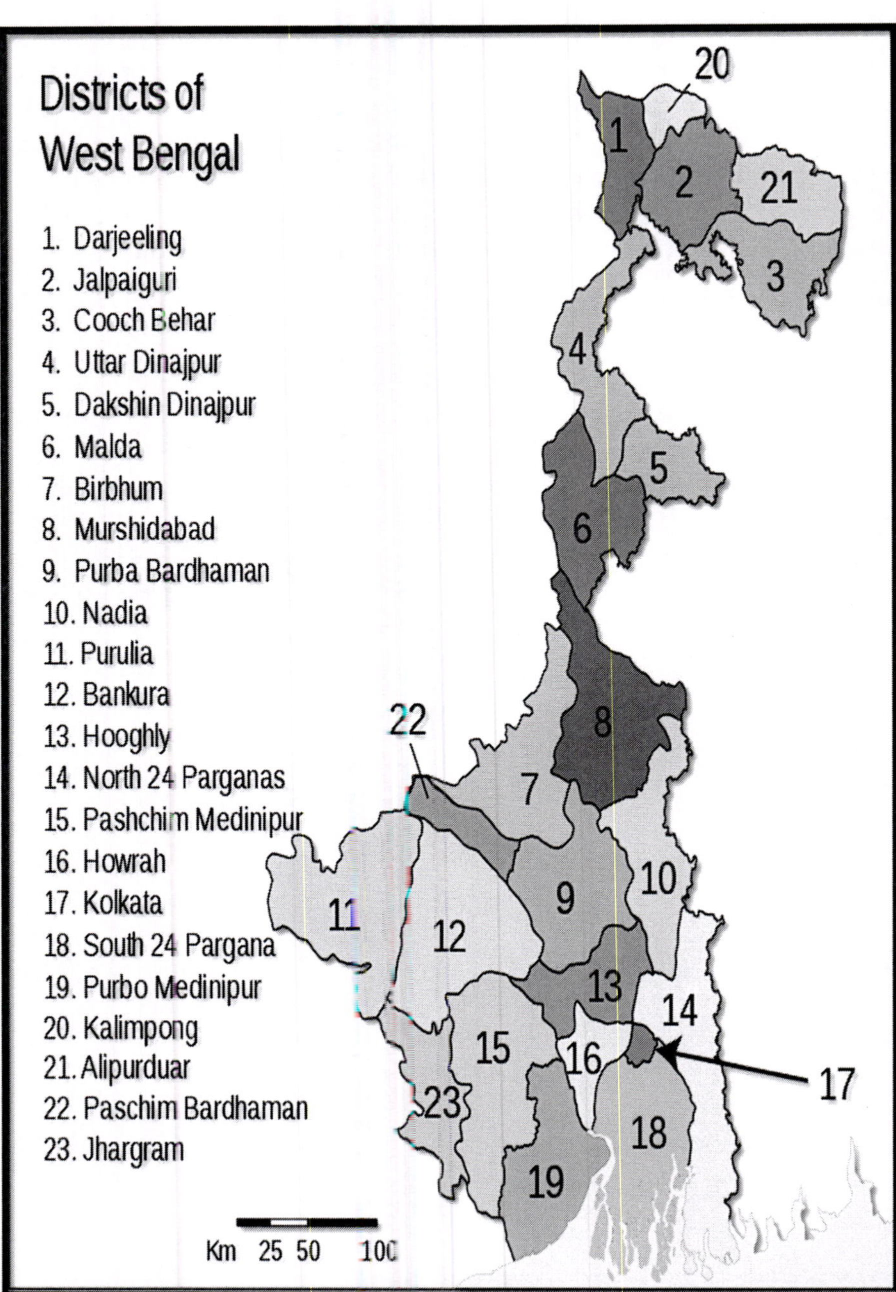

Source: https://www.mapsofindia.com/maps/westbengal/

Table 4.2. Physical parameters and arsenic affected areas of West Bengal

Parameters	
Area in sq. km.	88,750
Population in million	80.2
Total number of districts (no. of district surveyed)	19(19)
Total number of water samples analyzed	1,40,150
% of samples having arsenic > 10 gg L'1	48.1
% of samples having arsenic > 50 gg L4	23.8
No. of severely arsenic affected districts	9
No. of mildly arsenic affected districts	5
No. of arsenic safe districts	5
Total population of severely arsenic affected 9 districts in million	50.4
Total area of severely arsenic affected 9 districts in sq. km.	38,861
Total number of blocks/ police station	341
Total number of blocks/ police station surveyed	241
Number of blocks / police station having arsenic >50ggL4	111
Number of blocks / police station having arsenic >10ggL4	148
Total number of village	37910
Total number of village surveyed	7823
Number of villages/Paras having arsenic above 50 ggL4	3417
People at risk of drinking arsenic contaminated water >10 pgL4 (in million)	9.5
People at risk of drinking arsenic contaminated water >50 pgL4 (in million)	4.6
No. of districts surveyed for arsenic patients	9
No. of districts where arsenic patients found	7
People screened for arsenic patients including children (preliminary7 survey)	96,000
No. of adults screened for arsenic patient	82,000
Number of registered patients with clinical manifestations	9,356(11.0%)
No. of children screened for arsenic patient	14,000
No. of children showing arsenical manifestation	778 (5.6%)
Total hair, nail, urine analyzed	39624
Arsenic above normal/toxic level in hair, nail and urine samples	91%, 97% and 92%

***Source**: Mukherjee et. al (2009)*

Table 4.3. Distribution of hand tubewells against arsenic concentration ranges (pg/L) in all 19 districts of West-Bengal, India.

Districts	Area inkrn2	Population	Total no. of blocks	No. of blocks surveved	No. of blocks with As >10 Mg/L	No. of blocks with As >50pg/L	No. of samples anaivzed	Distribution of total samples in different arsenic concentration (iig/ Lj ranges									Maximum concent-ration found (ug/L)
								<3	4-10	11-50	51-100	101-200	201-300	301-500	501-1000	> 1000	
North-24-PGS	4094	8934286	22	22	22	21	54368	22221	3129	13001	6403	5531	2249	1308	477	49	2830
South-24- PGS	9960	6906689	29	17	12	11	8333	4407	427	1141	743	741	327	305	212	30	3700
Murshidabad	5324	5866569	26	26	25	24	29668	11471	2244	8042	3267	2366	941	884	382	71	3003
Nadia	3927	4604827	17	17	17	17	28794	11431	2613	9810	2265	1520	630	360	152	13	3200
Maldah	3733	3290468	15	14	13	9	4449	1754	373	810	488	559	183	163	97	22	1904
Haora	1467	4273099	14	12	12	7	1471	889	226	192	87	41	22	12	1	1	1333
Hugh	3147	5041976	18	17	16	11	2212	1469	346	251	77	52	14	2	1		600
Kolkata *	185	4572876			-	-	3626	2224	855	345	85	75	27	10	3		825
Bardhaman	7024	6895514	31	24	12	7	2634	2091	79	244	86	89	27	11	6	1	2230
Sub Total	38861	50386304	172	149	129	107	13555 5	57957	10292	33836	13501	10974	4420	3055	1333	187	
Koch Bihar	3387	2479155	12	5	4	1	474	403	57	13	1	-	-	-	-	-	54
Darjiling	3149	1609172	12	4	3	0	562	502	50	10		-	-	-	-	-	19
Dinajpur (N)	3140	2441794	9	7	6	2	990	817	57	112	4	-		-	-	-	68
Dinajpur (SI	2219	1503178	8	6	2	1	452	398	47	6	1	-	-	-	-	-	51
Jalpaiguri	6227	3401173	13	7	4	0	445	355	74	16		-	-		r	r	27
Sub Total	18122	11434472	54	29	19	4	2923	2475	285	157	6	-	-	-	-	-	
Bankura	6882	3192695	22	17	0	0	279	279				-					<3
Birbhum	4545	3015422	19	11	0	0	718	718				-					<3
Puruliya	6259	2536516	20	14	0	0	314	314				-					<3
Medinipur IE)	14081	9610788	54	10	0	0	182	182				-					<3
Medinipur (W|	NA	NA	-	8	0	0	179	179				-					<3
Sub Total	31767	18355421	115	60	0	0	1672	1672					-				
Grand Total	88750	80176197	341	241	148	111	14015 0	62104	10577	33993	13507	10974	4420	3055	1333	187	

Kolkata has no Blocks (It is an urban area)

Source: Mukherjee et al (2009)

4.5. Profile of Nadia District

Nadia is a district in the state of West Bengal, India. Purba Bardhaman to the west, Murshidabad to the north, North 24 Parganas and Hooghly districts to the south, and Bangladesh to the east are its neighbours. the central region of the Bengal lowlands. The district was divided at Partition, and a sizable portion of Nadia was transferred to East Bengal (Bangladesh). Nadia district is highly influential in the cultural history of Bengal. The standard version of Bengali, developed in the 19th century, is based on the dialect spoken around Nadia. The district is still largely agricultural. "Nadia" is a shortened form of Nabadwip, the name for a historic city in the district. Nabadwip, literally "new island", was formerly an island created by alluvial deposits of the Ganga. Nabadwip is an ancient town within the Nadia district. Chaitanya Mahaprabhu was born in Nabadwip. In 1202, Nabadwip was captured by Bakhtiyar Khilji. This victory paved the way for Muslim rule in Bengal. The British defeated Siraj ud-Daulah, Nawab of Bengal, at Palashi in this district. The 1859 revolt against European Indigo planters started from the village of Chaugacha in Krishnanagar, Nadia. Nadia is thought to have had trade relations with Tibet, Nepal and Bhutan.

4.5.1. Geography

The west-central Bengal region, in southern West Bengal, contains the Nadia district. The majority of the district is an alluvial plain created by the ongoing movement of the Ganges Delta's many rivers. To the west of the district is the Bhagirathi (or Hooghly) river, which was once the main distributary of the Ganga towards the Bay of Bengal, and is still considered to be the continuation of the Ganga for Hindus. As the main flow of the Ganga flowed east into the Padma, the Bhagirathi largely dried up. Most of the rivers that once flowed through Nadia are now mostly dry. The district has almost entirely been transformed into agriculture.

4.5.2. Rivers

Nadia district is home to many rivers. The district's north-eastern border is touched by the Padma, currently the Ganga's primary tributary. Before flowing south into Nadia district, the Jalangi, which originates in Murshidabad district, forms a significant portion of the district's northern border with Murshidabad. Around Krishnanagar, it turns west and flows into the Bhagirathi near Nabadwip. The Mathabhanga originates in the far northeast of the district and forms part of the border with Bangladesh. At Maijdia, it splits into the Churni and Ichamati.

4.6. General Information of Nadia District

The district has a land area of 3927 sq km and a population of 5,167,601. The 187 Gram Panchayats and 8 Municipalities make up the district's 17 Panchayat Samities. The total number of Police Stations in the district is 19. The density of population in this district is 1173 persons per sq km. Nadia district has 950 females as against 1000 males. The majority of the people of the district speak Bengali followed by Hindi, Santali and others. Religion wise about 73.75% are Hindus and 25.42% are Muslims. In the district of Nadia, the percentage of literacy by sex is 78.75% (Male) & 70.98% (Female) as per Census 2011.

Table 4.4: General information of Nadia district

Total population	5,167,601
Male	2,653,768
Female	2,513,832
Population Growth	12.22%
Sex ratio (per 1000)	950
Population density	1300/ km2
Literates Rate	74.97%
Male	78.75%
Female	70.98%
No of Sub divisions	5
No of the Development blocks	17
No of Gram Panchayats	187
Inhabited Villages	2639
Area (sq. Km)	3,927

Source: Office of the Register General and Census Commissioner 2011 Ministry of Home Affairs, GOI.

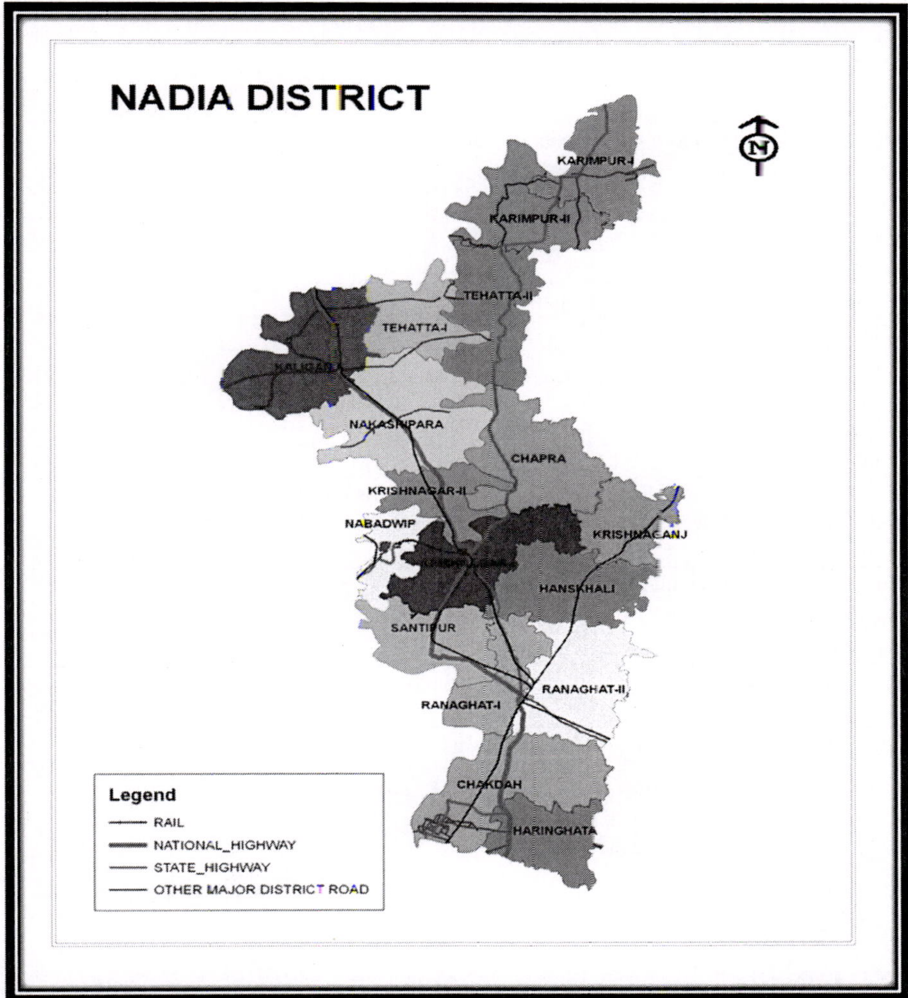

Fig. 4.2: Map of Nadia district

Source: https://www.researchgate.net/figure/Map-of-Nadia-District_fig1_344162873

Table: 4.5: Details of Nadia District, no. of blocks and affected villages along with other relevant data on contaminated and uncontaminated tube wells, number of arsenicosis cases in the exposed population, etc and estimated disease burden in the district of Nadia.

Parameters	**Total**
The population of villages having contamination (>50μg/l)	3220182
% of exposure (% of arsenic contaminated public tube wells in the blocks)	6.36-45.96
Total population of villages exposed to arsenic	839675
No. of villages studied	37
The total population in the study villages	635728
Total Households in the study villages	65905
No. of households surveyed	2297
Total no. of public tube wells in the district: (PHED water test report of Nadia)	29640
Total no. of public tube well water above 50 μg/l	7662 (25.85)
No of water samples which are collected from tube wells (Govt., Pvt.) and dug wells used by villagers for drinking and cooking purposes and tested for arsenic	2116
No. of water samples contaminated with arsenic more than 50 μg/l	809 (38.23)
No. of persons examined	10469
No. of patients with arsenical skin lesion Identified	1616
The proportion of the exposed population of the blocks affected	0.04781 - 0.32249
Population derived as probable affected with arsenicosis	141592

Source:*Indian J Community Med. 2010 Apr; 35(2): 331–338*

4.7. Profile of Haringhata block

Haringhata is a town in the Kalyani subdivision of Nadia district in the state of West Bengal of India.

4.7.1. Location

Haringhata is located at 22^0 95'N 88^0 57'E.

It has an average elevation of 10 meters. Haringhata community development block has an area of 168.59 Km^{-2}.

4.7.2. Gram panchayats

Gram panchayats of Haringhata block/ panchayat samiti are: Birohi I, Birohi II, Fatepur, Haringhata I, Haringhata II, Kastodanga I, Kestodanga II, Mollabelia, Nagarukhra I, Nagarukhra.

4.7.3. Language

Bengali is the local language in these areas.

Table 4.6: General Information about Haringhata block

Block area	170.32 sq.
No. of gram panchayats	10
No. of gram sabhas	129
No. of Mouza	87

Source: *Office of the Register General and Census Commissioner 2011 Ministry of Home Affairs, GOI.*

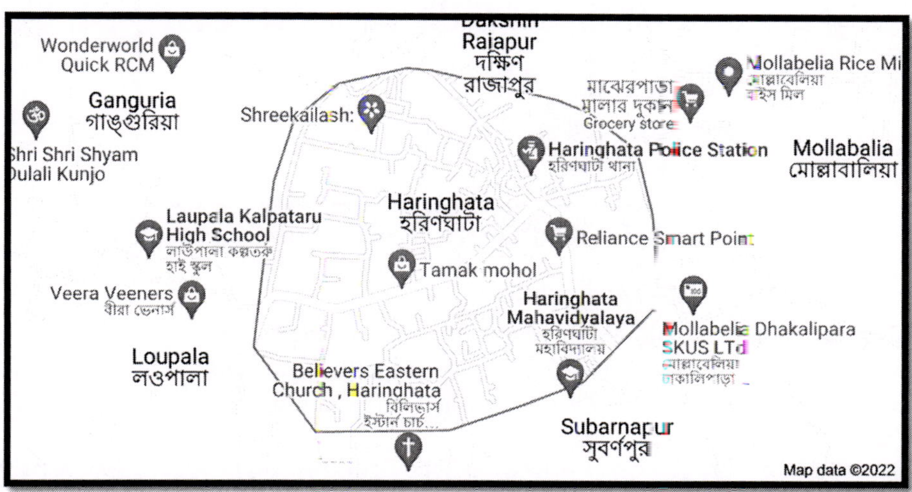

Fig. 4.3: Map of Haringhata block

4.8. Profile of Mollabelia Gram Panchayat

Mollabelia is a large village located in the Haringhata Block of Nadia district. There are 12 villages under the mollabelia gram panchayat. Those are 1) Bamunbaria 2) Basantapur 3) Dasdia 4) Hazrabelia 5) Jalkarampur 6) Kurambelia 7) Madhpur 8) Mitrapur 9) Mollabalia 10) Nischintapur 11) Panpur 12) Uttar Duttapara

Table 4.7: General information of Mollabelia Gram Panchayat

Particulars	**Total**
Total House Hold	8534
Total Population	34089
Male Population	18720
Female Population	15369
Total No of Sansad	24
Area (Square kilometre)	28.94

Source: Office of the Register General and Census Commissioner 2011 Ministry of Home Affairs, GOI.

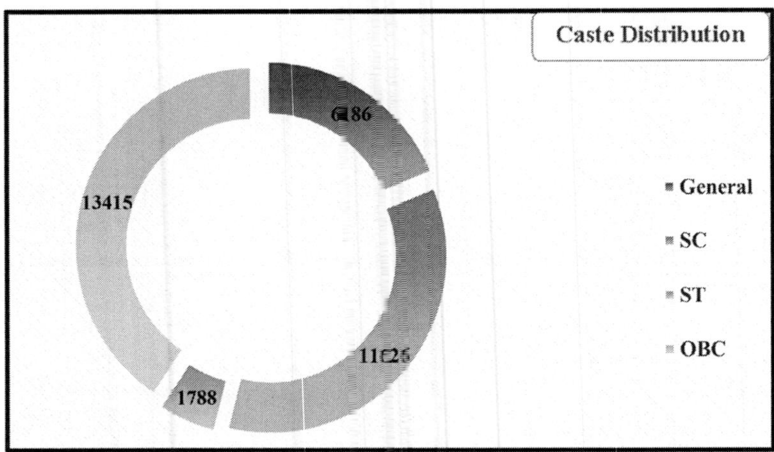

Fig. 4.4: Caste distribution of Mollabelia gram panchayat

4.9. Profile of Mitrapur Village

Mitrapur is a large village located in the Haringhata Block of Nadia district, West Bengal with a total of 707 families residing. According to the Population Census of 2011, the Mitrapur village has a population of 3048, of which 1593 are men and 1455 are women.

The number of children in Mitrapur village between the ages of 0 and 6 is 352, which represents 11.55 percent of the hamlet's overall population. The Mitrapur village's average sex ratio is 913, which is lower than the 950 average for West Bengal as a whole. According to the census, Mitrapur's child sex ratio is 778, which is lower than the 956 average for West Bengal. Mitrapur village has a higher literacy rate compared to West Bengal. In 2011, the literacy rate of Mitrapur village was 76.97 % compared to 76.26 % in West Bengal. In Mitrapur Male literacy stands at 82.72 % while the female literacy rate was 70.79 %. The main occupation of the villagers is cultivation.

Table 4.8: General information of Mitrapur village

Particulars	Total	Male	Female
Total No. of Houses	707	-	-
Population	3,048	1,593	1,455
Child (0-6)	352	198	154
Schedule Caste	814	416	398
Schedule Tribe	16	9	7
Literacy	76.97 %	82.72 %	70.79 %
Total Workers	1,125	945	180
Main Worker	739	-	-
Marginal Worker	386	243	143

Source: *Office of the Register General and Census Commissioner 2011 Ministry of Home Affairs, GOI.*

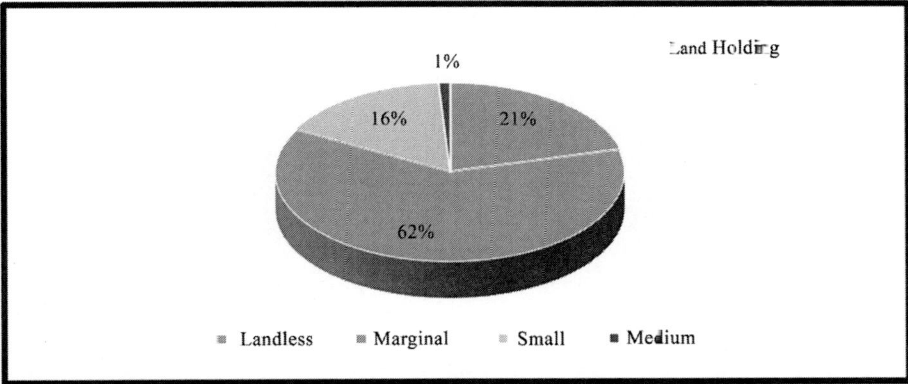

Fig. 4.5: Land holding of Mitrapur village

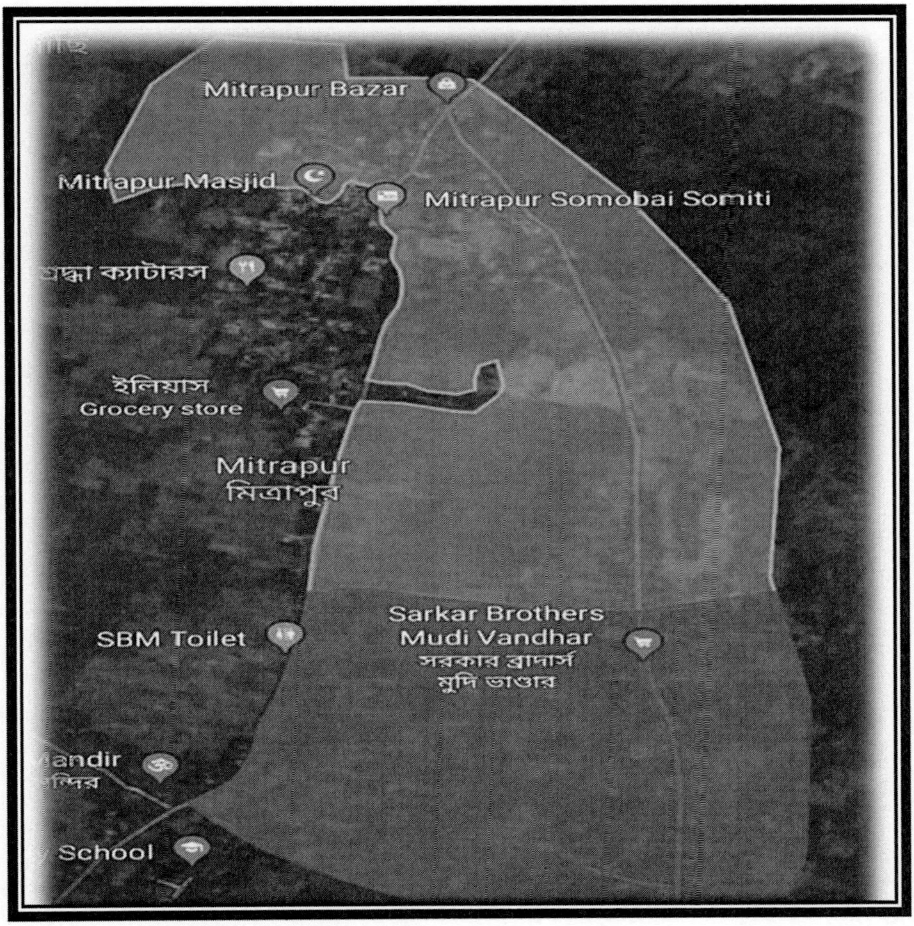

Fig. 4.6: Maps of Mitrapur village

5

Research Approaches and Methodology

Research methodology is a way of explaining how a researcher intends to carry out their research. It's a logical, systematic plan to resolve a research problem. A methodology details a researcher's approach to the research to ensure reliable, valid results that address their aims and objectives. It encompasses what data they're going to collect and where from, as well as how it's being collected and analyzed. The deliberation on the methodology has been made to understand the concept, methods and techniques which are utilized to design the study, collection of information, analysis of data and interpretation of the findings for the revelation of truths and formulation of theories. This chapter deals with the method and procedure used in the study and consists of eight main parts

- Locale of research.
- Pilot study.
- Sampling Design.
- Empirical measurement of the Independent and Dependent variable
- Preparation of interview Schedule.
- Pre-testing of Interview Schedule.
- Techniques of Data collection.
- Statistical Tools used for Analysis of Data.

5.1. Locale of Research

Mitrapur village of the Haringhata block of Nadia district in West Bengal was randomly selected for the study. The area has been selected for the study because of: (a) There is a sample scope for collecting relevant data for the present study (b) Acquaintance with the local people as well as local language (c) The concerned area was very easily accessible to the researcher in terms of place of residence (d) The area was very easily accessible to the researcher in terms of transportation (e) The closer familiarities of the student researchers with the area, people, officials and local dialects.

5.2. Pilot Study

A pilot study is the preliminary study conducted on a limited scale before the original study is carried out in order to gain some primary information, based on which the main project would be planned and formulated.

A pilot study enables the researcher to gain some systematic knowledge of the universe and its population. It plays a vital role in identifying the variables involved in research, the nature of the problem, the nature of the respondents, the kind of responses that they are likely to be available etc. gained during the study and the schedule is then framed.

5.3. Sampling Design

In West Bengal Mitrapur village under Haringhata Block of Nadia district is selected in purposive sampling approach, but the respondents within the village are selected in non-random snowball sampling approaches.

Table 5.1: Sampling techniques and Sampling design

Step	Items	Level	Approach
1	State	West Bengal	Purposive
2	District	Nadia	Purposive
3	Subdivision	Kalyani	Purposive
4	Block	Haringhata	Purposive
5	Gram Panchayat	Mollabelia	Purposive
6	Village	Mitrapur	Purposive
7	Respondents	70	Non-Random snowballing

5.4. Variables and Their Measurements

A variable is something that varies. More specifically variables are those attributes of objects, events, things and beings that vary and can be measured. By followings, the specific objectives of the study, the theoretical basis for the selection of variables and empirical measures are presented.

5.4.1. Independent Variables

An independent variable is a variable you manipulate or vary in an experimental study to explore its effects. It's called "independent" because it's not influenced by any other variables in the study. Independent variables are also called: Explanatory variables (they explain an event or outcome), Predictor variables (they can be used to predict the value of a dependent variable), and Right-hand-side variables (they appear on the right-hand side of a regression equation).

Table 5.2: Independent variables and their notations

Sl. No.	Variables	Notation
1	Age	X_1
2	Education	X_2
3	Family Size	X_3
4	Income	X_4
5	Size of land holding	X_5
6	Homestead land	X_6
7	Land under irrigation	X_7
8	No of source of irrigation	X_8
9	Cropping intensity	X_9
10	Depth of ground water table for tube well	X_{10}
11	Depth of groundwater table for shallow pump	X_{11}

- **Age(X_1):** In all societies, is one of the most important determinants of social status and the social role of the individual. Age here refers to the chronological age of the respondents measured in terms of full years they completed during the time age here refers to the chronological age of the respondents measured in terms of full years they completed during the time of interview.
- **Education(X_2):** Education may refer to the level of formal schooling attained/literacy acquired by the respondent at the time of the interview. Education is instrumental in building personality and helps in changing one's behaviour in social life.
- **Family Size(X_3):** It refers to whether the farmer's family is single or joint. To quantify the family type of farmers.
- **Income(X_4):** Annual income is the economic measurement of the farmer's status. Net income from farming in a single year. It was measured in terms of rounded rupees.
- **Size of Land Holding(X_5):** The amount of land owned by a person is an important parameter to access the economic status of the person in society.
- **Homestead Land(X_6):** Amount of land acquired by the home building and surrounding. This total amount is divided by the family size.
- **Land under irrigation(X_7):** Amount of irrigated land acquired by the home building and surrounding. This total amount is divided by the family size.
- **Source of irrigation(X_8):** Irrigation Sources from where water is used for cultivation purposes.

- **Cropping intensity(X):** It has been conceptualizing the proportion of total annual cropped area to the size of holding expressed in percentage. It's calculated as-

  Cropping Intensity = Total annual cropped area/size of holding x 100

- **Depth of groundwater table for tube well (X_{10}):** The depth of the groundwater table for tube well is measured. This is measured by the total no. of pipes and filters used and estimated their length for setting the tube well.

- **Depth of groundwater table for shallow pump(X_{11}):** The depth of the groundwater table for shallow pump is measured. This is measured by the total no. of pipes and filters used and estimated their length for setting the shallow pump.

5.4.2. Dependent Variables

A dependent variable is what you measure in the experiment and what is affected during the experiment. The dependent variable responds to the independent variable. It is called dependent because it "depends" on the independent variable. In a scientific experiment, you cannot have a dependent variable without an independent variable.

Table 5.3: Dependent variables and their notations

Sl. No.	Variables	Notation
1	Disease Severity (1-10 Scale) Stimuli items have been selected through expert rating and quality responses have undergone a data normalisation process	Y_1
2	Impact on Crop (1-10 Scale) (Z transformation)	Y_2
3	Impact on Livestock (1-10 Scale) (Z transformation)	Y_3
4	Impact on Health (1-10 Scale) (Z transformation)	Y_4
5	Expenditure on Treatment (Rs)	Y_5
6	Overall perception on arsenic (1-10 Scale)	Y_6

5.5. Preparation of Interview Schedule

Since the researcher is familiar with the local people and the dialect of the people, an interview schedule has been prepared for the collection of data with the help of researchers and experts of the department. The interview method was selected for the purpose of information.

5.6. Pre-testing of Interview Schedule

It's the process of advance testing of the study design after the schedule has been prepared. The object of pretesting is to detect the discrepancies that have emerged and to remove them after necessary modifications in the schedule. After conducting pretesting appropriate changes and modifications to the interview schedule have been made. The individuals who responded in pretesting have been excluded from the final sample selected for the study.

5.7. Techniques of Field Data Collection:

Relevant data from the farmers were collected based on the personal interview method with the help of an interview schedule. The respondents were personally interviewed during weekends in the month of December 2021 to May 2022. The items were asked in the local language, Bengali.

5.8. Statistical Tools Used for Analysis of Data

The data collected in the interview schedule have been processed and analyzed in accordance with the outline laid down for the purpose at the time of developing the research plan. Processing implies editing, coding, classification and tabulation of collected data.

Table 5.4: Approaches and statistical tools

Sl. No	Approaches	Statistical tools
1	Descriptive visualization	Mean, minimum, maximum, range, Standard deviation, Coefficient of variation
2	Influence and direction analysis	Coefficient of correlation and path analysis
3	Dimension reduction	Factor analysis
4	Functionality	Multiple and stepwise regression analysis
5	Conjoint	Canonical covariate analysis
7	Multilayer perception	Artificial neural networking

5.8.1. Mean

It is a measure of central tendency (or statistical averages) that tells us the point about which items have a tendency to cluster. Such a measure is considered as the most representative figure for the entire mass of data. The measure of central tendency is also known as the statistical average. Mean, median and mode are the most popular averages. Mean, also known as arithmetic average, is the most common measure of central tendency and may be defined as the value, which we get by dividing the total of the values of various given items in a series by the total number of items. We can work it out as follows

$$\text{Mean, } \bar{x} = \frac{\sum x_{ij}}{N}$$

Mean is the simplest measurement of central tendency and is a widely used measure. Its chief use consists in summarizing the essential features of a series and in enabling data to be compared. It is a relatively stable measure of central tendency, but it suffers from some limitations of being unduly affected by extremes; it may not coincide with the actual value of an item in a series, and it may lead to strong impressions, particularly when the item values are not given with the average. However, mean is better than other averages, especially in economic and social studies where direct quantitative measurements are possible.

5.8.2. Standard Deviation

Standard deviation is the most widely used measure of dispersion of a series and is commonly denoted by the symbol σ (pronounced as sigma). Standard deviation is the square root of the arithmetic mean of the square of deviations, the deviations being measured from the arithmetic mean of the distribution. It is less affected by sampling errors and is more stable measure of dispersion. It is worked out as follows,

Standard deviation, $\sigma = \frac{\sqrt{\Sigma(x-\overline{x})}}{N-1}$

5.8.3. Coefficient of Variation

A measure of variation which is independent of the unit of measurement is provided by the coefficient of variation. Being unit free, this is useful for the computation of variability between different populations. The coefficient of variation is the standard deviation expressed as a percentage of the mean and is measured by the formula,

5.8.4. Coefficient of correlation

When an increase or decrease in one variable is accompanied by an increase or decrease in other variables, the two are said to be correlated and the phenomenon is known as correlation. The correlation coefficient (r) is a measure of the relationship between two variables, which are at the interval or ratio level of measurement and are linearly related. A Spearman's coefficient of correlate

The value of 'r' lies between +1 to -1. Positive values of r indicate that positive correlation between the two variables (i.e. change in both variables takes place in the same direction), whereas negative values of 'r' indicate negative correlation i.e. changes in the two variables taking place in the opposite direction. A zero value of 'r' indicates that there is no association between the two variables. When 'r' (+) 1, it indicates a perfect positive correlation and when it is (-) 1, it indicates a perfect negative correlation, meaning thereby that variations in the independent variable (x) explain 100 percent of the variations in the dependent variable (y). We can also say that for a unit change in the independent variable if there happens to be a constant change in the dependent variable in the same direction, the correlation will be termed as perfect positive. But if such change occurs in the opposite direction, the correlation will be termed as perfect negative. The values of 'r' nearer to =1 or -1 indicate a high degree of correlation between the two variables.

5.8.5. Stepwise Multiple Regression

Stepwise regression is a variation of multiple regressions which provides a means of choosing independent variables that in the best prediction possible with the fewest independent variables. It permits the user to solve a sequence of one or more multiple linear regression problems by stepwise application of the least square method. At each step in the analysis, a variable is added or removed which results in the greatest production of the error sum of square (Burroghs corporation, 1975).

According to Drapper and Smith (1981), the method of stepwise multiple regression analysis is to insert variables in turn until the regression equation is satisfactory. The order of insertion is determined by using the partial correlation coefficient as the measure of the importance of variables not yet in the equation.

The program, according to Burrogh's Corporation (1975), first forms a correlation matrix, finds the best predictor (the independent variable having the highest correlation with the criterion variable) and performs a regression analysis with this predictor. Then, the second best predictor (independent), and so on. At any given stage, the group of predictors being used is not necessarily the best group of that size (i.e. the particular group of independent variables does not necessarily have the highest multiple correlations with the criterion that any group of this size does). Rather, this group contains the variables that have the highest individual correlation with the criterion.

Stepwise regression is the step-by-step iterative construction of a regression model that involves the selection of independent variables to be used in a final model. It involves adding or removing potential explanatory variables in succession and testing for statistical significance after each iteration. Stepwise regression is a method that iteratively examines the statistical significance of each independent variable in a linear regression model.

The underlying goal of stepwise regression is, through a series of tests (e.g. F-tests, t-tests) to find a set of independent variables that significantly influence the dependent variable. Stepwise regression can be achieved either by trying out one independent variable at a time and including it in the regression model if it is statistically significant or by including all potential independent variables in the model and eliminating those that are not statistically significant. There are three approaches to stepwise regression

1. **Forward selection-** It begins with no variables in the model, tests each variable as it is added to the model, then keeps those that are deemed most statistically significant—repeating the process until the results are optimal.
2. **Backward elimination-** This starts with a set of independent variables, deleting one at a time, then testing to see if the removed variable is statistically significant.
3. **Bidirectional elimination-** This is a combination of the first two methods that test which variables should be included or excluded.

5.8.6. Path Analysis

The term 'path analysis' was first introduced by the biologist, Sewall Wright in 1934 in connection with decomposing the total correlation between any two variables in a causal system. The technique of path analysis is based on a series of multiple regression analysis with the added assumption of a causal relationship between independent and dependent variables. Path analysis makes use of standardized partial regression coefficients (known as beta weights) as effect coefficients. In linear additive effects are assumed, and then through path analysis, a simple set of equations can be built up showing how each variable depends on preceding variables The main principle of path analysis is that any correlation coefficient between two variables, or a gross or overall measure of empirical relationship can be decomposed into a series of paths: separate paths of influence leading through chronologically intermediate variables to which both the correlated variables have linked.

The merit of path analysis in comparison to correlation analysis is that it makes possible the assessment of the relative influence of each antecedent or explanatory variable on the consequent or criterion variables by first making explicit the assumptions underlying the causal connections and then by elucidating the indirect effect of the explanatory variables.

5.8.7 Canonical Covariate Analysis

A canonical correlation is the correlation of two canonical variables, one representing a set of independent variables, the other a set of dependent variables. Each set may be considered a latent variable based on measured indicator variables in its set. The canonical correlation is optimized such that the linear correlation between the two latent variables is maximized. Canonical correlation is used for many relationships. There may be more than one such linear correlation relating the two sets of variables, with each such correlation representing a different dimension by which the independent set of variables is related to the dependent set. The purpose of canonical correlation is to explain the relation of the two sets of variables not to model the individual variables. In addition to asking how strong the relationship is between two latent variables, canonical correlation is useful in determining how many dimensions are needed to account for that relationship. Canonical correlation finds the linear combination of variables that produces the largest correlation with the second set of variables.

Canonical correlation is a member of the multiple general linear hypothesis (MLGH) family and shares many assumptions of multiple regression such as linearity of relationship, homoscedasticity (same level of relationship for the full range of data), interval or near interval data, untruncated variables, proper specification of the model, lack of high multicollinearity and multivariate normality for purpose of hypothesis testing.

5.8.8. Artificial Neural Network Analysis

ANNs are computing systems inspired by the biological neural network that constitute animal brains. Such systems learn (progressively improve performance on) tasks by considering examples, generally without task specific programming. An ANN is based on a collection of connected units or nodes called artificial neurons (analogous to biological neurons in an animal brain). Each connection (analogous to a synapse) between artificial neurons can transmit a signal from one to another. The artificial neuron that receives the signal can process it and then signal artificial neurons connected to it. In common ANN implementation, the signal at a connection between artificial neurons is a real number, and the output of each artificial neuron is calculated by a non-linear function of the sum of its inputs. Artificial neurons and connections typically have a weight that adjusts as learning proceeds. The weight increases or decreases the strength of the signal at a connection. Artificial neurons may have a threshold such that only if the aggregate signal crosses that threshold is the signal sent. Typically, artificial neurons are organized in layers. Different layers may perform different kinds of transformations on their inputs. Signals travel from the first (input) to the last (output) layer, possibly after traversing the layers multiple times. The original goal of the ANN approach was to solve problems in the same way that a human brain would. Over time, attention focused on matching specific mental abilities, leading to deviations from biology. ANNs have been used on a variety of tasks, including computer vision, speech recognition, machine translation, social network filtering, playing board and video games and medical diagnosis.

6

Ground Research Results Discussion

This chapter deals with the findings of the study along with the discussion based on the analysis. This study tried to represent its findings as per the objectives of the study. At the end of this chapter, interrelation has been made, an explanation has been tried to put out down and an attempt has been done to reveal the cause behind it. The results and their pertaining discussion are presented according to the specific objectives of the study.

Table 6.1: Descriptive Statistics of Variable with respect to Minimum, Maximum, Range, Mean, Standard Deviation of Values, Variance, and Coefficient of Variance

Variable	Minimum	Maximum	Mean	Std. Deviation	Variance	CV (%)
Age(x_1)	28.000	71.000	50.557	10.946	119.816	21.651
Education(x_2)	1.000	15.000	8.929	3.755	14.096	42.050
Family size(x_3)	3.000	9.000	6.229	1.406	1.976	22.569
Income(x_4)	64400.000	399000.000	194549.286	76999.676	5928950035.714	39.578
Size of land holding(x_5)	72.600	313.500	112.719	40.769	1662.109	36.169
Homestead land(x_6)	4.130	12.380	6.979	2.005	4.019	28.725
Land under irrigation(x_7)	68.480	303.600	116.174	48.958	2396.839	42.142
No of source of irrigation(x_8)	2.000	4.000	2.871	0.721	0.519	25.100
Cropping intensity(x_9)	168.800	300.000	233.478	27.493	755.862	11.775
Depth of ground water table for tube well(x_{10})	28.060	35.380	30.666	2.397	5.748	7.818
Depth of groundwater table for shallow pump(x_{11})	19.520	29.280	22.422	2.946	8.676	13.137
Disease severity (y_1)	3.000	9.000	5.400	1.172	1.374	21.706
Impact on crop (y_2)	3.000	7.000	5.129	1.034	1.070	20.171
Impact on livestock (y_3)	2.000	7.000	4.686	1.257	1.581	26.834
Impact on health (y_4)	2.000	7.000	4.886	1.378	1.900	28.211
Expenditure on treatment (y_5)	400.000	1000.000	714.286	139.394	19430.642	19.515
Overall perception on arsenic (y_6)	3.000	8.000	5.471	1.139	1.296	20.809

This table presents the distribution of 11 independent variables(x_1-x_{11}) and 6 dependent variables (y_1-y_6) in terms of their Minimum, Maximum, Mean, Variance, Standard deviation, and Coefficient of variation

Here, it is depicted that the average **age (x_1)** of the respondents is 50.557 and the standard deviation is 10.946. The C.V. value for the variable is 21.651. So, it indicates that the distribution of the variable is highly consistent.

From the table, it is depicted that the average **education (x_2)** of the respondents is 8.929. So, almost on average, the individual education is 9 for every respondent and standard deviation is 3.755. The C.V. value for the variable is 42.050. So, it indicates that the distribution of the variable is highly consistent.

From the table, it is depicted that the mean **family size (x_3)** of the respondents is 6.229. So, almost on an average, the family size is 6 for every respondent. The value of standard deviation is 1.406. The C.V. value for the variable is 22.569. So, it indicates that the distribution of the variable is highly consistent.

From the table, it is depicted that the mean **income (x_4)** of the respondents is 194549.286. So, almost on an average, the income is 195000 for every respondent. The value of standard deviation is 76999.676. The C.V. value for the variable is 39.578. So, it indicates that the distribution of the variable is highly consistent.

Here, it is depicted that the mean **size of land holding (x_5)** of the respondents is 112.719. The value of standard deviation is 40.769. The C.V. value for the variable is 36.169. So, it indicates that the distribution of the variable is highly consistent.

From the table, it is depicted that the mean **homestead land (x_6)** of the respondents is 6.979. So, almost on an average, the homestead land is 7 for every respondent. The value of standard deviation is 2.005. The C.V. value for the variable is 28.725. So, it indicates that the distribution of the variable is highly consistent.

Here, it is depicted that the mean total **irrigated land (x_7)** of the respondents is 116.174. The value of the standard deviation is 48.958. The C.V. value for the variable is 42.142. So, it indicates that the distribution of the variable is highly consistent.

From the table, it is depicted that the mean **number of irrigation sources (x_8)** of the respondents is 2.871. So, almost on average, the irrigation source is 3 for every respondent. The value of standard deviation is 0.721. The C.V. value for the variable is 25.100. So, it indicates that the distribution of the variable is highly consistent.

From the table, it is depicted that the mean **cropping intensity (x_9)** of the respondents is 233.476. The value of standard deviation is 27.493. The C.V. value for the variable is 11.775. So, it indicates that the distribution of the variable is highly consistent.

From the table, it is depicted that the mean **depth of the groundwater table for tube well (x_{10})** of the respondents is 30.666. So, almost on an average, the depth is 31 for every respondent. The value of standard deviation is 2.397. The C.V. value for the variable is 7.818. So, it indicates that the distribution of the variable is highly consistent.

From the table, it is depicted that the mean **depth of groundwater table for shallow pump (x_{11})** of the respondents is 22.422. So, almost on an average the depth is 22 for every respondent. The value of standard deviation is 2.946. The C.V. value for the variable is 13.137. So, it indicates that the distribution of the variable is highly consistent.

From the table, it is depicted that the mean **disease severity (y_1)** of the respondents is 5.400. So, almost on an average the disease severity is 5 for every respondent. The value of standard deviation is 1.172. The C.V. value for the variable is 21.706. So, it indicates that the distribution of the variable is highly consistent.

From the table, it is depicted that the mean **impact on crop (y_2)** of the respondents is 5.129. So, almost on an average impact on crop is 5 for every respondent. The value of standard deviation is 1.034. The C.V. value for the variable is 20.171. So, it indicates that the distribution of the variable is highly consistent.

From the table, it is depicted that the mean **impact on livestock (y_3)** of the respondents is 4.686. So, almost on an average the impact on livestock is 5 for every respondent. The value of standard deviation is 1.257. The C.V. value for the variable is 26.834. So, it indicates that the distribution of the variable is highly consistent.

From the table, it is depicted that the mean **impact on health (y_4)** of the respondents is 4.886. So, almost on an average the impact on health is 5 for every respondent. The value of standard deviation is 1.378. The C.V. value for the variable is 28.211. So, it indicates that the distribution of the variable is highly consistent.

Here, it is depicted that the mean **expenditure on treatment (y_5)** of the respondents is 714.286. The value of standard deviation is 139.394. The C.V. value for the variable is 19.515. So, it indicates that the distribution of the variable is highly consistent.

From the table it is depicted that the mean **overall perception on arsenic (y_6)** of the respondents is 5.471. So, almost on an average the perception on arsenic is 5.5 for every respondent. The value of standard deviation is 1.139. The C.V. value for the variable is 20.809. So, it indicates that the distribution of the variable is highly consistent.

Table 6.2: Coefficient of Correlation (r): Disease severity (y_1) Vs. 11 Independent Variables (x_1-x_{11})

Sl. No.	Independent Variables	'r' Value	Remarks
1	Age(x_1)	0.390	**
2	Education(x_2)	0.366	**
3	Family size(x_3)	-0.004	
4	Income(x_4)	0.359	**
5	Size of land holding(x_5)	0.512	**
6	Homestead land(x_6)	0.230	
7	Land under irrigation(x_7)	0.295	*
8	No of source of irrigation(x_8)	0.268	*
9	Cropping intensity(x_9)	0.383	**
10	Depth of ground water table for tube well(x_{10})	-0.128	
11	Depth of groundwater table for shallow pump(x_{11})	0.297	*

**Correlation is significant at the 0.01 level

*Correlation is significant at the 0.05 level

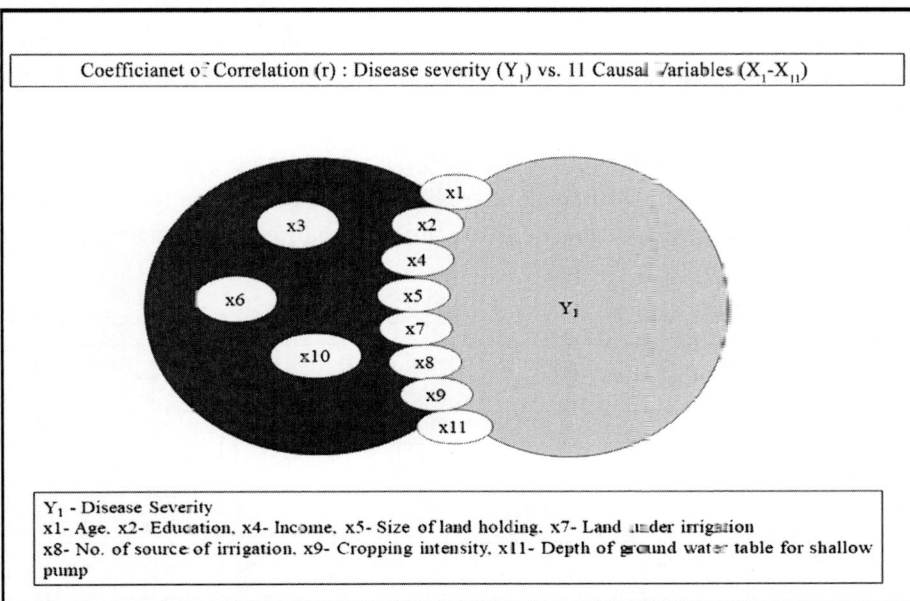

Fig. 6.1: Coefficient of Correlation (r): Disease severity (y_1) Vs. 11 Independent Variables (x_1-x_{11})

Results

The table presents the coefficient of correlation between Disease severity(y_1) and 11 independent variables(x_1-x_{11}). It has been found that the following variables viz. Age(x_1), Education(x_2), Income(x_4), Size of land holding(x_5), Land under irrigation(x_7), No of source of irrigation(x_8), Cropping intensity(x_9) and Depth of groundwater table for shallow pump(x_{11}) have recorded significant correlation with Disease Severity(y_1).

Revelation

The higher age group of the respondents are more vulnerable to the severity of the disease. Education indicates that the respondents falling victim to arsenic contamination are also having higher education level. Similar cases have also occurred to the respondents having higher income level. It may be due to the fact that those who are intensively engaged in agriculture enterprises, are more prone to arsenic contamination.

The higher size of land holding implicates wider and intensive involvement in an agricultural operation. Marginal farmers of agricultural wage labourers for their survival they need to migrate. On the other way atleast for a certain span of time, they are away from the danger zone. On the other hand, big farmers with large size of land holding are stuck of in their own village. So, they are more vulnerable to arsenic contamination. Groundwater itself with the source of arsenic pollution. That is how more is the hectors under irrigated agro-ecosystem in the part of West Bengal, the worst has been the burnt on arsenic contamination on human health, agriculture and livestock.

When the frequency of irrigation is growing up the arsenic contamination also increases. If the cropping intensity increases arsenic contamination also grows up. However, the depth of groundwater table for shallow pump has recorded a positive correlation.

Table 6.3: Multiple Regression Analysis: Disease severity (y_1) vs. 11 Causal Variables (x_1-x_{11})

Sl. No.	Variables	Reg. Coef. B	S.E. B	Beta	t Value
1	Age(x_1)	0.618	0.169	0.618	3.658
2	Education(x_2)	0.104	0.123	0.104	0.848
3	Family size(x_3)	-0.236	0.104	-0.236	-2.277
4	Income(x_4)	0.207	0.130	0.207	1.594
5	Size of land holding(x_5)	0.210	0.123	0.210	1.708
6	Homestead land(x_6)	-0.109	0.128	-0.109	-0.848
7	Land under irrigation(x_7)	0.259	0.122	0.259	2.122
8	No of source of irrigation(x_8)	0.013	0.125	0.013	0.100
9	Cropping intensity(x_9)	0.318	0.098	0.318	3.233

Sl. No.	Variables	Reg. Coef. B	S.E. B	Beta	t Value
10	Depth of ground water table for tube well(x_{10})	-0.005	0.112	-0.005	-0.040
11	Depth of groundwater table for shallow pump(x_{11})	-0.189	0.153	-0.189	-1.233

R square: 58.30%

The standard error of the estimate: 0.705

Result

In multiple regression analysis (full model) found that age is the strongest determinant in estimating disease severity(y_1). The highest beta value is **0.618.**

Revelation

The table presents the multiple regression analysis between Disease severity(y_1) and 11 causal variables(x_1-x_{11}). The full model of multiple regression analysis portraits with the combination of 11 causal variables, 58.30% variance in Disease severity(y_1) has been explained.

Table 6.4: Stepwise Regression Analysis: Disease severity (y_1) vs. 11 Causal Variables (x_1-x_{11})

Sl. No	Variables	Reg. coef. B	S.E. B	Beta	t value
1	Size of land holding(x_5)	0.425	0.091	0.425	4.689
2	Cropping intensity(x_9)	0.347	0.089	0.347	3.878
3	Age(x_1)	0.325	0.090	0.325	3.603

R square: 47.70%

The standard error of the estimate: 0.739

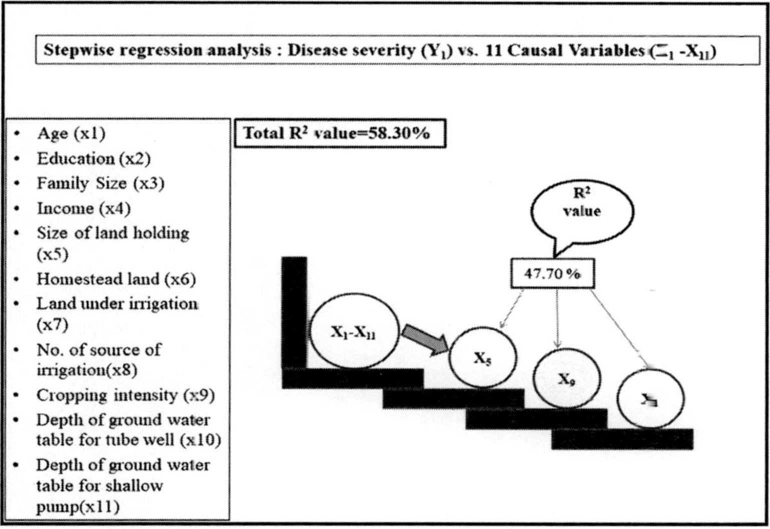

Fig. 6.2: Stepwise Regression Analysis: Disease severity (y_1) vs. 11 Causal Variables (x_1-x_{11})

Results

The table isolates the causal dominant variables contributing to the level of Disease severity (y_1). It has been found that Size of land holding(x_5), Cropping intensity(x_9) and Age(x_1) were written at the last step and have substantially contributed to the consequent variable Disease severity(y_1). These three variables together have contributed 47.70% of the variance.

Revelation

The impact of arsenic contamination has been dominant in the cases where the size of land holding has gone bigger. So, the severity of arsenic contamination has been precarious for large farmers. Higher is the cropping intensity, more has been the number of crops that acts as the sink for arsenic contamination. So, for the high intensity cropping the impact of arsenic has become precarious. Age has been written because the most vulnerable echelon of the population has been the higher age category. Among the 11 causal variables, the following variables size of land holding(x_5), cropping intensity(x_9) and age(x_1) were found most significant with the consequent variable disease severity (y_1). Those three variables are contributing 81.81% of the total 58.30% of the variance.

Table 6.5: Path Analysis: Decomposition of Total Effect into Direct, Indirect and Residual Effect: Disease severity (y_1) vs. 11 exogenous variables (x_1-x_{11})

Sl.No	Variables	Total Effect	Direct Effect	Indirect Effect	Highest Indirect Effect
1	Age(x_1)	0.390	0.615	-0.225	-0.14(x11)
2	Education(x_2)	0.366	0.104	0.262	0.221 (x1)
3	Family size(x_3)	-0.004	-0.236	0.232	0.236 (x1)
4	Income(x_4)	0.359	0.206	0.153	0.095 (x5)
5	Size of land holding(x_5)	0.512	0.210	0.302	0.137 (x7)
6	Homestead land(x_6)	0.230	-0.108	0.338	0.146 (x4)
7	Land under irrigation(x_7)	0.295	0.258	0.037	-0.119(x1)
8	No of source of irrigation(x_8)	0.268	0.013	0.255	0.173(x1)
9	Cropping intensity(x_9)	0.383	0.318	0.065	0.041(x4)
10	Depth of ground water table for tube well(x_{10})	-0.128	-0.005	-0.123	0.22 (x1)
11	Depth of groundwater table for shallow pump(x_{11})	0.297	-0.186	0.483	0.462(x1)

Residual effect: 0.497

Highest indirect individual effect: x_1 (6

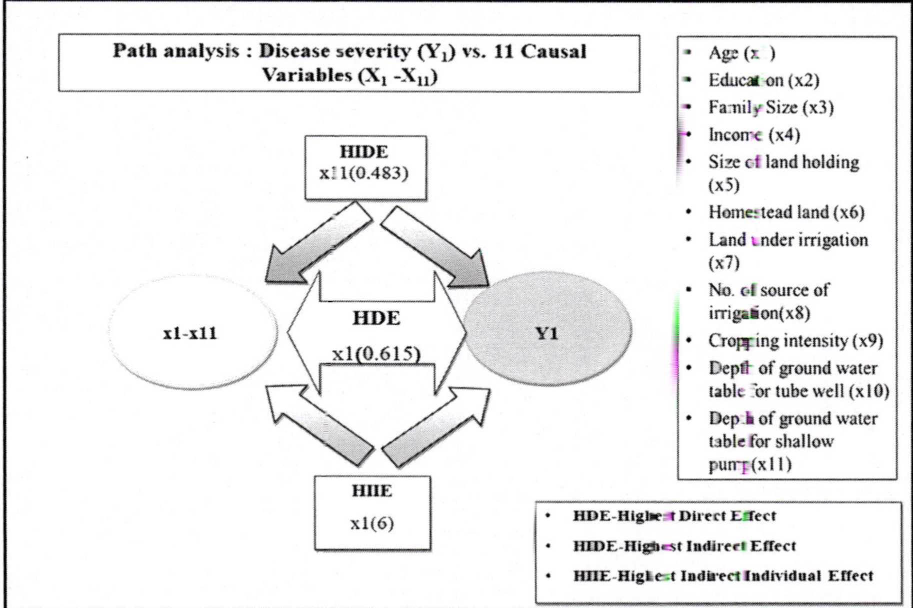

Fig. 6.3: Path Analysis: Decomposition of Total Effect into Direct, Indirect and Residual Effect: Disease severity (y_1) vs. 11 exogenous variables (x_1-x_{11})

Results

The table represents the path analysis by decomposing the total effect (r-value) into direct, indirect and residual effects. It has been found that the highest direct effect is exerted by the causal variable Age(x_1) and the highest indirect effect is exerted by the causal variable Depth of groundwater table for shallow pump(x_{11}). Interestingly the variable age(x_1) has routed the highest indirect effect of 6 other causal variables to ultimately characterised disease severity(y_1).

Revelation

Age has come out as the most influencing exogeneous variable to estimate the severity of diseases due to arsenic contamination. So, there should be an age specific intervention plan to deal with arsenic contamination and its precarious impact on health.

Due to some geo morphological factors, the quality and quantity of water retained in the sub lithospheric layer is exerting a discernible impact on disease severity.

If groundwater depletion grows high, the solubility of arsenic will be simmered. The anaerobic condition in the groundwater raising enhances the transformation of recalcitrant arsenic into labile one. So, we need to have long-term planning to deal with the arsenic problems by managing groundwater with proper technological and scientific intervention.

Table 6.6: Coefficient of Correlation (r): Impact on crop (y_2) Vs. 11 Independent Variables (x_1-x_{11})

Sl. No.	Independent Variables	'r' Value	Remarks
1	Age(x_1)	0.303	*
2	Education(x_2)	0.234	
3	Family size(x_3)	-0.050	
4	Income(x_4)	0.358	**
5	Size of land holding(x_5)	0.429	**
6	Homestead land(x_6)	0.201	
7	Land under irrigation(x_7)	0.200	
8	No of source of irrigation(x_8)	0.236	*
9	Cropping intensity(x_9)	0.398	**
10	Depth of ground water table for tube well(x_{10})	-0.265	*
11	Depth of groundwater table for shallow pump(x_{11})	0.253	*

**Correlation is significant at the 0.01 level

*Correlation is significant at the 0.05 level

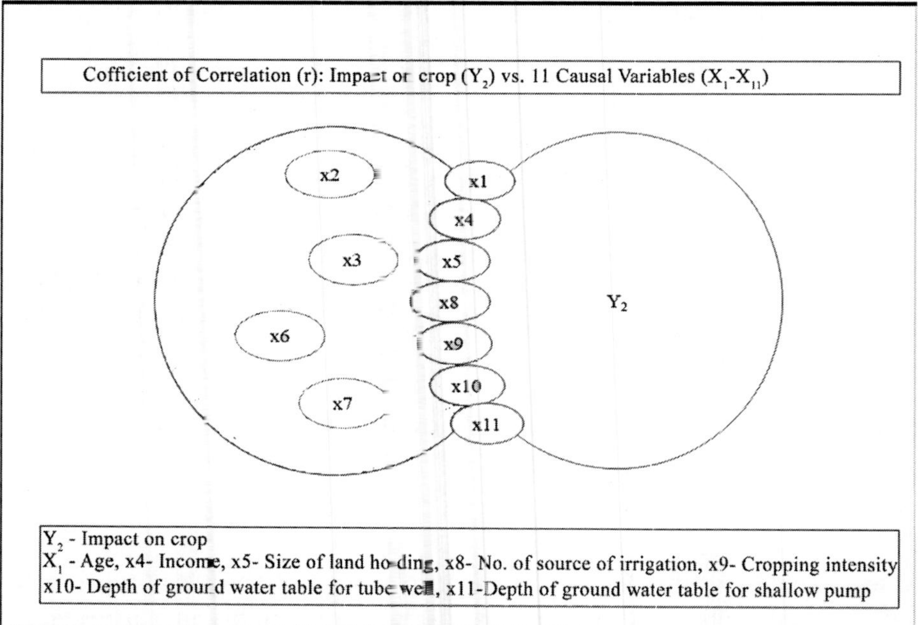

Fig. 6.4: Coefficient of Correlation (r): Impact on crop (y_2) Vs. 11 Independent Variables (x_1-x_{11})

Results: The table presents the coefficient of correlation between Impact on crop(y_2) and 11 independent variables(x_1-x_{11}). It has been found that the following variables viz. age(x_1), income(x_4), size of land holding(x_5), no. of source of irrigation(x_8), cropping intensity(x_9), depth of groundwater table(x_{10}) and depth of groundwater table for shallow pump(x_{11}) have recorded significant correlation with impact on crop(y_2).

Revelation: The higher age group of the respondents have recorded a positive correlation with the impact on crop of arsenic. Similar cases have also occurred in the respondents having higher income level. It may be due to the fact that those who are intensively engaged more time in the crop field, are more prone to arsenic contamination. The higher size of land holding implicates wider and intensive involvement in agricultural operation. That's why they are more vulnerable to arsenic contamination. Groundwater itself with the source of arsenic pollution. That is how more is the hectors under irrigated agro-ecosystem in the part of West Bengal, the worst has been the burnt on arsenic contamination on human health, agriculture and livestock. When the frequency of number of irrigation grows up the arsenic contamination on crops also increases. If the cropping intensity increases arsenic contamination in crops also grows up. The depth of the groundwater table for tube well decreases, the impact of crops on arsenic grows up because in the upper layer of groundwater the volume of arsenic is more. However, the depth of groundwater table for shallow pump has recorded a positive correlation with the impact of arsenic on crops.

Table 6.7: Multiple Regression Analysis: Impact on crop (y_2) Vs. 11 Causal Variables (x_1-x_{11})

Sl. No.	Variables	Reg. Coef. B	S.E. B	Beta	t Value
1	Age(x_1)	0.525	0.189	0.525	2.783
2	Education(x_2)	-0.009	0.137	-0.009	-0.066
3	Family size(x_3)	-0.201	0.116	-0.201	-1.736
4	Income(x_4)	0.211	0.145	0.211	1.457
5	Size of land holding(x_5)	0.167	0.137	0.167	1.214
6	Homestead land(x_6)	-0.110	0.143	-0.110	-0.768
7	Land under irrigation(x_7)	0.096	0.136	0.096	0.705
8	No of source of irrigation(x_8)	0.143	0.139	0.143	1.025
9	Cropping intensity(x_9)	0.282	0.110	0.282	2.570
10	Depth of ground water table for tube well(x_{10})	-0.207	0.125	-0.207	-1.652
11	Depth of groundwater table for shallow pump(x_{11})	-0.130	0.171	-0.130	-0.759

R square: 48.00%

The standard error of the estimate: 0.787

Results

In multiple regression analysis (full model) found that age is the strongest determinant in estimating the impact on crop(y_2). The highest beta value is 0.525.

Revelation

The table presents the multiple regression analysis between the impact on the crop(y_2) and 11 causal variables(x_1-x_{11}). The full model of multiple regression analysis portraits with the combination of 11 causal variables, 48.00% variance in impact on crop (y_2) has been explained.

Table 6.8. Stepwise Regression Analysis: Impact on crop (y_2) Vs. 11 Causal Variables (x_1-x_{11})

Sl. No	Variables	Reg. coef. B	S.E. B	Beta	t value
1	Size of land holding(x_5)	0.353	0.099	0.353	3.554
2	Cropping intensity(x_9)	0.368	0.098	0.368	3.761
3	Age(x_1)	0.251	0.099	0.251	2.538

R square: 37.40%

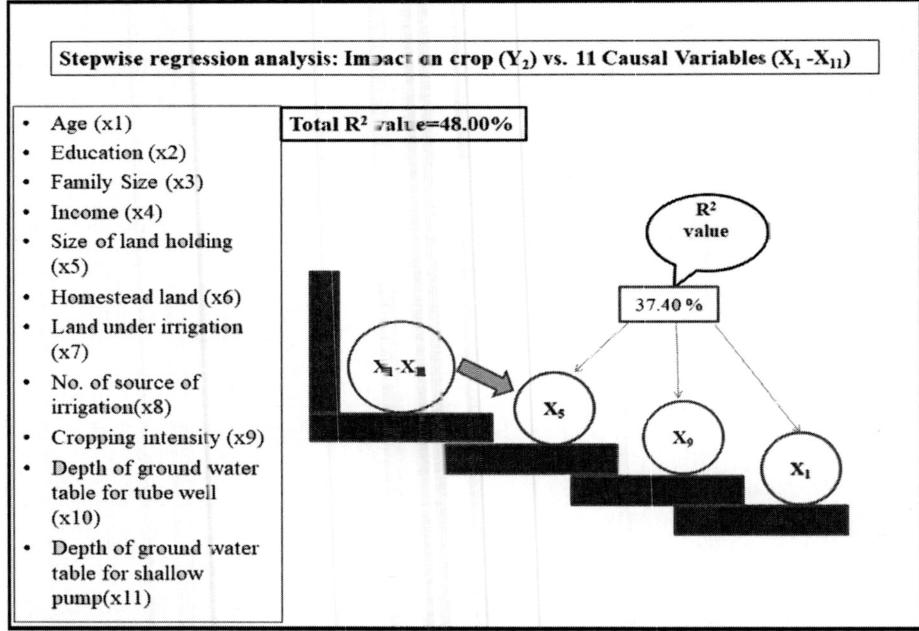

Fig. 6.5: Stepwise Regression Analysis: Impact on crop (y_2) Vs. 11 Causal Variables (x_1-x_{11})

Results

The table isolates the causal dominant variables contributing to the level of Impact on crop (y_2). It has been found that Size of land holding(x_5), Cropping intensity(x_9) and Age(x_1) were written at the last step and have substantially contributed to the consequent variable Impact on crop(y_2). These three variables together have contributed 37.40% of variance.

Revelation

The impact of arsenic contamination has been dominant in the cases where the size of land holding has gone bigger. So, the impact on crops of arsenic contamination has been precarious for farmers. Higher cropping intensity more has been the number of crops which acts as the sink for arsenic contamination. So, for high intensity cropping the impact of arsenic on crop has become precarious. Age has been written because the most vulnerable echelon of the population has been the higher age category. Among the 11 causal variables, the following variables size of land holding(x_5), cropping intensity(x_9) and age(x_1) were found most significant with the consequent variable impact on crop (y_2). Those three variables are contributing 77.92% of the total 48% of variance.

Table: 6.9. Path Analysis: Decomposition of Total Effect into Direct, Indirect and Residual Effect: Impact on crop (y_2) vs. 11 exogenous variables (x_1-x_{11})

Sl.No	Variables	Total Effect	Direct Effect	Indirect Effect	Highest Indirect Effect
1	Age(x_1)	0.303	**0.523**	-0.220	-0.096 (x11)
2	Education(x_2)	0.234	-0.009	0.243	0.188 (x1)
3	Family size(x_3)	-0.050	-0.200	0.150	0.2 (x1)
4	Income(x_4)	0.358	0.212	0.146	-0.078 (x6)
5	Size of land holding(x_5)	0.429	0.166	0.263	0.096 (x4)
6	Homestead land(x_6)	0.201	-0.110	0.311	0.15 (x4)
7	Land under irrigation(x_7)	0.200	0.097	0.103	-0.101 (x1)
8	No of source of irrigation(x_8)	0.236	0.142	0.094	0.147 (x1)
9	Cropping intensity(x_9)	0.398	0.282	0.116	0.064 (x10)
10	Depth of ground water table for tube well(x_{10})	-0.265	-0.207	-0.058	0.187 (x1)
11	Depth of groundwater table for shallow pump(x_{11})	0.253	-0.128	**0.381**	0.393 (x1)

Residual effect: 0.521

Highest indirect individual effect: x_1(6)

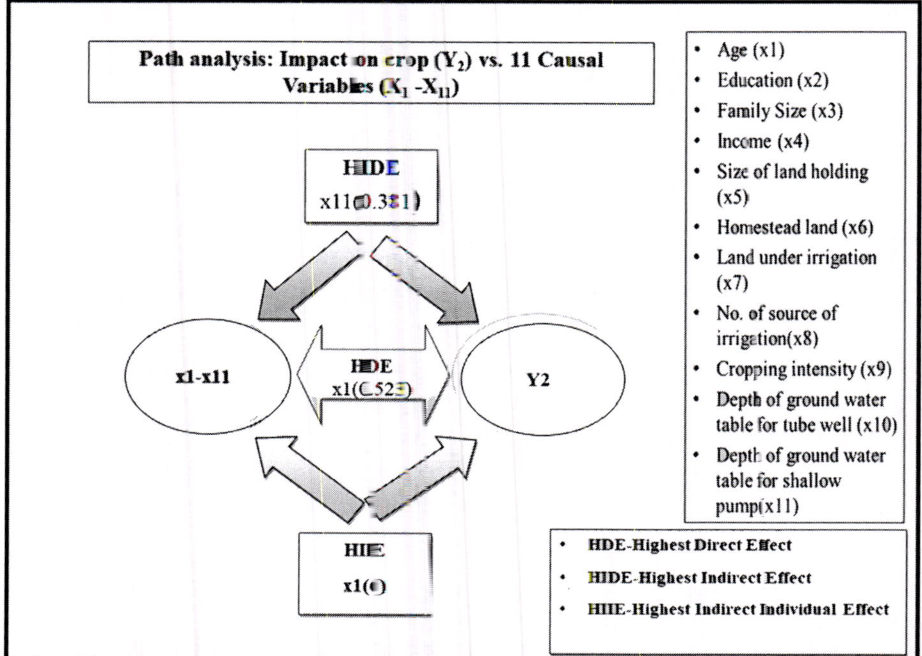

Fig. 6.6: Path Analysis: Decomposition of Total Effect into Direct, Indirect and Residual Effect: Impact on crop (y_2) vs. 11 exogenous variables (x_1-x_{11})

Results

The table represents the path analysis by decomposing the total effect (r-value) into direct, indirect and residual effects. It has been found that the highest direct effect is exerted by the causal variable Age(x_1) and the highest indirect effect is exerted by causal variable Depth of groundwater table for shallow pump(x_{11}). Interestingly the variable Age(x_1) has routed the highest indirect effect of 6 other causal variables to ultimately characterised Impact on the crop (y_2).

Revelation

Age has come out as the most influencing exogeneous variable to estimate the impact on crops due to arsenic contamination. So, there should be an age specific intervention plan to deal with arsenic contamination and its precarious impact on the crop.

Due to some geo morphological factors, the quality and quantity of water retained in the sub lithospheric layer is exerting discernible impact on crops.

If groundwater depletion grows high, the solubility of arsenic will be simmered. The anaerobic condition in the groundwater raising enhances the transformation of recalcitrant arsenic into labile one. So, we need to have a long-term plan to deal with the arsenic problem by managing groundwater with proper technological and scientific intervention.

Table:6.10: Coefficient of Correlation (r): Impact on livestock (y_3) Vs. 11 Independent Variables (x_1-x_{11})

Sl. No.	Independent Variables	'r' Value	Remarks
1	Age(x_1)	-0.460	**
2	Education(x_2)	-0.290	*
3	Family size(x_3)	0.000	
4	Income(x_4)	-0.312	**
5	Size of land holding(x_5)	-0.455	**
6	Homestead land(x_6)	-0.185	
7	Land under irrigation(x_7)	-0.108	
8	No of source of irrigation(x_8)	-0.221	
9	Cropping intensity(x_9)	-0.356	**
10	Depth of ground water table for tube well(x_{10})	0.135	
11	Depth of groundwater table for shallow pump(x_{11})	-0.361	**

**Correlation is significant at the 0.01 level

*Correlation is significant at the 0.05 level

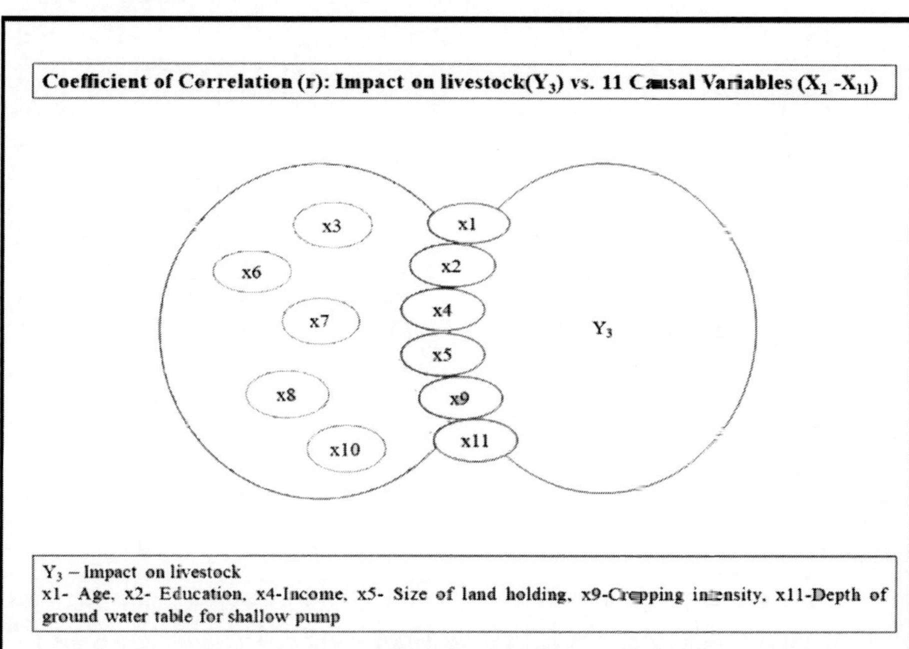

Fig. 6.7: Coefficient of Correlation (r): Impact on livestock (y_3) Vs. 11 Independent Variables (x_1-x_{11})

Results: The table presents the coefficient of correlation between Impact on livestock (y_3) and 11 independent variables(x_1-x_{11}). It has been found that the following variables viz. Age(x_1), Education(x_2), Income(x_4), Size of land holding(x_5), Cropping intensity(x_9) and Depth of groundwater table for shallow pump(x_{11}) have recorded significant correlation with Impact on livestock(y_3).

Revelation: The livestock population with lower age level have been more prone to arsenic contamination. The young livestock are extensively exposed to the threats and bites of arsenic contamination. Similar cases have also occurred in the respondents having lower income level. Those who have less land holding, have more livestock to run their family. So, livestock contamination is high. If the cropping intensity decreases, arsenic contamination on livestock also grows up. The depth groundwater table has become a strong determinant in estimating the impact and intensity of arsenic contamination. The area of study has been unique for intensive depletion of groundwater, high intensity cropping and density of population per unit area. All of these are correlated to invite the problem of arsenic contamination and contributed to the impact on livestock on arsenic.

Table 6.11. Multiple Regression Analysis: Impact on livestock (y_3) Vs. 11 Causal Variables (x_1-x_{11})

Sl. No.	Variables	Reg. Coef. B	S.E. B	Beta	t Value
1	Age(x_1)	-0.600	0.178	-0.600	-3.370
2	Education(x_2)	-0.063	0.129	-0.063	-0.489
3	Family size(x_3)	0.203	0.109	0.203	1.857
4	Income(x_4)	-0.177	0.137	-0.177	-1.293
5	Size of land holding(x_5)	-0.258	0.130	-0.258	-1.988
6	Homestead land(x_6)	0.104	0.135	0.104	0.769
7	Land under irrigation(x_7)	-0.019	0.129	-0.019	-0.149
8	No of source of irrigation(x_8)	-0.030	0.131	-0.030	-0.231
9	Cropping intensity(x_9)	-0.271	0.104	-0.271	-2.619
10	Depth of ground water table for tube well(x_{10})	0.120	0.118	0.120	1.010
11	Depth of groundwater table for shallow pump(x_{11})	0.098	0.162	0.098	0.605

R square: 53.60%

The standard error of the estimate: 0.743

Results

In multiple regression analysis (full model) found that age is the strongest determinant in estimating the impact on livestock(y_3). The highest beta value is 0.600.

Revelation

The table presents the multiple regression analysis between the impact on livestock(y_3) and 11 causal variables(x_1-x_{11}). The full model of multiple regression analysis portraits with the combination of 11 causal variables, 53.60% variance in Impact on livestock (y_3) has been explained.

Table 6.12: Stepwise Regression Analysis: Impact on livestock (y_3) vs. 11 Causal Variables(x_1-x_{11})

Sl. No	Variables	Reg. coef. B	S.E. B	Beta	t value
1	Age(x_1)	-0.485	0.096	-0.485	-5.048
2	Size of land holding(x_5)	-0.341	0.089	-0.341	-3.823
3	Cropping intensity(x_9)	-0.348	0.088	-0.348	-3.968
4	Family size(x_3)	0.199	0.095	0.199	2.099

R square: 50.60%

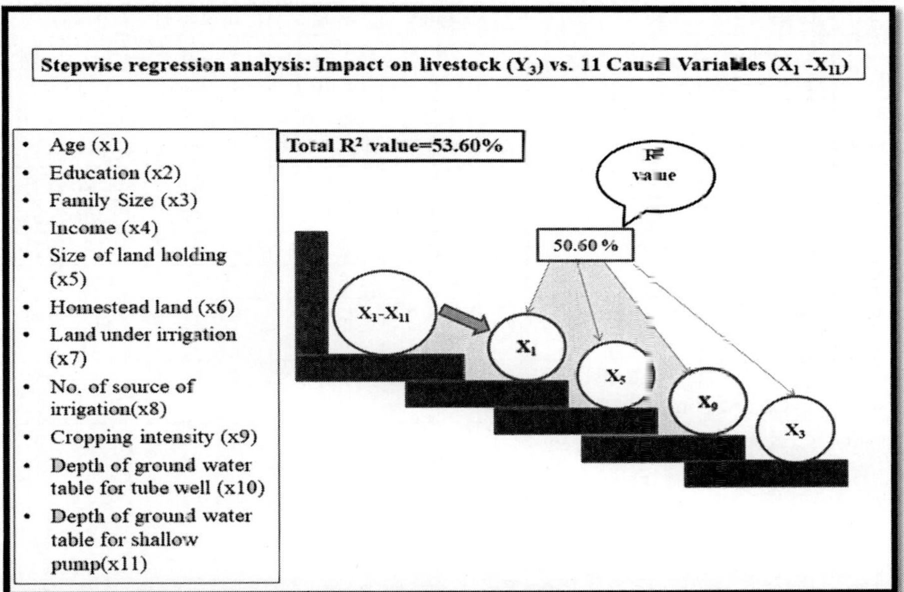

Fig. 6.8: Stepwise Regression Analysis: Impact on livestock (y_3) vs. 11 Causal Variables(x_1-x_{11})

Results

The table isolates the causal dominant variables contributing to the level of Impact on livestock(y_3). It has been found that Age(x_1), Size of land holding(x_5), Cropping intensity(x_9) and Family size(x_3) were written at the last step and have substantially contributed to the consequent variable Impact on livestock(y_3). These four variables together have contributed 50.60% of variance.

Revelation

Among the 11 causal variables, the following variables age(x_1), size of land holding(x_5), cropping intensity(x_9) and family size(x_3) were found most significant with the consequent variable impact on livestock (y_3). Those four variables are contributing 94.40% of the total 53.60% of variance. So, while dealing with the problem of arsenicosis suffer by the ill-fated responses, proper intervention should be made at the proper age. Farmers also should be aware of the size of land holding, cropping intensity and size of their family.

Table 6.13: Path Analysis: Decomposition of Total Effect into Direct, Indirect and Residual Effect: Impact on livestock (y_3) vs. 11 exogenous variables (x_1-x_{11})

Sl.No	Variables	Total Effect	Direct Effect	Indirect Effect	Highest Indirect Effect
1	Age(x_1)	**-0.460**	**-0.600**	0.140	0.078 (x3)
2	Education(x_2)	-0.290	-0.061	-0.229	-0.215 (x1)
3	Family size(x_3)	0.000	0.202	-0.202	-0.23 (x1)
4	Income(x_4)	-0.312	-0.178	-0.134	-0.116 (x5)
5	Size of land holding(x_5)	-0.455	-0.257	-0.198	-0.099 (x1)
6	Homestead land(x_6)	-0.185	0.104	-0.289	-0.126 (x4)
7	Land under irrigation(x_7)	-0.108	-0.020	-0.088	-0.136 (x5)
8	No of source of irrigation(x_8)	-0.221	-0.031	-0.190	-0.169 (x1)
9	Cropping intensity(x_9)	-0.366	-0.272	-0.094	-0.037 (x10)
10	Depth of ground water table for tube well(x_{10})	0.135	0.119	0.016	-0.215 (x1)
11	Depth of groundwater table for shallow pump(x_{11})	-0.361	0.098	**-0.459**	-0.451 (x1)

Residual effect: 0.464

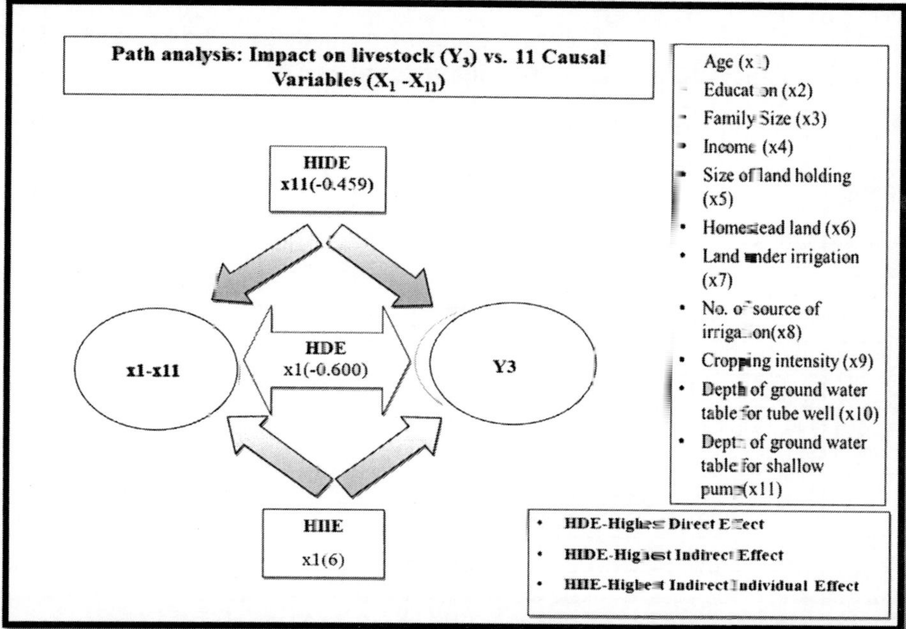

Fig. 6.9: Path Analysis: Decomposition of Total Effect into Direct, Indirect and Residual Effect: Impact on livestock (y_3) vs. 11 exogenous variables (x_1-x_{11})

Results

The table presents the path analysis by decomposing the total effect (r-value) into direct, indirect and residual effects. It has been found that the highest direct effect is exerted by the causal variable Age(x_1) and the highest indirect effect exerted by the causal variable Depth of groundwater table for shallow pump(x_{11}). Interestingly the variable Age(x_1) has routed the highest indirect effect of 6 other causal variables to ultimately charactersed Impact on livestock (y_3).

Revelation

Age has come out as the most influencing exogeneous variable to estimate the impact on livestock due to arsenic contamination. So, there should be an age specific intervention plan to deal with arsenic contamination and its precarious impact on the health of livestock.

Due to some geo morphological factors, the quality and quantity of water retained in the sub lithospheric layer is exerting discernible impact on livestock.

If groundwater depletion grows high, the solubility of arsenic will be simmered. The anaerobic condition in the groundwater raising enhances the transformation of recalcitrant arsenic into labile one. So, we need to have long-term plan to deal with the arsenic problem by managing groundwater with proper technological and scientific intervention.

Table 6.14: Coefficient of Correlation (r): Impact on health (y_4) Vs. 11 Independent Variables (x_1-x_{11})

Sl. No.	Independent Variables	'r' Value	Remarks
1	Age(x_1)	0.243	*
2	Education(x_2)	0.253	*
3	Family size(x_3)	-0.106	
4	Income(x_4)	0.414	**
5	Size of land holding(x_5)	0.400	**
6	Homestead land(x_6)	0.219	
7	Land under irrigation(x_7)	0.186	
8	No of source of irrigation(x_8)	0.204	
9	Cropping intensity(x_9)	0.431	**
10	Depth of ground water table for tube well(x_{10})	-0.321	**
11	Depth of groundwater table for shallow pump(x_{11})	0.205	

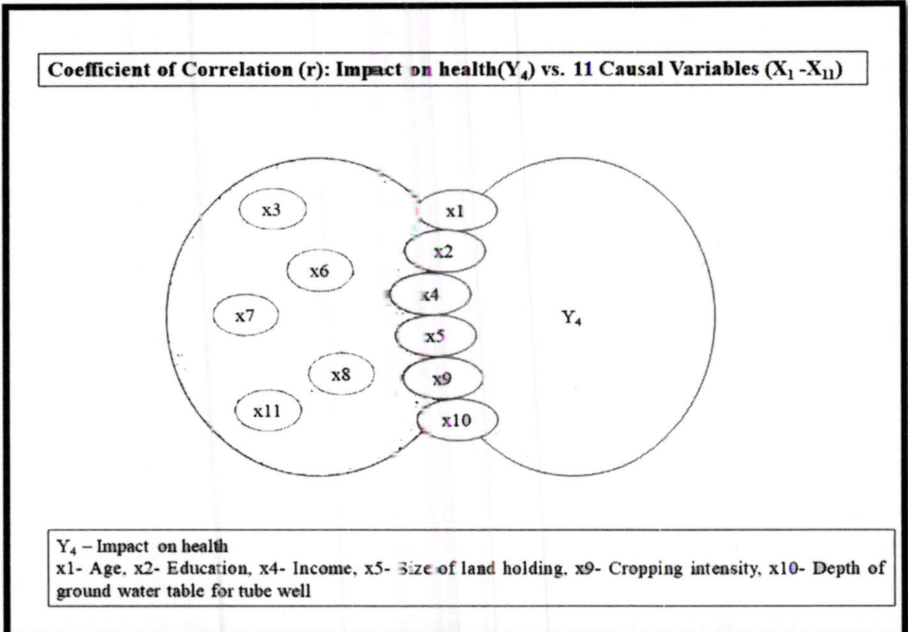

Fig. 6.10: Coefficient of Correlation (r): Impact on health (y_4) Vs. 11 Independent Variables (x_1-x_{11})

Results

The table presents the coefficient of correlation between Impact on health(y_4) and 11 independent variables(x_1-x_{11}). It has been found that the following variables viz. Age(x_1), Education(x_2), Income(x_4), Size of land holding(x_5), Cropping intensity(x_9) and Depth of groundwater table for tube well(x_{10}) have recorded significant correlation with Impact on health(y_4).

Revelation

The higher age group of the respondents are more vulnerable to the impact of arsenic on health. Those who have higher age are more affected by arsenic. Education indicates that the respondents falling victim to arsenic contamination are also having higher education level. Similar cases have also occurred to the respondents having higher income level. It may be due to the fact that those who are intensively engaged in agriculture enterprises, are more prone to arsenic contamination. So, the impact of arsenic on health is also higher. The higher size of land holding implicates wider and intensive involvement in agricultural operation. That's why they are more affected by arsenic contamination and impact on health is higher. If the cropping intensity increases, the impact of arsenic contamination on health also grows up. However, the depth of the groundwater table for tube well has recorded a negative correlation with the impact of arsenic on health. The depth of groundwater table has become strong determinant in estimating the impact and intensity of arsenic contamination. The area of study has been unique for intensive depletion of groundwater, high intensity cropping and density of population per unit area.

Table 6.15: Multiple Regression Analysis: Impact on health (y_4) vs. 11 Causal Variables (x_1-x_{11})

Sl. No.	Variables	Reg. Coef. B	S.E. B	Beta	t Value
1	Age(x_1)	0.484	0.184	0.484	2.634
2	Education(x_2)	0.098	0.133	0.098	0.738
3	Family size(x_3)	-0.234	0.113	-0.234	-2.078
4	Income(x_4)	0.279	0.141	0.279	1.976
5	Size of land holding(x_5)	0.114	0.134	0.114	0.854
6	Homestead land(x_6)	-0.121	0.139	-0.121	-0.872
7	Land under irrigation(x_7)	0.066	0.133	0.066	0.498
8	No of source of irrigation(x_8)	0.075	0.136	0.075	0.551
9	Cropping intensity(x_9)	0.276	0.107	0.276	2.583
10	Depth of ground water table for tube well(x_{10})	-0.264	0.122	-0.264	-2.162
11	Depth of groundwater table for shallow pump(x_{11})	-0.127	0.1[illegible]7	-0.127	-0.764

R square: 50.70%

The standard error of the estimate: 0.766

Result

In multiple regression analysis (full model) found that age is the strongest determinant in estimating impact on livestock(y_4). The highest beta value is 0.484.

Revelation

The table presents the multiple regression analysis between Impact on health(y_4) and 11 causal variables(x_1-x_{11}). The full model of multiple regression analysis portraits with the combination of 11 causal variables, 50.70% variance in Impact on health(y_4) has been explained.

Table 6.16: Stepwise Regression Analysis: Impact on health (y_4) Vs. 11 Causal Variables (x_1-x_{11})

Sl. No	Variables	Reg. coef. B	S.E. B	Beta	t value
1	Cropping intensity(x_9)	0.309	0.098	0.309	3.164
2	Income(x_4)	0.272	0.098	0.272	2.783
3	Age(x_1)	0.444	0.106	0.444	4.203
4	Depth of ground water table for tube well(x_{10})	-0.272	0.107	-0.272	-2.539
5	Family size(x_3)	-0.207	0.100	-0.207	-2.057

R square: 46.30%

The standard error of the estimate: 0.7[illegible]1

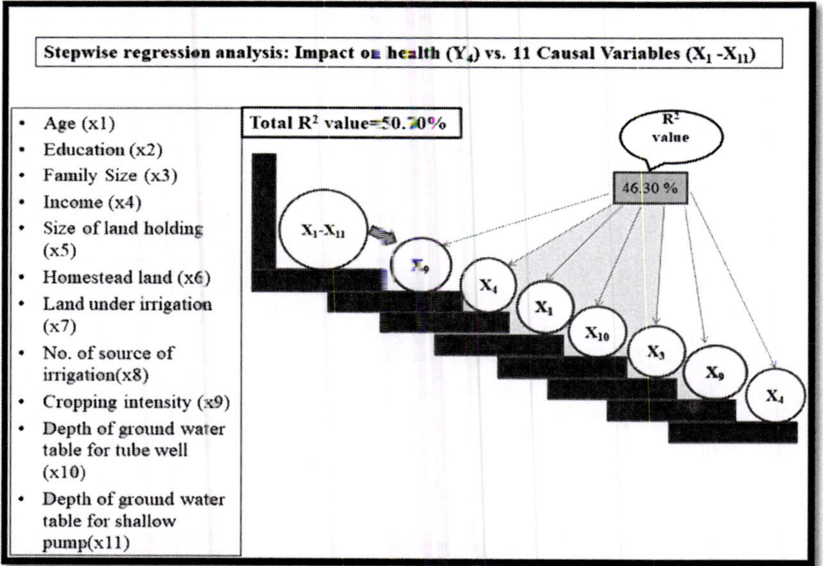

Fig. 6.11: Stepwise Regression Analysis Impact on health (y_4) Vs. 11 Causal Variables (x_1-x_{11})

Results

The table isolates the causal dominant variables contributing to the level of Impact on health(y_4). It has been found that Cropping intensity(x_9), Income(x_4), Age(x_1), Depth of groundwater table for tube well(x_{10}) and Family size(x_3) written at the last step and has substantially contributed to the consequent variable Impact on health(y_4). These five variables together have contributed 46.30% of variance.

Revelation

Among the 11 causal variables, the following variables cropping intensity(x_9), income(x_4), age(x_1), depth of groundwater table for tube well (x_{10}) and family size(x_3) were found most significant with the consequent variable impact on health (y_4). Those five variables are contributing 91.32% of the total 50.70% of variance. So, while dealing with the problem of arsenicosis suffer by the ill-fated responses, proper intervention should be made at the proper age. Farmers also should be aware from cropping intensity, size of their family, their income and depth of groundwater table for tube well.

Table 6.17: Path Analysis: Decomposition of Total Effect into Direct, Indirect and Residual Effect: Impact on health (y_4) vs. 11 exogenous variables (x_1-x_{11})

Sl. No	Variables	Total Effect	Direct Effect	Indirect Effect	Highest Indirect Effect
1	Age(x_1)	0.243	0.486	-0.243	-0.096 (x11)
2	Education(x_2)	0.253	0.096	0.157	0.174 (x1)
3	Family size(x_3)	-0.106	-0.235	0.129	0.186 (x1)
4	Income(x_4)	0.414	0.280	0.134	-0.086 (x6)
5	Size of land holding(x_5)	0.400	0.113	0.287	0.126 (x4)
6	Homestead land(x_6)	0.219	-0.122	0.341	0.198 (x4)
7	Land under irrigation(x_7)	0.186	0.067	0.119	-0.094 (x1)
8	No of source of irrigation(x_8)	0.204	0.076	0.128	0.137 (x1)
9	Cropping intensity(x_9)	0.431	0.276	0.155	0.082 (x10)
10	Depth of ground water table for tube well(x_{10})	-0.321	-0.265	-0.056	0.174 (x1)
11	Depth of groundwater table for shallow pump(x_{11})	0.205	-0.128	0.333	0.365 (x1)

Residual effect: 0.493

Highest indirect individual effect: x_1(6)

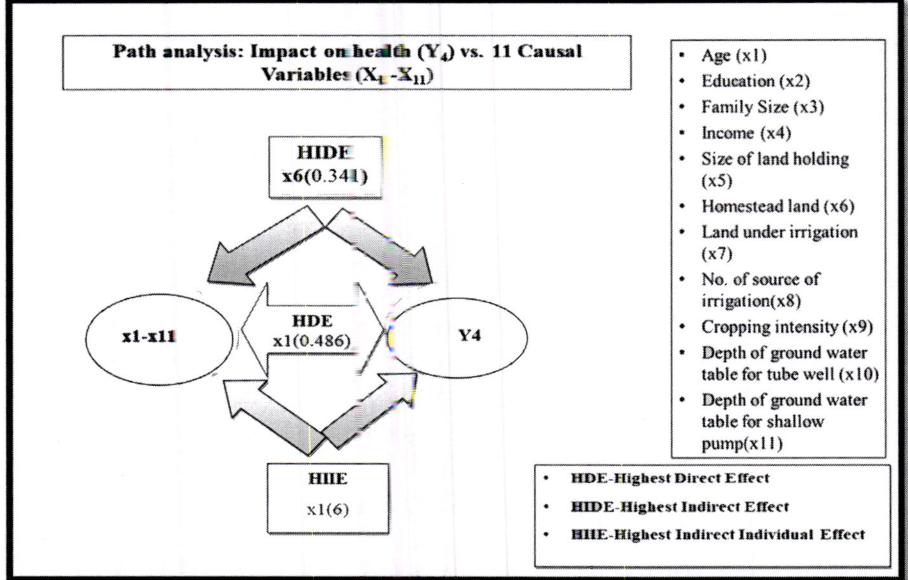

Fig. 6.12: Path Analysis: Decomposition of Total Effect into Direct, Indirect and Residual Effect: Impact on health (y_4) vs. 11 exogenous variables (x_1-x_{11})

Results

The table represents the path analysis by decomposing the total effect (r-value) into direct, indirect and residual effects. It has been found that the highest direct effect is exerted by the causal variable Age(x_1) and the highest indirect effect is exerted by causal variable Homestead land (x_6). Interestingly the variable Age(x_1) has routed the highest indirect effect of 6 other causal variables to ultimately characterised Impact on health (y_4).

Revelation

Age has come out as the most influencing exogeneous variable to estimate the impact on health of diseases due to arsenic contamination. So, there should be an age specific intervention plan to deal with arsenic contamination and its precarious impact on health.

The higher size of homestead land implicates wider and intensive involvement in agricultural operation. Marginal farmers of agricultural wage labourers for their survival they need to migrate. On the other way at least for a certain span of time, they are away from the danger zone. On the other hand, big farmers with large size of homestead land are stuck in their own village. So, they are more vulnerable to arsenic contamination, the worst has been burnt of arsenic contamination on human health.

Table 6.18: Coefficient of Correlation (r): Expenditure on treatment (y_5) Vs. 11 Independent Variables (x_1-x_{11})

Sl. No.	Independent Variables	'r' Value	Remarks
1	Age(x_1)	0.390	**
2	Education(x_2)	0.240	*
3	Family size(x_3)	-0.028	
4	Income(x_4)	0.362	**
5	Size of land holding(x_5)	0.475	**
6	Homestead land(x_6)	0.175	
7	Land under irrigation(x_7)	0.177	
8	No of source of irrigation(x_8)	0.177	
9	Cropping intensity(x_9)	0.414	**
10	Depth of ground water table for tube well(x_{10})	-0.249	*
11	Depth of groundwater table for shallow pump(x_{11})	0.300	*

**Correlation is significant at the 0.01 level

*Correlation is significant at the 0.05 level

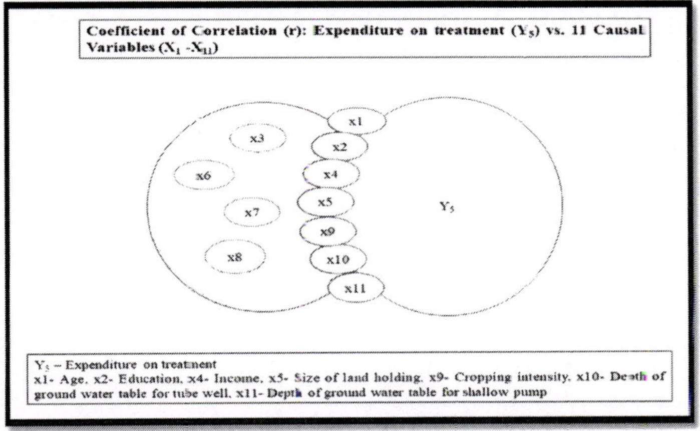

Fig. 6.13: Coefficient of Correlation (r): Expenditure on treatment (y_5) Vs. 11 Independent Variables (x_1-x_{11})

Results: The table presents the coefficient of correlation between Expenditure on treatment(y_5) and 11 independent variables(x_1-x_{11}). It has been found that the following variables viz. Age(x_1), Education(x_2), Income(x_4), Size of land holding(x_5), Cropping intensity(x_9), Depth of groundwater table for tube well(x_{10}) and Depth of groundwater table for shallow pump(x_{11}) have recorded significant correlation with Expenditure on treatment(y_5).

Revelation: For those who have higher age, the amount of treatment for the arsenic contaminated disease is also higher. Education has recorded a positive correlation with expenditure on treatment for the arsenic related disease. For those who have higher income, the expenditure for disease treatment is also higher. If the size of land holding increases, the expenditure on treatment for disease also increases. Cropping intensity and depth of groundwater table for shallow pump have recorded a positive correlation with expenditure on treatment for arsenic related disease. The depth of groundwater table has become a strong determinant in estimating the impact and intensity of arsenic contamination. The area of study has been unique for intensive depletion of groundwater, high intensity cropping and density of population per unit area. All of these are correlated to invite the problem of arsenic contamination and contributed to the expenditure on the treatment of arsenic.

Table 6.19: Multiple Regression Analysis: Expenditure on treatment (y_5) Vs. 11 Causal Variables (x_1-x_{11})

Sl. No.	Variables	Reg. Coef. B	S.E. B	Beta	t Value
1	Age(x_1)	0.619	0.171	0.619	3.627
2	Education(x_2)	0.031	0.124	0.031	0.254
3	Family size(x_3)	-0.219	0.105	-0.219	-2.092
4	Income(x_4)	0.255	0.131	0.255	1.945
5	Size of land holding(x_5)	0.226	0.124	0.226	1.816
6	Homestead land(x_6)	-0.183	0.130	-0.183	-1.411
7	Land under irrigation(x_7)	0.097	0.123	0.097	0.790
8	No of source of irrigation(x_8)	0.017	0.126	0.017	0.136
9	Cropping intensity(x_9)	0.294	0.099	0.294	2.963
10	Depth of ground water table for tube well(x_{10})	-0.184	0.114	-0.184	-1.621
11	Depth of groundwater table for shallow pump(x_{11})	-0.124	0.155	-0.124	-0.797

R square: 57.40%

The standard error of the estimate: 0.71[illegible]

Results

In multiple regression analysis (full model) found that age is the strongest determinant in estimating expenditure on treatment(y_5). The highest beta value is 0.619.

Revelation

The table presents the multiple regression analysis between Expenditure on treatment(y_5) and 11 causal variables(x_1-x_{11}). The full model of multiple regression analysis portraits with the combination of 11 causal variables, 57.40% variance in Expenditure on treatment(y_5) has been explained.

Table 6.20: Stepwise Regression Analysis: Expenditure on treatment (y_5) Vs. 11 Causal Variables (x_1-x_{11})

Sl. No	Variables	Reg. coef. B	S.E. B	Beta	t value
1	Size of land holding(x_5)	0.302	0.093	0.302	3.258
2	Cropping intensity(x_9)	0.335	0.090	0.335	3.710
3	Age(x_1)	0.498	0.103	0.498	4.856
4	Family size(x_3)	-0.201	0.093	-0.201	-2.167
5	Depth of ground water table for tube well(x_{10})	-0.215	0.103	-0.215	-2.090

R square: 53.50%

The standard error of the estimate: 0.7

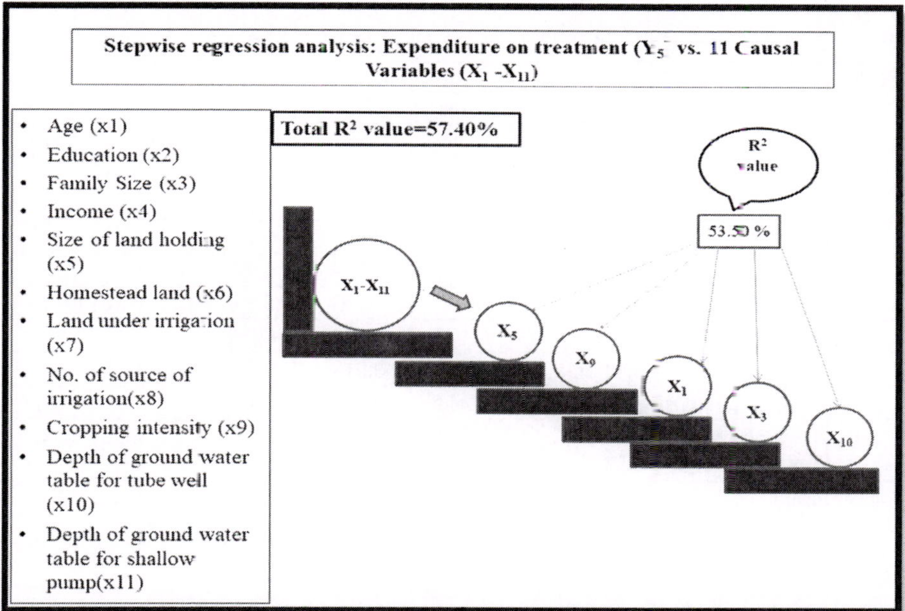

Fig. 6.14: Stepwise Regression Analysis: Expenditure on treatment (y_5) Vs. 11 Causal Variables (x_1-x_{11})

Results

The table isolates the causal dominant variables contributing to the Expenditure on treatment (y_5). It has been found that the Size of land holding(x_5), Cropping intensity(x_9), Age(x_1), Family size(x_3) and Depth of groundwater table for tube well(x_{10}) were written at the last step and has substantially contributed to the consequent variable Expenditure on treatment(y_5). These five variables together have contributed 53.50% of variance.

Revelation

Among the 11 causal variables, the following variables cropping intensity(x_9), income(x_4), age(x_1), depth of groundwater table for tube well (x_{10}) and family size(x_3) were found most significant with the consequent variable expenditure on treatment (y_5). Those five variables are contributing 93.20% of the total 57.40% of variance. So, while dealing with the problem of arsenicosis suffer by the ill-fated responses, proper intervention should be made at the proper age. Farmers also should be aware of their land holding size, cropping intensity, size of their family, and depth of groundwater table for tube well and improve the status of their family.

Table 6.21: Path Analysis: Decomposition of Total Effect into Direct, Indirect and Residual Effect: Expenditure on treatment (y_5) vs. 11 exogenous variables (x_1-x_{11})

Sl. No	Variables	Total Effect	Direct Effect	Indirect Effect	Highest Indirect Effect
1	Age(x_1)	0.390	0.619	-0.229	-0.093 (x11)
2	Education(x_2)	0.240	0.030	0.210	0.222 (x1)
3	Family size(x_3)	-0.028	-0.219	0.191	0.237 (x1)
4	Income(x_4)	0.362	0.255	0.107	-0.129 (x6)
5	Size of land holding(x_5)	0.475	0.227	0.248	0.115 (x4)
6	Homestead land(x_6)	0.175	-0.183	0.358	0.181 (x4)
7	Land under irrigation(x_7)	0.177	0.097	0.080	0.12 (x5)
8	No of source of irrigation(x_8)	0.177	0.018	0.159	0.174 (x1)
9	Cropping intensity(x_9)	0.414	0.295	0.119	0.057 (x10)
10	Depth of ground water table for tube well(x_{10})	-0.249	-0.183	-0.066	0.221 (x1)
11	Depth of groundwater table for shallow pump(x_{11})	0.300	-0.124	0.424	0.465 (x1)

Residual effect: 0.426

Highest indirect individual effect: x_1(5)

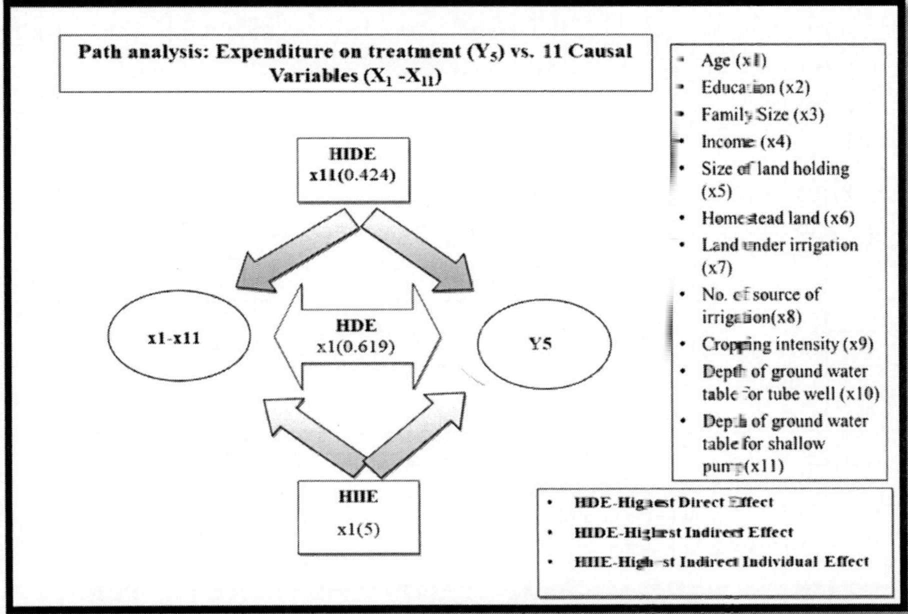

Fig. 6.15: Path Analysis: Decomposition of Total Effect into Direct, Indirect and Residual Effect: Expenditure on treatment (y_5) vs. 11 exogenous variables (x_1-x_{11})

Results

The table represents the path analysis by decomposing the total effect (r-value) into direct, indirect and residual effects. It has been found that the highest direct effect is exerted by the causal variable Age(x_1) and the highest indirect effect is exerted by the causal variable Depth of groundwater table for shallow pump(x_{11}). Interestingly the variable Age(x_1) has routed the highest indirect effect of 5 other causal variables to ultimately characterised Expenditure on treatment(y_5).

Revelation

Age has come out as the most influencing exogeneous variable to estimate the expenditure on treatment of diseases due to arsenic contamination. So, there should be an age specific intervention plan to deal with arsenic contamination and its precarious impact on expenditure.

Due to some geo morphological factors, the quality and quantity of water retained in the sub lithospheric layer is exerting a discernible impact on expenditure for treatment.

If groundwater depletion grows high, the solubility of arsenic will be simmered. The anaerobic condition in the groundwater raising enhances the transformation of recalcitrant arsenic into labile one. So, we need to have a long-term plan to deal with the arsenic problem by managing groundwater with proper technological and scientific intervention.

Table 6.22: Coefficient of Correlation (r): Overall perception on arsenic (y_6) Vs. 11 Independent Variables (x_1-x_{11})

Sl. No.	Independent Variables	'r' Value	Remarks
1	Age(x_1)	0.345	**
2	Education(x_2)	0.306	**
3	Family size(x_3)	-0.059	
4	Income(x_4)	0.345	**
5	Size of land holding(x_5)	0.469	**
6	Homestead land(x_6)	0.209	
7	Land under irrigation(x_7)	0.239	*
8	No of source of irrigation(x_8)	0.199	
9	Cropping intensity(x_9)	0.390	**
10	Depth of ground water table for tube well(x_{10})	-0.204	
11	Depth of groundwater table for shallow pump(x_{11})	0.258	*

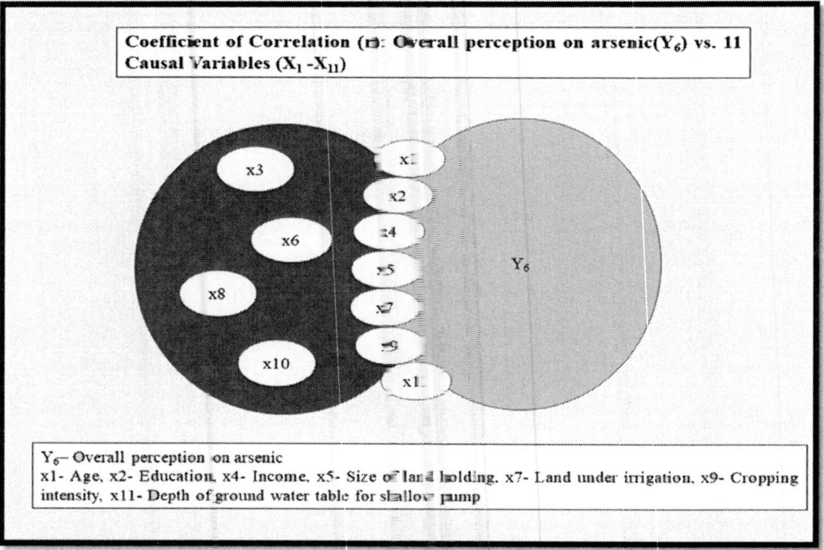

Fig. 6.16: Coefficient of Correlation (r): Overall perception on arsenic (y_6) Vs. 11 Independent Variables (x_1-x_{11})

Results: The table presents the coefficient of correlation between the overall perception on arsenic (y_6) and 11 independent variables(x_1-x_{11}). It has been found that the following variables viz. Age(x_1), Education(x_2), Income(x_4), Size of land holding(x_5), Land under irrigation(x_7), Cropping intensity(x_9) and Depth of groundwater table for shallow pump(x_{11}) have recorded significant correlation with Overall perception on arsenic(y_6).

Revelation

Age is one of the strongest determinants to estimate arsenic on human health. That is how it has been associated with the overall perception on arsenic. It has been found that a certain category of the age gap has been followed vulnerable to arsenic contamination.

Education brings a scientific perception on arsenic contamination, its cause and its impact on human and livestock health, economy and ecology.

Arsenic contamination has jeopardized the prospect of income, market is not that responsive, rather it is enough conservative to dispose of vegetables grown from the arsenic contaminated village. An arsenic contaminated person cannot exert his/her fullest potential to contribute to the income earning activity.

Land and crop are the sinks for arsenic contamination, that is how the size of holding, land under irrigation and cropping intensity have been found closely associated with the problem of arsenic contamination.

The depth of the groundwater table has become a strong determinant in estimating the impact and intensity of arsenic contamination. The area of study has been unique for intensive depletion of groundwater, high intensity cropping and density of population per unit area. All of these are correlated to invite the problem of arsenic contamination and contributed to the overall perception on arsenic.

Table 6.23: Multiple Regression Analysis: Overall perception on arsenic (y_6) Vs. 11 Causal Variables (x_1-x_{11})

Sl.No.	Variables	Reg. Coef. B	S.E. B	Beta	t Value
1	Age(x_1)	0.611	0.180	0.611	3.403
2	Education(x_2)	0.123	0.130	0.123	0.941
3	Family size(x_3)	-0.277	0.110	-0.277	-2.508
4	Income(x_4)	0.179	0.138	0.179	1.294
5	Size of land holding(x_5)	0.184	0.131	0.184	1.408
6	Homestead land(x_6)	-0.098	0.136	-0.098	-0.717
7	Land under irrigation(x_7)	0.191	0.130	0.191	1.470
8	No of source of irrigation(x_8)	-0.026	0.133	-0.026	-0.193
9	Cropping intensity(x_9)	0.301	0.105	0.301	2.874
10	Depth of ground water table for tube well(x_{10})	-0.113	0.120	-0.113	-0.949
11	Depth of groundwater table for shallow pump(x_{11})	-0.178	0.163	-0.178	-1.090

R square: 52.80%

The standard error of the estimate: 0.750

Results

In multiple regression analysis (full model) found that age is the strongest determinant in estimating the overall perception on arsenic(y_6). The highest beta value is 0.611.

Revelation

The table presents the multiple regression analysis between the overall perception on arsenic (y_6) and 11 causal variables(x_1-x_{11}). The full model of multiple regression analysis portraits with the combination of 11 causal variables, 52.80% variance in Overall perception on arsenic(y_6) has been explained.

Table 6.24: Stepwise Regression Analysis: Overall perception on arsenic (y_6) Vs. 11 Causal Variables (x_1-x_{11})

Sl. No	Variables	Reg.coef. B	S.E. B	Beta	t value
1	Size of land holding(x_5)	0.371	0.093	0.371	3.980
2	Cropping intensity(x_9)	0.369	0.092	0.369	4.020
3	Age(x_1)	0.372	0.101	0.372	3.698
4	Family size(x_3)	-0.215	0.099	-0.215	-2.172

R square: 45.90%

The standard error of the estimate: 0.7[illegible]8

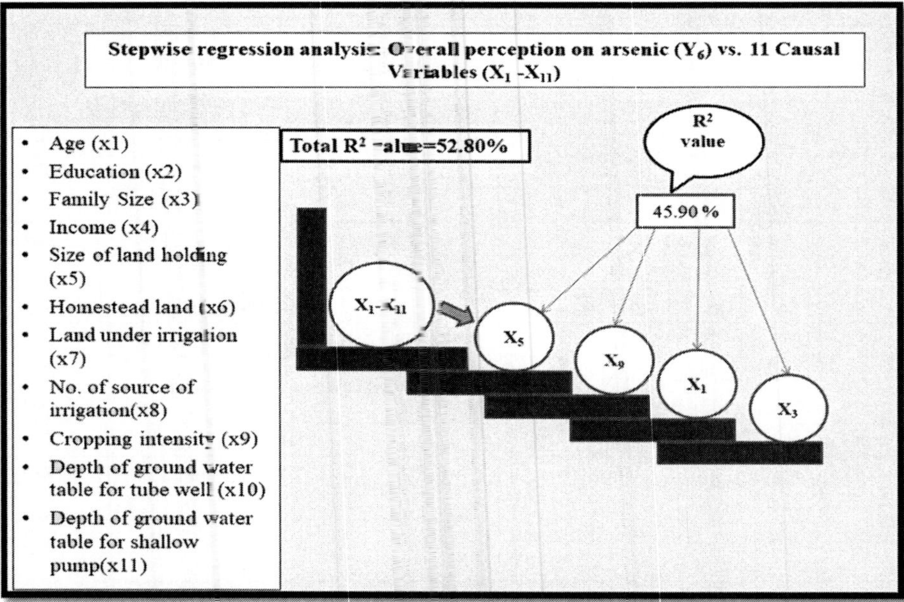

Fig. 6.17: Stepwise Regression Analysis: Overall perception on arsenic (y_6) Vs. 11 Causal Variables (x_1-x_{11})

Results

The table isolates the causal dominant variables contributing to the Overall perception on arsenic(y_6). It has been found that Size of land holding(x_5), Cropping intensity(x_9), Age(x_1) and Family size(x_3) were written at the last step and has substantially contributed to the consequent variable overall perception on arsenic(y_6). These four variables together have contributed 45.90% of variance.

Revelation

Among the 11 causal variables, the following variables size of land holding(x_5), cropping intensity(x_9), age(x_1) and family size(x_3) were found most significant with the consequent variable expenditure on treatment (y_5). Those four variables are contributing 86.93% of the total 52.80% of variance. So, while dealing with the problem of arsenicosis suffer by the ill-fated responses, proper intervention should be made at the proper age. Farmers also should be aware of their land holding size, cropping intensity and size of their family and improve the status of their family.

Table 6.25: Path Analysis: Decomposition of Total Effect into Direct, Indirect and Residual Effect: Overall perception on arsenic (y_6) vs. 11 exogenous variables (x_1-x_{11})

Sl. No	Variables	Total Effect	Direct Effect	Indirect Effect	Highest Indirect Effect
1	Age(x_1)	0.345	0.611	-0.266	-0.134 (x11)
2	Education(x_2)	0.306	0.121	0.185	0.219 (x1)
3	Family size(x_3)	-0.059	-0.276	0.217	0.234 (x1)
4	Income(x_4)	0.345	0.179	0.166	0.084 (x5)
5	Size of land holding(x_5)	0.469	0.185	0.284	0.101 (x1)
6	Homestead land(x_6)	0.209	-0.098	0.307	0.127 (x4)
7	Land under irrigation(x_7)	0.239	0.189	0.050	-0.118 (x1)
8	No of source of irrigation(x_8)	0.199	-0.023	0.222	0.172 (x1)
9	Cropping intensity(x_9)	0.390	0.300	0.090	0.036 (x4)
10	Depth of ground water table for tube well(x_{10})	-0.204	-0.114	-0.090	0.219 (x1)
11	Depth of groundwater table for shallow pump(x_{11})	0.258	-0.179	0.437	0.459 (x1)

Residual effect: 0.473

Highest indirect individual effect: x_1(7)

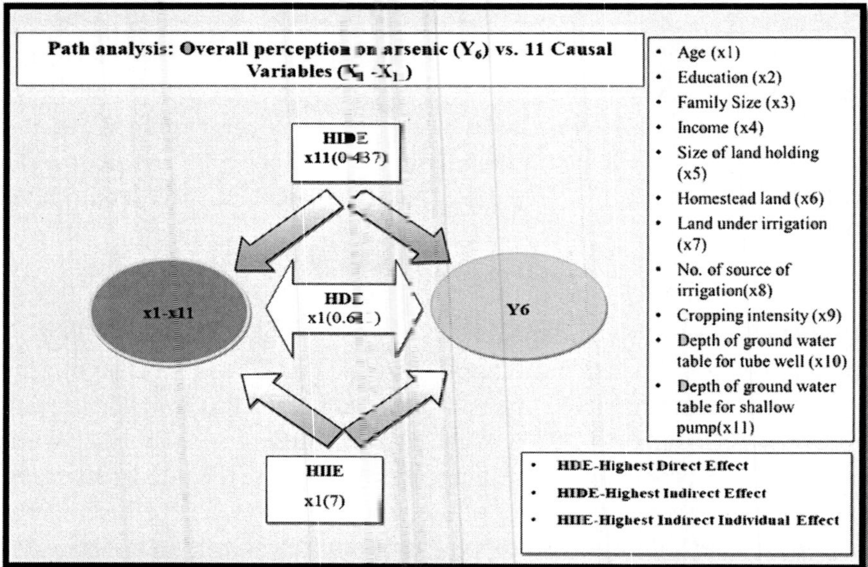

Fig. 6.18: Path Analysis: Decomposition of Total Effect into Direct, Indirect and Residual Effect: Overall perception on arsenic (y_6) vs. 11 exogenous variables (x_1-x_{11})

Results

The table presents the path analysis by decomposing the total effect (r-value) into direct, indirect and residual effects. It has been found that the highest direct effect is exerted by the causal variable Age(x_1) and the highest indirect effect is exerted by the causal variable Depth of groundwater table for shallow pump(x_{11}). Interestingly the variable Age(x_1) has routed the highest indirect effect of 7 other causal variables to ultimately characterised Overall perception on arsenic (y_6).

Revelation

Age has come out as the most influencing exogeneous variable to estimate the overall perception on arsenic due to arsenic contamination. So, there should be an age specific intervention plan to deal with arsenic contamination and its precarious impact on overall perception.

Due to some geo morphological factors, the quality and quantity of water retained in the sub lithospheric layer is exerting a discernible impact on overall perception.

If groundwater depletion grows high, the solubility of arsenic will be simmered. The anaerobic condition in the groundwater raising enhances the transformation of recalcitrant arsenic into labile one. So, we need to have a long-term plan to deal with the arsenic problem by managing groundwater with proper technological and scientific intervention.

Table 6.26: Canonical Covariate Analysis: Interaction between Left Side Dependent and Right Side Independent Variables

Dependent Variable	Coefficients
Disease severity (y_1)	0.180
Impact on crop (y_2)	0.116
Impact on livestock (y_3)	-0.127
Impact on health (y_4)	0.103
Expenditure on treatment (y_5)	0.021
Overall perception on arsenic (y_6)	0.081
Independent Variable	**Coefficients**
Age(x_1)	-0.001
Education(x_2)	-0.067
Family size(x_3)	0.001
Income(x_4)	0.029
Size of land holding(x_5)	0.020
Homestead land(x_6)	-0.022
Land under irrigation(x_7)	-0.012
No of source of irrigation(x_8)	-0.046
Cropping intensity(x_9)	0.078
Depth of ground water table for tube well(x_{10})	0.374
Depth of groundwater table for shallow pump(x_{11})	-0.222

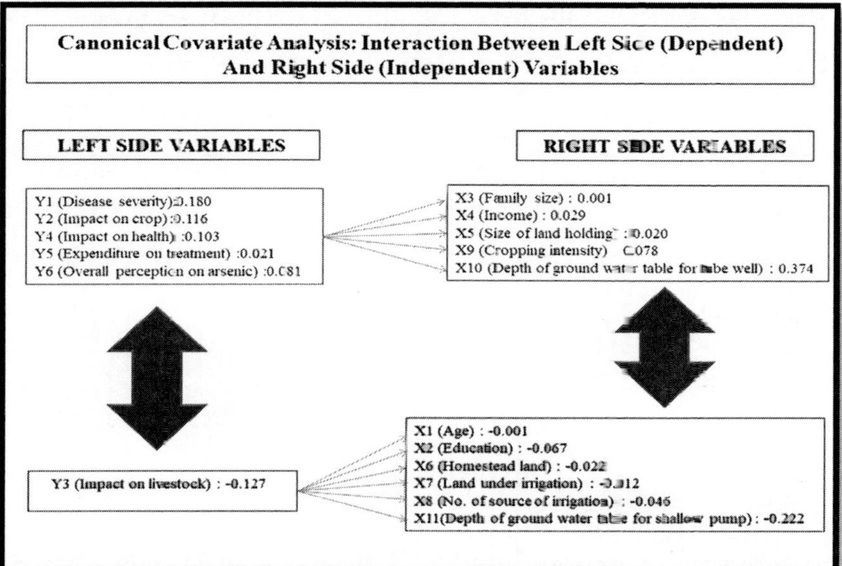

Fig. 6.19: Canonical Covariate Analysis: Interaction between Left Side Dependent and Right Side Independent Variables

Revelation

Canonical covariate analysis is a splendid statistical tool which can delineate and interact between two sets of variables. The table represents the canonical covariate analysis into two sets of variables (y and x) is in clandestine interactions. 'y' sets of variables are generally called left side variables and 'x' sets of variables are generally called right side variables. It has been found that the 'y' variables have got two subgroups.

Through this canonical covariate analysis, it is revealed that the following five dependent variables are disease severity (y_1), impact on crop (y_2), impact on health(y_4), expenditure on treatment(y_5) and overall perception on arsenic (y_6) are directly and strongly impacted by following independent variables viz. family size (x_3), income(x_4), size of land holding (x_5), cropping intensity (x_9) and depth of groundwater table for tube well(x_{10}).

On the other hand, the other predicted variable Impact on livestock (y_3) has been conglomerated together and impacted by the following predictor variables viz. age (x_1), education (x_2), homestead land (x_6), land under irrigation (x_7), no. of the source of irrigation (x_8) and depth of groundwater table for shallow pump (x_{11}).

The entire spectrum of canonical covariate analysis suggests that without dealing with 'y' variables in isolation, we can deal with them into two groups of 'y' variables and accordingly we can go concerned with the selected 'x' variables on the right-side variables.

Table 6.27: Factor Analysis: Strategic Conglomeration of Variables into Factor

Factors	Variables	Factor loading	% of variance	Cumulative %	Factor renamed
1	Size of land holding (x_5) Disease severity (y_1) Impact on crop (y_2) Impact on livestock (y_3) Impact on health (y_4) Expenditure on treatment (y_5) Overall perception on arsenic (y_6)	0.579 0.963 0.938 -0.948 0.941 0.961 0.950	40.873	40.873	Health ecology
2	Age (x_1) Education(x_2) Depth of ground water table for tube well (x_{10}) Depth of groundwater table for shallow pump (x_{11})	0.737 0.500 0.716 0.709	16.212	57.084	Water ecology

Factors	Variables	Factor loading	% of variance	Cumulative %	Factor renamed
3	Homestead land (Decimal) (x_6) Land under irrigation (x_7)	0.614 0.543	10.528	57.612	Resource ecology
4	No of source of irrigation(x_3) Income (x_4) Cropping intensity (x_9)	-0.488 -0.473 0.428	6.804	74.416	Crop eco-dynamics
5	Family size(x_3)	0.783	6.470	80.886	Family size

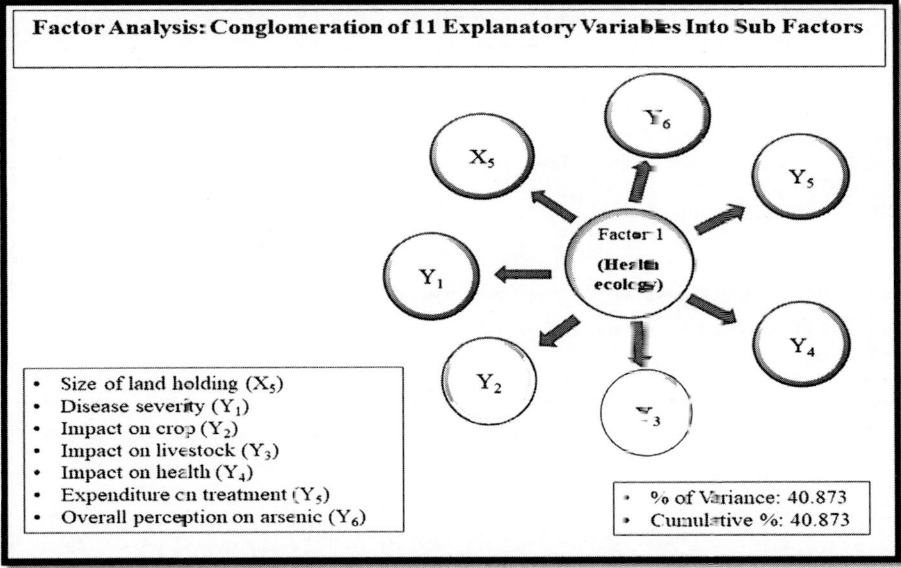

Fig. 6.20: Factor Analysis (Health ecology): Conglomeration of 11 Explanatory variables into sub factors

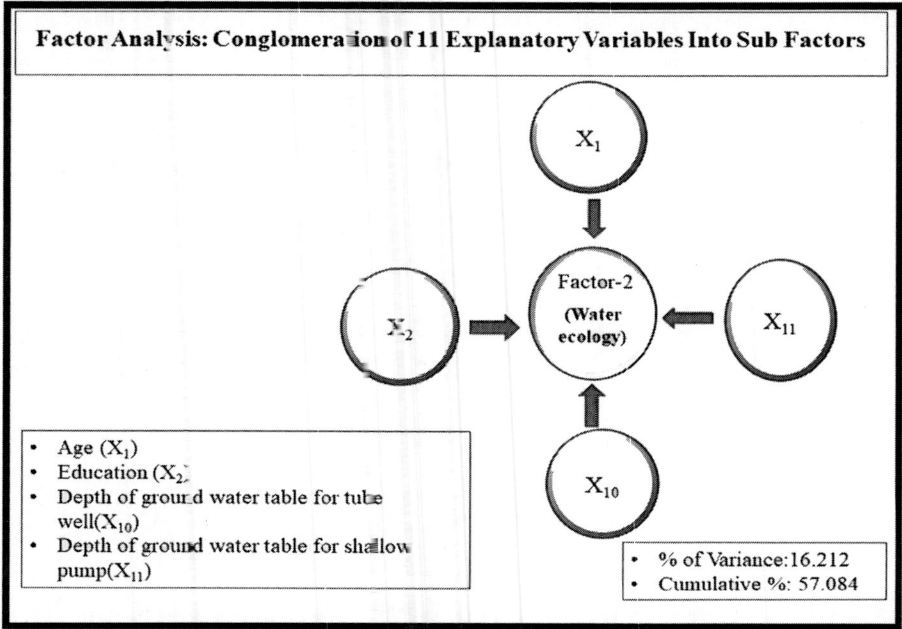

Fig. 6.21: Factor Analysis (Water ecology): Conglomeration of 11 Explanatory variables into sub factors

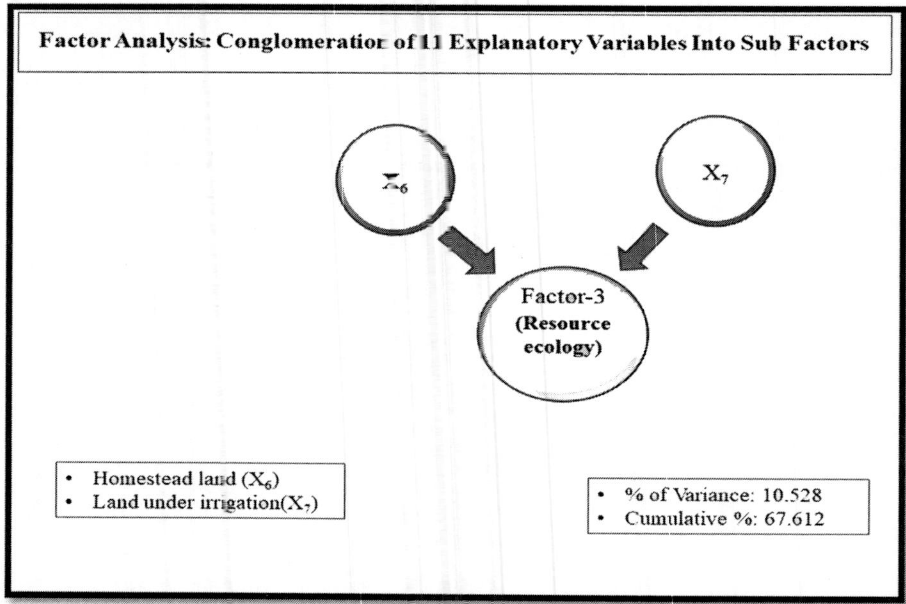

Fig. 6.22: Factor Analysis (Resource ecology):Conglomeration of 11 Explanatory variables into sub factors

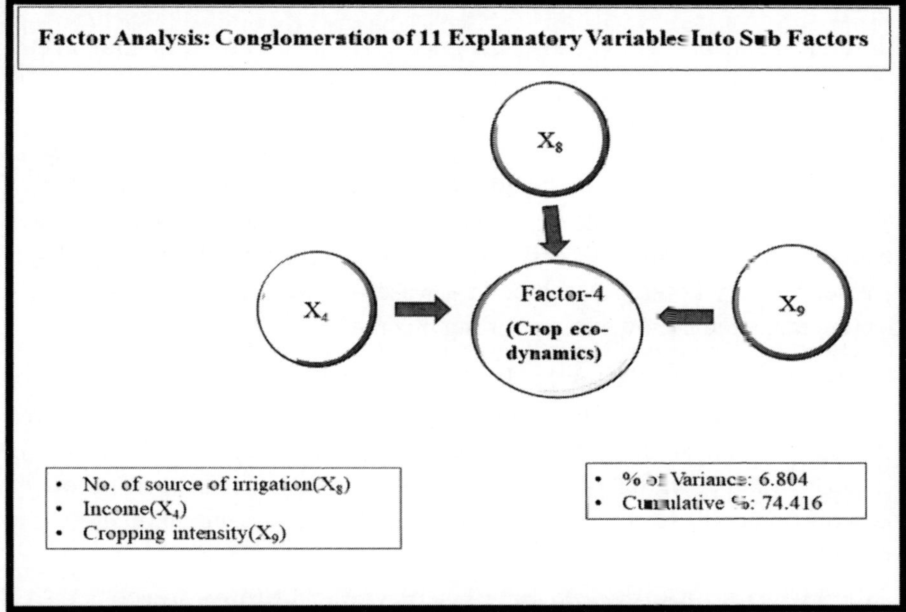

Fig. 6.23: Factor Analysis (Crop eco-dynamics):Conglomeration of 11 Explanatory variables into sub factors

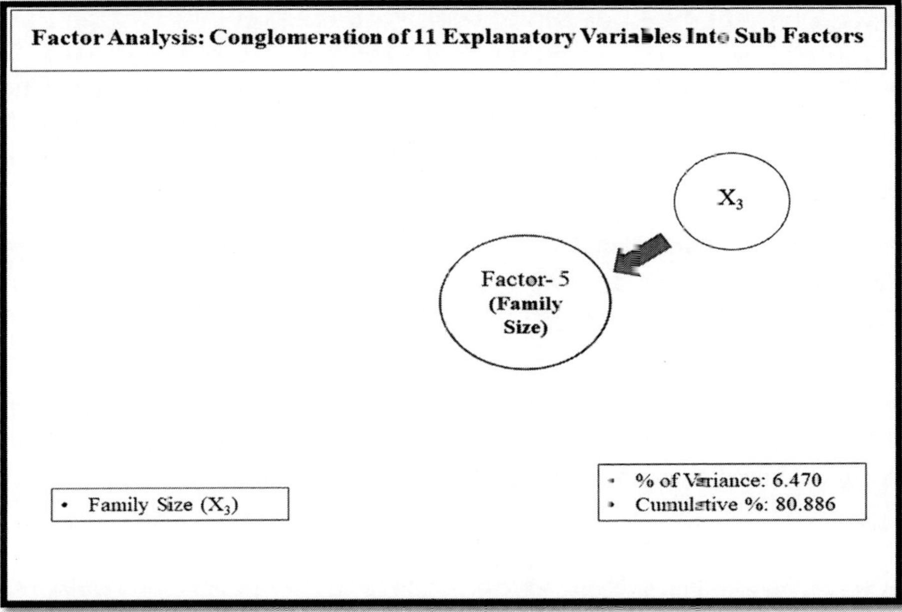

Fig. 6.24: Factor Analysis (Family size):Conglomeration of 11 Explanatory variables into sub factors

Results

This table presents the Strategic Conglomeration of Variables into Factor through factor analysis.

Revelation

Factor analysis is a data reduction approach. Here the total number of interacting variables, based on factor loading and eigen values, has been grouped into identifiable factors. Each factor is retaining the homogeneous variable, strongly bound to each other towards taking a strategic decision for effective management of resource or organization.

It has been found that **Factor 1:** Has accommodated the following variables, viz., Size of land holding(decimal)(x_5), disease severity (y_1), impact on crop (y_2), impact on livestock (y_3), impact on health (y_4), expenditure on treatment (y_5) and overall perception on arsenic (y_6) has been renamed as **health ecology.** Ithas contributed to 40.873 percent alone and 40.873 percent cumulatively to explain the variance.

Factor 2: Has accommodated the following variables, viz., age(x_1), education(y_4), depth of groundwater table for tube well(x_{10}) and depth of groundwater table for shallow pump(x_{11}) have been renamed as **water ecology.** Ithas contributed to 16.212 percent alone and 57.084 percent cumulatively to explainthe variance.

Factor 3: Has accommodated the following variables, viz., homestead land (x_6)and land under irrigation($x_{7)}$ have been renamed as **resource ecology**. It hascontributed to 10.528 percent alone and 67.612 percent cumulatively to explain thevariance.

Factor 4 hasaccommodatedthefollowingvariables,viz., number of sources of irrigation(x_8), income(x_4), and cropping intensity(x_9) have been renamedas **crop eco-dynamics**. It has contributed to 6.804 percent alone and 74.416 percentcumulatively to explain the variance.

Factor 5 hasaccommodatedtheonlyvariable family size(x_3) has been renamedas **family size**. It has contributed to 6.470 percent alone and 80.886 percentcumulatively to explain the variance.

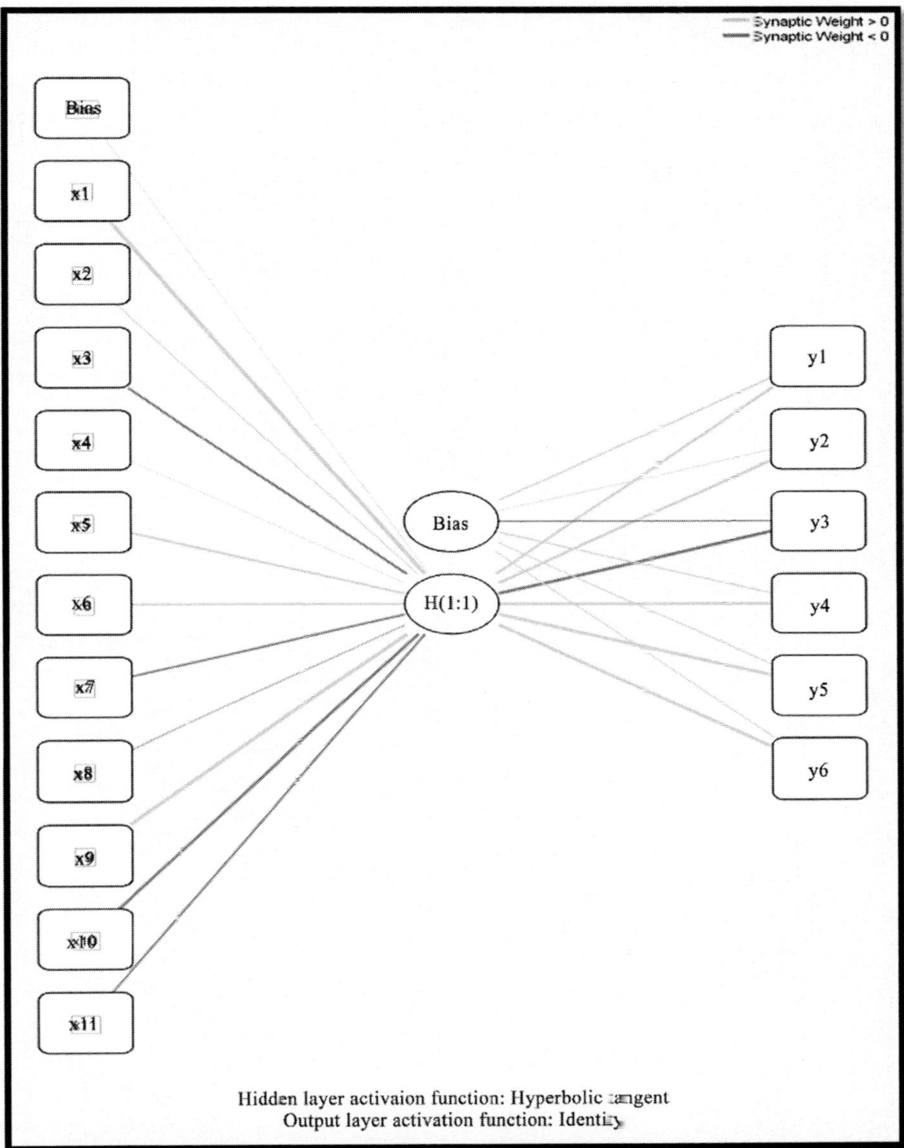

Fig. 6.25: Artificial Neural Networking

Revelation

Artificial Neural Network is an advance and statistical tool for screening out for dominant variables out of total pool of input variables for any kind of ANN there would be Input variables, hidden layers and output variables.The following input variables family size (x_3), land under irrigation(x_7), depth of groundwater table for tube well (x_{10}) and depth of groundwater table for shallow pump (x_{11}) being displayed through blue or bold lines are extracting strong and dominant impact on the output variables, whenever the input variables are pursuing through the hidden layers, their respective error will be minimized by the application of the bias value.

Table 6.28: Model summary

Model Summary			
Training	Sum of Squares Error		80.405
	Average Overall Relative Error		0.558
	Relative Error for Scale Dependents	Disease severity (yl)	0.615
		Impact on crop (y2)	0.557
		Impact on livestock (y3)	0.597
		Impact on health (y4)	0.528
		Expenditure on treatment (y5)	0.489
		Overall perception on arsenic (y6)	0.563
Testing	Sum of Squares Error		31.542
	Average Overall Relative Error		0.479
	Relative Error for Scale Dependents	Disease severity (yl)	0.552
		Impact on crop (y2)	0.461
		Impact on livestock (y3)	0.521
		Impact on health (y4)	0.407
		Expenditure on treatment (y5)	0.416
		Overall perception on arsenic (y6)	0.562

7

Conclusion

7.1. Summary

The present research had been carried out to study the social ecology of arsenic contamination in groundwater; factors, perception and impact on life and livelihood of the selected village of West Bengal and in the end the comparison is made. The study had been conducted at village Mitrapur in the Haringhata block of West Bengal. A pilot study was conducted to understand the area, its people, institution, communication and extension system of these states.

An exhaustive list of respondents was prepared critically with the help of some villagers. From the list, 70 respondents from the village were selected for the study through a non-random snowball sampling method. The primary data were collected with the help of a structured interview schedule by following the personal interview method. The secondary data were collected from our departmental library, internet, Department of Agriculture, West Bengal, Bidhan Chandra Krishi Viswavidyalaya etc. for establishing the conceptual framework of the study.

Three types of Issues are Covered in this Study

- The influence of social, cultural, and/or economic variables on how drinking water poisoned with arsenic affects people's life, notably their health (for example socio-economic patterns of illness and social),
- Social factors influencing people's response: especially in relation to other priorities, perceptions of arsenic, and social roles and relationships.
- Institutional and project/program initiatives to address issues related to arsenic.

Some professionals consider that mental health problems, such as depression, may also result from intense social isolation or ostracism of arsenicosis patients. Such problems, however, have not been scientifically studied.

The general **objective** of the study is to estimate and analyze the **"Social Ecology of Arsenic Contamination in Groundwater; Factors, Perception and Impact on Life and Livelihood"** and the **specific objectives** are:

- To know the health impact of arsenic contaminated water
- To calculate the sociological factor-based impact of arsenic poisoning on the health of rural residents
- To elucidate the effect of a set of socio-ecological variables on the perceived impact of arsenic contamination on health, agriculture and allied sectors
- To extract policy implications based on empirical research for a micro - sociological perspectives
- To highlight the efforts made by the Government to combat the problem

Coefficient of Correlation (r) between Disease severity (Y_1) and 11 Causal Variables (X_1 -X_{11}):

It is found that the coefficient of correlation between Disease severity(y_1) and 11 independent variables. It has been found that the following variables are Age (x_1), Education(x_2), Income (x_4), Size of land holding(x_5), Land under irrigation (x_7), No. of source of irrigation (x_8), Cropping intensity(x_9) and Depth of groundwater table for shallow pump(x_{11}) have recorded a significant correlation with Disease severity(y_1).

Coefficient of Correlation (r) between Impact on crop(Y_2) and 11 Independent Variables (X_1 -X_{11}):

It has been found that the following variables Age(x_1), Income(x_4), Size of land holding(x_5), No. of source of irrigation(x_8), Cropping intensity (x_9), Depth of groundwater table for tube well (x_{10}) and Depth of groundwater table for shallow pump (x_{11}) have recorded significant and positive correlation with impact on crop.

Coefficient of Correlation (r) between Impact on livestock(Y_3) and 11 Independent Variables (X_1 -X_{11}):

This presents the coefficient of correlation between Impact on livestock(y_3) and 11 independent variables. It has been found that Age(x_1), Education (x_2), Income(x_4), Size of land holding(x_5), Cropping intensity(x_9) and Depth of groundwater table for shallow pump (x_{11}) have recorded significant and positive correlation with Impact on livestock(y_3).

Coefficient of Correlation (r) between Impact on health(Y_4) and 11 Independent Variables (X_1 -X_{11}):

It has been found that the following variables are Age(x_1), Education (x_2), Income(x_4), Size of land holding(x_5), Cropping intensity(x_9) and Depth of groundwater table for tube well(x_{10}) have recorded significant correlation with impact on health.

Coefficient of Correlation (r) between Expenditure on treatment(Y_5) and 11 Independent Variables (X_1 -X_{11}):

It has been found that the following variables are Age(x_1), Education(x_2), Income(x_4), Size of land holding(x_5), Cropping intensity(x_9), Depth of groundwater table for tube well(x_{10}) and Depth of groundwater table for shallow pump(x_{11}) have recorded significant correlation with Expenditure on treatment(y_5).

Coefficient of Correlation (r) between Overall perception on arsenic(Y_6) and 11 Independent Variables (X_1 -X_{11}):

This presents the coefficient of correlation between the overall perception on arsenic(y_6) and 11 independent variables(x_1-x_{11}). It has been found that the following variables viz. Age(x_1), Education(x_2), Income(x_4), Size of land holding(x_5), Land under irrigation(x_7), Cropping intensity(x_9) and Depth of groundwater table for shallow pump(x_{11}) have recorded significant correlation with Overall perception on arsenic(y_6).

Stepwise Regression Analysis between Disease severity(y_1) and 11 Causal Variables(x_1-x_{11}):

The three variables Size of land holding(x_5), cropping intensity(x_9) and Age(x_1) have been retained at the last step. The r^2 value is 47.70%. These three variables have together contributed to 81.81% of the total 58.30% of explicated variables. The Size of land holding(x_5), Cropping intensity(x_9) and Age(x_1) were come up as strong determinants for the disease severity.

Stepwise Regression Analysis between Impact on crop(y_2) and 11 Causal Variables(x_1-x_{11})

The three variables Size of land holding(x_5), Cropping intensity(x_9) and Age(x_1) have been retained at the last step. The r^2 value is 37.40%. These three variables have together contributed to 77.92% of the total 48.00% of explicated variables. The three variables Size of land holding(x_5), Cropping intensity(x_9) and Age(x_1) were come up as strong determinants for the impact on crops due to arsenic.

Stepwise Regression Analysis between Impact on livestock(y_3) and 11 Causal Variables(x_1-x_{11}):

The four variables Age(x_1), Size of land holding(x_5), Cropping intensity(x_9) and Family size(x_3) have been retained at the last step. The r^2 value is 50.60%. These four variables have together contributed to 94.40% of the total 53.60% of explicated variables. The four variables Age(x_1), Size of land holding(x_5), Cropping intensity(x_9) and Family size(x_3) were come up as strong determinants for accessibility to impact on livestock.

Stepwise Regression Analysis between Impact on health(y_4) and 11 Causal Variables(x_1-x_{11}):

The five variables Cropping intensity(x_9), Income(x_4), Age(x_1), Depth of groundwater table for tube well(x_{10}) and Family size(x_3) have been retained at the last step. The r^2 value is 46.30%. These five variables have together contributed to 91.32% of the total 50.70% of explicated variables. TheCropping intensity(x_9), Income(x_4), Age(x_1), Depth of groundwater table for tube well(x_{10}) and Family size(x_3) come up as strong determinants of impact on health due to arsenic contamination.

Stepwise Regression Analysis between Expenditure on treatment(y_5) and 11 Causal Variables(x_1-x_{11}):

The five variables Size of land holding(x_5), Cropping intensity(x_9), Age(x_1), Family size(x_3) and Depth of groundwater table for tube well(x_{10}) have been retained at the last step. The r^2 value is 53.50%. These five variables have together contributed to 93.20% of the total 57.40% of explicated variables. The Size of land holding(x_5), Cropping intensity(x_9), Age(x_1), Family size(x_3) and Depth of groundwater table for tube well(x_{10}) were come up as strong determinants for expenditure on treatment due to arsenic.

Stepwise Regression Analysis between Overall perception on arsenic(y_6) and 11 Causal Variables(x_1-x_{11}):

The four variables Size of land holding(x_5), Cropping intensity(x_9), Age(x_1) and Family size(x_3) have been retained at the last step. The r^2 value is 45.90%. These four variables have together contributed to 86.93% of the total 52.80% of explicated variables. The Size of land holding(x_5), Cropping intensity(x_9), Age(x_1) and Family size(x_3) were come up as strong determinants for characterizing the overall perception on arsenic.

Path Analysis: Decomposition of Total Effect into Direct, Indirect and Residual Effect: Disease severity(y_1) vs. 11 exogenous variables(x_1-x_{11})

The highest direct effect was exerted by the causal variable Age(x_1) and the highest indirect effect was exerted by the causal variable Depth of groundwater table for shallow pump(x11). Interestingly the variable age(x_1) has routed the highest indirect effect of 6 other causal variables to ultimately characterised disease severity (y_1).

Path Analysis: Decomposition of Total Effect into Direct, Indirect and Residual Effect: Impact on crop(y_2) vs. 11 exogenous variables(x_1-x_{11})

The highest direct effect was exerted by the causal variable Age(x_1) and the highest indirect effect was exerted by the causal variable Depth of groundwater table for shallow pump(x_{11}). Interestingly the variable Age(x_1) has routed the highest indirect effect of 6 other causal variables to ultimately characterised Impact on crop (y_2).

Path Analysis: Decomposition of Total Effect into Direct, Indirect and Residual Effect: Impact on livestock (y_3) vs. 11 exogenous variables(x_1-x_{11})

The highest direct effect was exerted by the causal variable Age(x_1) and the highest indirect effect was exerted by the causal variable Depth of groundwater table for shallow pump(x_{11}). Interestingly the variable Age(x_1) has routed the highest indirect effect of 6 other causal variables to ultimately characterised Impact on livestock (y_3).

Path Analysis: Decomposition of Total Effect into Direct, Indirect and Residual Effect: Impact on health(y_4) vs. 11 exogenous variables(x_1-x_{11})

The highest direct effect was exerted by the causal variable Age(x_1) and the highest indirect effect was exerted by the causal variable Homestead land (x_6). Interestingly the variable Age(x_1) has routed the highest indirect effect of 6 other causal variables to ultimately characterised Impact on health(y_4).

Path Analysis: Decomposition of Total Effect into Direct, Indirect and Residual Effect: Expenditure on treatment(y_5) vs. 11 exogenous variables(x_1-x_{11})

The highest direct effect was exerted by the causal variable Age(x_1) and the highest indirect effect was exerted by the causal variable Depth of groundwater table for shallow pump(x_{11}). Interestingly the variable Age(x_1) has routed the highest indirect effect of 5 other causal variables to ultimately characterised Expenditure on treatment(y_5).

Path Analysis: Decomposition of Total Effect into Direct, Indirect and Residual Effect: Overall perception on arsenic(y_6) vs. 11 exogenous variables(x_1-x_{11})

It has been found that the highest direct effect is exerted by the causal variable Age(x_1) and the highest indirect effect is exerted by the causal variable Depth of groundwater table for shallow pump(x_{11}). Interestingly the variable Age(x_1)

has routed the highest indirect effect of 7 other causal variables to ultimately characterised Overall perception on arsenic(y_6).

Canonical Covariate Analysis: Interaction between Dependent and Independent Variables

Through this canonical covariate analysis, it is revealed that the following five dependent variables are disease severity (y_1), impact on crop (y_2), impact on health(y_4), expenditure on treatment(y_5) and overall perception on arsenic (y_6) are directly and strongly impacted by following independent variables viz. family size (x_3), income(x_4), size of land holding (x_5), cropping intensity (x_9) and depth of groundwater table for tube well(x_{10}).

On the other hand, the other predicted variable Impact on livestock (y_3) has been conglomerated together and impacted by the following predictor variables viz. age (x_1), education (x_2), homestead land (x_6), land under irrigation (x_7), no. of the source of irrigation (x_8) and depth of groundwater table for shallow pump (x_{11}).

Factor Analysis by Conglomeration of 11 independent variables (x_1-x_{11}) ofvillages Mitrapur, Haringhata, West Bengal, into 5 factors, based on factorloadingand renamingofall thefactors.

Factor analysis is a data reduction approach. Here the total number of interacting variables, based on factor loading and Eigen values, has been grouped into identifiable factors. Each factor is retaining the homogeneous variable, strongly bound to each other towards taking a strategic decision for effective management of resource or organization.

It has been found that, Factor 1(Health ecology) has contributed to 40.873 percent alone and 40.873 percent cumulatively to explain the variance.

Factor 2(Water ecology) has contributed to 16.212 percent alone and 57.084 percent cumulatively to explain the variance.

Factor 3(Resource ecology) has contributed to 10.528 percent alone and 67.612 percent cumulatively to explain the variance.

Factor4 (Crop eco-dynamics)has contributed to 6.804 percent alone and 74.416 percent cumulatively to explain the variance.

Factor5(Family size)has contributed to 6.470 percent alone and 80.886 percent cumulatively to explain the variance.

7.2. Future Scope of the Study

The followings are the considerations

- Gender can be included in future studies to see whether there are any gender differences in terms of vulnerability and disasters caused by this heavy metal contamination.
- More variables, particularly those of a perceptual and biomedical character, ought to have been included in the study; if not, they can be in future studies.
- Rain water harvesting, use of surface water and consumptive use of irrigation water merit issues of opportunities and possible interventions. So, a more robust intervention is required in the future.
- There is also a scope for studying arsenic in the food chain, which is very important for the ecosystem.
- Future research may take into account other affected locations. On the eastern side of the Hooghly River in west Bengal, this threat has claimed 120 blocks. The magnitude and current scope of the respondents are like the tip of an iceberg.

7.3. Limitations of the Study

- More variables, taking into account applicability and relevance, should have been used.
- Research and control groups should have been included to isolate the critical factors which are really responsible for the menace.
- More sophisticated statistical methods, such as discriminate function analysis, and LDF (Linear Discrimination Function), would have been useful for modelling the empirical research and for identifying the marker variables.
- In order to identify the trend and pattern of this crisis, the temporal fluctuations may have been investigated using time series analysis or ARIMA (Autoregressive Integrated Moving Average)
- Girls are frequently absent from patient surveys, and little is known about their struggles or other social aspects of the arsenic issue.
- More research is needed to determine how diet and nutrition affect illnesses linked to toxicity from arsenic.

- More research is needed on how men and women play distinct roles as well as how men and women view women who seek medical attention.
- Epidemiological investigations will reveal gender disparities in arsenic-related sickness; special attention should be paid to age-adjusted prevalence odds ratios and distinct risk factors.

7.4. Recommendation

- A specific policy for livestock management is needed to protect them from exposure to arsenic contamination. Cow milk is also found to retain a higher amount of arsenic and which subsequently enters the food chain and undergoes a biomagnification process to make the problem extensive
- Vegetables, livestock and orchard crops deserve a unique techno-managerial approach to get them a deterrent from arsenic contamination. Also, their marketability will be absent and the farm ecology will be disrupted severely
- Raising public awareness and changing water use behaviour
- Building self-help and community mobilisation
- Alternatives that are secure, inexpensive, and practical, and how communities and organisations should identify them
- Developing comprehensive, coordinated approaches to project implementation
- Change from the dependable tube well that is at or near the home to community-based water sources
- Including women and poor people in planning
- Staff training to develop the necessary abilities for mobilising and educating the public

7.5. Conclusion

Arsenicosis is a very complicated issue, and both the source of drinking water and the source of irrigation water are crucial. Approximately 120 blocks in West Bengal, nearly all of which are located on the eastern bank of the Ganges and include the districts of Malda, Murshidabad, Nadia, North 24 Parganas, and South 24 Parganas, are at risk from this social and environmental threat. The present study has well identified some of the important factors (age, education, cropping intensity, depth of groundwater, size of land holding) that are mainly responsible for this type of deceptive level arsenicosis. The safest and most abundant water supply should not be the ground. For residential supplies and irrigation purposes, groundwater is crucial in the basin's rural sections. Although there are aquifers of different depths, the majority are shallow. The dragon in sleep (As.) will awaken if groundwater is drawn without regard for the environment and will obliterate humans, animals, and the ecosystem as a whole. Therefore, to make the lives of the unfortunate victims safe secure, and of course, enjoyable, policymakers, leaders, NGO activists, government departments, and benefactors should band together. Before using groundwater for drinking or cooking, it should first be tested for potability.

The entire empirical study reveals the following facts

(i) Big size farmers are more vulnerable to arsenic

(ii) The higher is the cropping intensity, the more has been the impact

(iii) There should be an age specific intervention plan to deal with arsenic contamination and its precarious impact on health

(iv) Groundwater management should be focused on higher priority

(v) It is advisable to assess the potability of groundwater before utilising it for drinking or cooking purpose

(vi) To diagnose the problem regarding the range of arsenic contamination in the water-soil-plant-animal system. It is imperative to create genetically modified crop types that are arsenic resistant.

(vii) Community mobilization for enhancing surface water consumtion after proper treatment and promoting rain water harvesting.

Bibliography

Ahmad, S. A., Khan, M. H., and Haque, M. (2018). Arsenic contamination in groundwater in Bangladesh: implications and challenges for healthcare policy. Risk management and healthcare policy, 11, 251.

Alaerts, G. J., and Khouri, N. (2004). Arsenic contamination of groundwater: Mitigation strategies and policies. Hydrogeology Journal, 12(1), 103-114.

Alam, M. G. M., Allinson, G., Stagnitti, F., Tanaka, A., and Westbrooke, M. (2002). Arsenic contamination in Bangladesh groundwater: a major environmental and social disaster. International journal of environmental health research, 12(3), 235-253.

Barringer, J.L., Mumford, A, Young, L.Y., Reilly, P.A., Bonin, J.L. and Rosman, R. (2010). Pathways for arsenic from sediments to groundwater to streams: Biogeochemical processes in the Inner Coastal Plain, New Jersey, USA. Water Research, 44(19), 5532-5544.

Berg, M., Tran, H. C., Nguyen, T. C., Pham, H. V., Schertenleib, R., and Giger, W. (2001). Arsenic contamination of groundwater and drinking water in Vietnam a human health threat. Environmental science & technology, 35(13), 2621-2626.

Chakrabarti, K. B. (2017). Arsenic Contamination in Bengal Basin: Reinventing Mitigation through Participatory Social Innovations. Washington State University.

Charlson, J. (2010). "Phytoremediation of Arsenic Joe Charlson ABSTRACT Arsenic Is Listed as the Number One Hazardous Substance by The U.S Agency.

Chatterjee, R., Acharya, S. K., and Mitra, S. (2018). Distributive Family Expenditure of Farmers due to Arsenic Contamination: The Interpretation and Inferences. Bangladesh Journal of Extension Education ISSN, 1011, 3916.

Chatterjee, R., Acharya, S. K., and Mitra, S. (2018). Estimation of impact of arsenic contamination: The severity and contextuality.

Chowdhury, A. M. R., Hossain, M., Nickson, R., Rahman, M., Jakariya, M., and Sharnimuddin, M. (2000) "Combating a Deadly Menace: Early Experiences with a community-based arsenic mitigation project in Bangladesh."

Curry, A., Carrin, G., Bartram, J., Yamamura, S., Sims, J., Hueb J., ... and World Health Organization. (2000). Towards an assessment of the socioeconomic impact of arsenic poisoning in Bangladesh (No. WHO/SDE/WSH/00.4). World Health Organization.

Das, A., and Mondal, S. (2021). Geomorphic controls on shallow groundwater arsenic contamination in Bengal basin, India. Environmental Science and Pollution Research, 28(31), 42177-42195.

Das, D., Samanta, G., Mandal, B. K., Roy Chowdhury, T., Chanda, C. R. Chowdhury, P. P., ... and Chakraborti, D. (1996). Arsenic in groundwater in six districts of West Bengal, India. Environmental Geochemistry and Health, 18(1), 5-15

Delhi, N. (2011). Office of the Registrar General and Census Commissioner, Ministry of Home Affairs, Government of India; 2014. Census of India.

Dhillon, A. K. (2020). Arsenic contamination of India's groundwater: a review and critical analysis. Arsenic Water Resources Contamination, 177-205

Fendorf, S., Michael, H. A., and van Geen, A. (2010). Spatial and temporal variations of groundwater arsenic in South and Southeast Asia. Science, 328(5982), 1123-1127.

Garelick, H., Dybowska, A., Valsami-Jones, E., and Priest, N. (2005). Remediation technologies for arsenic contaminated drinking waters (9 pp). Journal of Soils and Sediments, 5(3), 182-190.

Harvey, C.F., Swartz, C.H., Badruzzaman, A.B.M., Keon-Blute, N., Yu, W., Ali, M.A.l, Jay, J., Beckie, R., Niedan, V., Brabander, D., Oates, R.M., Ashfaque, K.N., Islam, S., Hemond, H.F. and Ahmed, M.F. (2002). Arsenic mobility and groundwater extraction in Bangladesh. Science, 298(5695), 1602-1606.

Islam, M. A., Shiragi, M. H. K., Baque M. A., Rahman, M. S., and Khanam, T. (2009). Arsenic contamination in food chain due to arsenic contaminated groundwater irrigation. International Journal of Sustainable Agricultural Technology, 5(9), 40-45.

Jain, C. K., Sharma, S. K., and Singh, S. (2018). Physico-chemical characteristics and hydrogeological mechanisms in groundwater with special reference to arsenic contamination in Barpeta District, Assam (India). Environmental monitoring and assessment, 190(7), 1-16.

Kapaj, S., Peterson, H., Liber, K., and Bhattacharya, P. (2006). Human health effects from chronic arsenic poisoning–a review. Journal of Environmental Science and Health, Part A, 41(10), 2399-2428.

Kinniburgh, D. G ,and Smedley, P. (2001). Arsenic contamination of groundwater in Bangladesh.

Lear, G., Song B., Gault, A.G., Polya, D.A. and Lloyd, J.R. (2007). Molecular analysis of arsenate reducing bacteria within Cambodian sediments following amendment with acetate. Applied and Environmental Microbiology, 73(4), 1041-1048.

Lena, Q. M., Hong-Jie, S., Bala, R. Bing, W., Jun, L., and Li-Ping, P. (2014). "Arsenic and selenium toxicity and their interactive effects in humans," Environment International, vol. 69, 148–158.

Lin, T. Y., Wei, C. C., Huang, C. W., Chang, C. H., Hsu, F. L., and Liao, V. H. C. (2016). Both phosphorus fertilizers and indigenous bacteria enhance arsenic release into groundwater in arsenic-contaminated aquifers. Journal of Agricultural and Food Chemistry, 64(11), 2214-2222.

Maharjan, M., Watanabe, C., Ahmad, S. A. and Ohtsuka, R. (2005). Arsenic contamination in drinking water and skin manifestations in lowland Nepal: the first community-based survey. The American journal of tropical medicine and hygiene, 73(2), 477-479.

Majumdar, P. K., Ghosh, N. C., and Chakravorty, B. (2002). Analysis of arsenic-contaminated groundwater domain in the Nadia district of West Bengal (India). Hydrological sciences journal, 47(S1), S55-S66.

Mandal, P. (2017). An insight of environmental contamination of arsenic on animal health. Emerging Contaminants, 3(1), 17-22

Mazumder, D. N. G., Ghosh, A., Majumdar, K. K., Ghosh, N., Saha, C., and Mazumder, R. N. G. (2010). Arsenic contamination of ground water and its health impact on population of district of Nadia, West Bengal, India. Indian Journal of Community Medicine: Official Publication of Indian Association of Preventive & Social Medicine, 35(2), 331.

McArthur, J.M., Banerjee, D.M., Hudson-Edwards, K.A., Mishra, R., Purohit, R., Ravenscroft, P., Cronin, A., Howarth, R.J., Chatterjee, A., Talukder, T., Lowry, D., Houghton, S.M and Chadha, D.K. (2004). Natural organic matter in sedimentary basins and its relation to arsenic in anoxic ground water: the example of West Bengal and its worldwide implications. Applied Geochemistry, 19(8), 1255-1293.

Mishra, S., Dwivedi, S., Kumar, A., Chauhan, R., Awasthi, S., Mattusch, J., and Tripathi, R. D. (2016). Current status of ground water arsenic contamination in India and recent advancements in removal techniques from drinking water. International Journal of Plant and Environment, 2(1 and 2), 01-15.

Muehe, E. M., and Kappler, A. (2014). Arsenic mobility and toxicity in South and South-east Asia–a review on biogeochemistry, health and socio-economic effects, remediation and risk predictions. Environmental Chemistry, 11(5), 483-495.

Mukherjee, A. B., and Bhattacharya, P. (2001). Arsenic in groundwater in the Bengal Delta Plain: slow poisoning in Bangladesh. Environmental Reviews, 9(3), 189-220.

Mukherji, A., and Shah, T. (2005). Groundwater socio-ecology and governance: a review of institutions and policies in selected countries. Hydrogeology Journal, 13(1), 328-345.

Nahar, N. (2009). Impacts of arsenic contamination in groundwater: case study of some villages in Bangladesh. Environment, development and sustainability, 11(3), 571-588.

Nordstrom, D. K. (2002). Worldwide occurrences of arsenic in ground water. Science, 296(5576), 2143-2145.

Rahman, M. H., Rahman, M. M., Watanabe, C., and Yamamoto, K. (2003). Arsenic contamination of groundwater in Bangladesh and its remedial measures. In Arsenic Contamination in Groundwater-Technical and Policy Dimensions. Proceedings of the UNU-NIES International Workshop, United Nations University, Tokyo, Japan (pp. 9-21).

Rajesh, S., Gond, D. P., and Amit, P. (2010). Assessment of ground water quality of Ghazipur district, Eastern Uttar Pradesh, India, special reference to arsenic contamination. Recent Research in Science and Technology, 2(3), 38-41.

Ravenscroft, P., McArthur, J. M., & Hoque, B. A. (2001). Geochemical and palaeohydrological controls on pollution of groundwater by arsenic (Vol. 5, pp. 1-2■). Elsevier Science Ltd.: Oxford.

Ravenscroft, P., Burgess, W. G., Ahmed, K. M., Burren, M., and Perrin, J. (2005). Arsenic in groundwater of the Bengal Basin, Bangladesh: Distribution, field relations, and hydrogeological setting. Hydrogeology Journal, 13(5), 727-75■.

Roy, M. (2019). Arsenic Contamination of Groundwater in West Bengal: A Human Health Threat. Page, 1(1).

Saha, D., and Sahu, S. (2016). A decade of investigations on groundwater arsenic contamination in Middle Ganga Plain, India. Environmental geochemistry and health, 38(2), 315-337.

Saikia, K. C., and Gupta, S. (2012). Assessment of surface water quality in an arsenic contaminated village. American Journal of Environmental Sciences, 8(5), 523.

Saldaña-Robles, A., Abraham-Juárez, M. R., Saldaña-Robles, A. L., Saldaña-Robles, N., Ozuna, C., and Gutiérrez-Chávez, A. J. (2018). The negative effect of arsenic in agriculture: irrigation water, soil and crops, state of the art. Appl. Ecol. Environ. Res, 16, 1533-1551.

Samal, A. C., Kar, S., Bhattacharya, P., and Santra, S. C. (2011). Human exposure to arsenic through foodstuffs cultivated using arsenic contaminated groundwater in areas of West Bengal, India. Journal of Environmental Science and Health, Part A, 46(11), 1259-1265.

Sanjrani, M. A., Zhou, B., Zhao, H., Bhutto, S. A., Muneer, A. S., and Xia, S. B. (2019). Arsenic contaminated groundwater in China and its treatment options, a review. Applied Ecology and Environmental Research, 17(2), 1655-1683.

Santini, J.M. and Ward, S.A. (eds.) (2012) Metabolism of Arsenite CRC Press.

Sarkar, A. (2010). Ecosystem perspective of groundwater arsenic contamination in India and relevance in policy. EcoHealth, 7(1), 114-126.

Sarkar, A., Panigrahi, S., Mehrotra, R., Anand, M., and Mukherjee, A. B. (2006). Toward finding a sustainable solution for arsenic contamination of ground water: A SWOT analysis. Epidemiology, 17(6), S218-S219.

Sen, P. K. (2016). Abundant environmental arsenic contamination: some statistical perspectives. Sankhya B, 78(2), 341-361.

Shrivastava, A., Barla, A., Yadav, H. and Bose, S. (2014). Arsenic contamination in shallow groundwater and agricultural soil of Chakdaha block, West Bengal, India. Frontiers in Environmental Science, 2, 50.

Smedley P.L., and Kinniburgh, D.G. (2002). A review of the source, behavior and distribution of arsenic in natural waters Applied Geochemistry, 17(5), 517-568.

Smith, A. H., Lingas, E. O., and Rahman, M. (2000). Contamination of drinking-water by arsenic in Bangladesh: a public health emergency. Bulletin of the World Health Organization, 78(9), 1093-1103.

Srivastava, M., Ma, L. Q., and Santos, J A. G. (2006). Three new arsenic hyperaccumulating ferns. Science of the Total Environment, 364(1-3), 24-31.

Sultana, F. (2007). Suffering for water, suffering from water: Political ecologies of arsenic, water and* development in Bangladesh. University of Minnesota.

Sultana, F. A. R. H. A. N. A. (2012). Political ecology of the arsenic crisis in Bangladesh. Diverting Flow Gender Equity Water South Asia.

Van Geen, A., Zheng, Y., Versteeg, R., Stute, M., Horneman, A., Dhar, R., Steckler, M., Gelman, A., Small, C., Ahsan, H., Graziano, J.H., Hussain, I. and Ahmed, K.M. (2003). Spatial variability of arsenic in 6000 tube wells in a 25 km^2 area of Bangladesh. Water Resources Research, 39, (HWC 3) 1-16.

WSP. (2002) Fighting Arsenic, listening to rural communities: findings from a study on willingness to pay for Arsenic-free, safe drinking water in rural Bangladesh. Field Note, Water and Sanitation Program — South Asia. Dhaka, Bangladesh.

Zheng, Y., Stute, M., Geen, A.V., Gavrieli, 1., Dhar, R., Simpson, H.J., Schlosser, P. and Ahmed, K.M., 2004. Redox control of arsenic mobilization in Bangladesh ground water: Applied Geochemistry, v. 19, p. 201-214.

Following websites were consulted

URL: http://nriwestbengal.gov.in/nri/common/stateprofile.aspx

URL: https://www.bharatonline.com/west-bengal/travel-tips/weather.html

URL: https://www.mapsofindia.com/maps/westbengal/

URL: https://www.researchgate.net/figure/Map-of-Nadia-District_fig1_344162873

URL: hltp://wwvs.naisu.intö/

URL :< http://www.greenfacts.org/en/arsenic/index.htm>.

URL: .<http://www.who.int/water_sanitation health/dwq/arsenic/en/>.

URL: .<http://www.who.int/water_sanitation_health/diseases/ars enicosis/en/>.

Appendix

General Information

Name of the respondent: ....

Age: ....

Particulars regarding households: ....

Name of the head of the family: ....

Name of the block: ....

Name of the village: ....

Cast and religion: ....

Family type: Single/Joint: ....

Family size: ....

Number of adult male members: ....

Number of adult female members: ....

Number of children: ....

Total number of member in the family: ....

Total income of the farmer in one year (Rs): ....

Educational level: Given tick mark () in the appropriate item regarding the formal education received by me.

Sl. No.	Educational Qualification	
1	**Illiterate**	
2	Can read-only	
3	Can read and write/Primary passed	
4	Middle school passed	
5	Madhyamik passed	
6	Higher Secondary passed	
7	Graduate/Diploma or above	

Operational Land Holding size (Katha): Please mention the area of land that allot for different enterprises in your farm

Enterprise	Area owned (Katha)	Area leased in (Katha)	Area leased out (Katha)	Total
Crop				
Livestock				
Fishery				
Others (Specify)				

1. (a) Homestead Land: Yes/No

   (b) Size of Homestead land (Katha): ....

2.. Land under irrigation (Katha): ....

3. (a) No. of source of irrigation: ....

   (b) Name of the irrigation source: ....

4. Cropping intensity:

   CI = (Gross cropped area/ Net cropped area) *100

5. Depth of ground water table for tube well (meter): ....

6. Depth of ground water table for shallow pump (meter): ....

7. Name of the crops and yield (which are cultivated):

A. Rice

Area under rice production (Katha):

Rice	Summer			Winter			Autumn		
	UL	ML	LL	UL	ML	LL	UL	ML	LL
HYV									
Hybrid									
Local									

Production of rice (Kg)

Rice	Summer			Winter			Autumn		
	UL	ML	LL	UL	ML	LL	UL	ML	LL
HYV									
Hybrid									
Local									

B. Vegetables

Vegetable Crops	Cultivated Area (Katha)	Yield (Kg/Katha)

C. Pulses

Pulse Crops	Cultivated Area (Katha)	Yield (Kg/Katha)

8. Effect of Arsenic in body and mind
   a) Arsenic related problems (Male)
   b) Arsenic related problems (Female)
9. Name of the disease:

Disease Symptoms	Local Name	Doctor's Name

10. Treatment (Cost and health management of the problem)

Medicine trade name	Cost (Rs)

11. Disease Severity (10 Point Scale)

Disease Severity	Category	Scale
1-3	Mild	
3-6	High	
6-9	Very High	
9+	Lethal	

12. Causes of Arsenic contamination
13. Impact on Crop (10 Point Scale)

Impact on Crop	Category	Scale
1-3	Mild	
3-6	High	
6-9	Very High	
9+	Lethal	

14. Impact on livestock (10 Point Scale)

Impact on Livestock	Category	Scale
1-3	Mild	
3-6	High	
6-9	Very High	
9+	Lethal	

15. Impact on Health (10 Point Scale)

Impact on Health	Category	Scale
1-3	Mild	
3-6	High	
6-9	Very High	
9–	Lethal	

16. Overall Perception on Arsenic (10 Point Scale):

Overall Perception	Category	Scale
Low	1-4	
Medium	5-7	
High	8-10	

Index

आधुनिक हिन्दी नाटक का अग्रदूत

मोहन राकेश

आधुनिक हिन्दी नाटक का अग्रदूत

मोहन राकेश

गोविन्द चातक

राधाकृष्ण प्रकाशन

ISBN : 978-81-7119-817-7

आधुनिक हिन्दी नाटक का अग्रदूत : मोहन राकेश

पहला संस्करण : 2003
तीसरा संस्करण : 2016

This book is printed on **Print on Demand** Technology : 2025

मूल्य : ₹895

प्रकाशक

राधाकृष्ण प्रकाशन प्राइवेट लिमिटेड
जी-17, जगतपुरी, दिल्ली-110 051

शाखाएँ : अशोक राजपथ, साइंस कॉलेज के सामने, पटना-800 006
पहली मंजिल, दरबारी बिल्डिंग, महात्मा गांधी मार्ग, प्रयागराज-211 001
1, अनमोल सोराबजी संतुक लेन, धोबी तलाव, मरीन लाइंस, मुम्बई-400 002
वेबसाइट : www.radhakrishnaprakashan.com
ई-मेल : info@radhakrishnaprakashan.com

ADHUNIK HINDI NATAK KA AGRADOOT : MOHAN RAKESH
Critical study of wellknown modern Hindi playwright
Mohan Rakesh by Dr. Govind Chatak

गुरुवर श्री शम्भुप्रसाद बहुगुणा को
सादर
और कृतज्ञतापूर्वक

तीसरे संस्करण की भूमिका

पिछले संस्करणों का स्वागत हुआ। उसके लिए धन्यवाद। बहुत-से लोगों को मोहन राकेश को आधुनिक हिन्दी नाटक का मसीहा करार देना अच्छा नहीं लगा। अच्छा तो शायद मुझको भी नहीं लगता। पर अच्छा लगने न लगने से क्या होता है ? मैंने पहले संस्करण की भूमिका में लिख दिया था कि राकेश का जीने और लिखने का अपना एक अन्दाज़ था। दुश्मनों की बात जाने दीजिए, उनके दोस्तों ने उन्हें अपने जीवन-काल में ही एक मिथ बना डाला था। मेरा इरादा उन्हें मसीहा बनाने का नहीं रहा है। मोहन राकेश ने पहली बार हिन्दी नाटक को आधुनिकता की नई दिशा दी है पर उनके कृतित्व पर समय के साथ फिर से मूल्यांकन की ज़रूरत है। पर इतना निश्चित है कि उन्होंने हिन्दी में पहली बार मानव अस्तित्व के मूल में स्थित प्रश्न उठाए हैं—एक नहीं, अनेक प्रश्न : नियति और स्थिति के प्रश्न; सत्ता-मोह और सृजन का प्रश्न; पार्थिव और अपार्थिव का प्रश्न; पुरुष मन की तृप्ति और नारी मन की आत्मरति का प्रश्न ! वह अपने नाटकों में निरन्तर प्रश्नों से उलझते रहे—कहीं उत्तरों के साथ, कहीं बिलकुल निरुत्तर। पर सब जगह एक जागरूक झेलनेवाले व्यक्ति की तरह, एक चिन्तक की मुद्रा में। एक ऐसे 'मसीहा' की तरह जिसने युग के प्रश्नों की चुनौती-भरी भंगिमा को महसूस किया और जो आधुनिक मानव की पीड़ा और उलझन का प्रवक्ता बना। राकेश और उनकी कृतियों में भावना का सम्बन्ध इतना सीधा और सच्चा है और अन्तर की आवाज़ इतनी तीखी और तल्ख़ है कि उनके तेवर मसीहाई-से लगने लगते हैं।

तब भी मेरा यह आग्रह नहीं था कि कोई उन्हें मसीहा कहे। पर मुझे एक शब्द चाहिए था; एक ऐसे साहित्यकार के लिए जिसने आज के टूटे, थके-हारे, भटके इंसान की नियति को उजागर किया हो; जिसने आधे-अधूरे और ज़िन्दगी के अन्ध-लोकों में भटकते मानव की चाहों की विडम्बना को महसूस किया हो और भँवर में पड़े जीवन का अर्थ खोजने का प्रयास किया हो। राकेश की कृतियों में मुझे जो एक विशेष अकुलाहट दिखाई दी उसमें हो सकता है ग़लत ही मैंने मसीहाईपन देखा हो। मेरी

दृष्टि उनकी भंगिमा पर रही है। पर आलोचक के लिए कोई मसीहा नहीं होता। इसीलिए आपको 'मसीहा' का कृतित्व सूली पर टँगा भी दिखाई देगा। सवाल असल में मेरे सामने फूल-मालाएँ चढ़ाने का नहीं रहा है। फूल-मालाएँ लोग चढ़ा चुके हैं। मैंने तो उनके नीचे दबे रचनाकार को उबारने की कोशिश मात्र की है।

कोई रचना या रचनाकार क्या है, यह बताना सरल कार्य नहीं है—यह एक प्रकार से रचना से जूझना है। इस संस्करण तक आते-आते मैं अब भी अपने को राकेश से जूझता पाता हूँ। सत्य की तलाश सभी वक्तव्यों से बड़ी कही जा सकती है। इस संस्करण को प्रेस में देने तक भी यह तलाश ज़ारी रही है। इसलिए मुझे आपसे यही कहना है कि मैं समीक्षा नहीं दे रहा हूँ, आपको मात्र देखने-समझने की दृष्टि देना चाहता हूँ। सोचने-समझने के कई पहलू हो सकते हैं। जहाँ तक मेरा सम्बन्ध है, मेरा अपना पहलू निवेदित है।

इस संस्करण में जहाँ-कहीं मुझे कुछ नया सूझा, कुछ बातें यहाँ-वहाँ जोड़ी गई हैं। 'मसीहा' के स्थान पर पुस्तक के नाम में 'अग्रदूत' शब्द रख दिया है। **पैर तले की ज़मीन** पर इसमें केवल चर्चा भर है। लिखना चाहता था, पर मन नहीं किया। सच्चाई यह है कि यह अधूरा नाटक पाठक-आलोचक को बेहद निराश करता है। एक और बात यह भी है कि सब-कुछ को लेकर सभी कुछ लिख देना मैं अनिवार्य भी नहीं समझता। राकेश पर लिखनेवाले और भी हैं। मैंने वे ही बातें लिखी हैं जो मेरी परिधि में आती हैं, क्योंकि अपेक्षाओं को ध्यान में रखकर लिखना मुझे पसन्द नहीं। मैं उनके प्रति कृतज्ञ हूँ। कृतज्ञता की बात चली है तो एक कर्त्तव्य और निभा दूँ। मैं उन सबका कृतज्ञ हूँ जिन्होंने इस पुस्तक के सम्बन्ध में मुझे अपने विचारों से अवगत कराया। इस दृष्टि से मैं डॉ. इन्द्रनाथ मदान का विशेष आभार मानता हूँ। पुस्तक के छपते ही उन्होंने एक पत्र लिखा था, "अगर मेरा इरादा लिखने का कायम रहा तो तुम्हारी पुस्तक मेरे लिए काम की होगी—खंडन और मंडन दोनों के लिए। बहरहाल तुमने काम जमकर किया है।" वे मोहन राकेश को एक उपलब्धि नहीं सम्भावना मानते हैं। मेरी यह पुस्तक उपलब्धि और सम्भावना दोनों की तलाश है।

—गोविन्द चातक

10 वाणी विहार, उत्तम नगर,
नई दिल्ली—110059

अनुक्रम

सृजन के स्रोत

एक यात्रा : पद-चिह्नों के आर-पार

हिन्दी नाट्य साहित्य में भारतेन्दु और प्रसाद के बाद यदि लीक से हटकर कोई नाम उभरता है तो मोहन राकेश का। हालाँकि बीच में और भी कई नाम आते हैं जिन्होंने आधुनिक हिन्दी नाटक की विकास-यात्रा में महत्त्वपूर्ण पड़ाव तय किए हैं; किन्तु मोहन राकेश का लेखन एक दूसरे ध्रुवान्त पर नज़र आता है। इसलिए ही नहीं कि उन्होंने अच्छे नाटक लिखे, बल्कि इसलिए भी कि उन्होंने हिन्दी नाटक को अँधेरे बन्द कमरों से बाहर निकाला और उसे युगों के रोमानी ऐन्द्रजालिक सम्मोहन से उबारकर एक नए दौर के साथ जोड़कर दिखाया। वस्तुतः मोहन राकेश के नाटक केवल हिन्दी के नाटक नहीं हैं। वे हिन्दी में लिखे अवश्य गए हैं, किन्तु वे समकालीन भारतीय नाट्य प्रवृत्तियों के द्योतक हैं। उन्होंने हिन्दी नाटक को पहली बार अखिल भारतीय स्तर ही नहीं प्रदान किया वरन् उसके सदियों के अलग-थलग प्रवाह को विश्व नाटक की एक सामान्य धारा की ओर भी अग्रसर किया।

'विश्व नाटक' शब्द का प्रयोग यहाँ जान-बूझकर किया गया है। दरअसल, जब हम आधुनिक नाटक की बात करते हैं तो नाटक न किसी भाषा का रह जाता है, न देश का। सारे आस्पदों और उपाधियों से मुक्त वह सिर्फ़ नाटक है—सार्वजनीन और सार्वदेशिक; अभिन्न और अविभाजित। इस अभेद के मूल में आज के युग की बढ़ती हुई अन्तर्राष्ट्रीयता है जो अब निरन्तर कलाओं का आधार बनती जा रही है। इसीलिए अनेक कलात्मक प्रवृत्तियाँ और चिन्तन धाराएँ निरन्तर आदान-प्रदान के माध्यम से एक विश्वजनीन रूप धारण करती जा रही हैं। आज का नाटक विश्व-भर में इसी आधार पर विकसित हो रहा है।

हिन्दी नाटक भारतेन्दु युग में पहली बार अपनी लक्ष्मण-रेखा से बाहर निकला। उसके बाद पुनरुत्थानवादी प्रवृत्तियों और स्वच्छन्दतावादी आग्रह के कारण पौर्वात्य और पाश्चात्य नाट्य शिल्प में समन्वय की प्रक्रिया प्रारम्भ हुई। किन्तु कालान्तर में जिस विश्वजनीन साहित्य-दौर ने हिन्दी नाटक को विश्व नाटक की समकालीन परिस्थितियों से सबसे अधिक जुड़ा पाया, वह था—यथार्थवाद। और फिर उसके बाद यथार्थवाद की प्रतिक्रिया में विश्व-भर में नाटक के जो विविध रूप उभरे और जो-जो तेवर बदले, हिन्दी नाटक ने उन भंगिमाओं को ग्रहण करने में देर ज़रूर लगाई, किन्तु वह उससे सर्वथा

अनजान नहीं रहा। इस दिशा में भी, कालक्रम से कई नाटककार आगे आए—उन आगे आनेवालों में अब तक सबसे आख़िरी नाम मोहन राकेश का कहा जा सकता है जिन्हें नए नाटक की पहल करने का श्रेय प्राप्त है।

राकेश ने हिन्दी नाटक को एक नई ज़मीन पर खड़ा किया। सोचने की बात इतनी है कि यह ज़मीन कहाँ तक उन्होंने स्वयं तोड़कर बनाई थी, और कहाँ तक उन्हें बनी-बनाई मिल गई थी ? वैसे हर प्रयोगधर्मा लेखक को अपनी पूर्ववर्ती परम्परा और चिन्तन का एक दाय अपने-आप ही हाथ लग जाता है जिस पर वह अपने लेखन को टिकाता है; और फिर उस बिन्दु से वह आगे बढ़ता है। राकेश के लिए वह बिन्दु था—हिन्दी नाटक का एक विश्वजनीन चेतना की ओर अग्रसर होता विकास-क्रम। इसमें लक्ष्मीनारायण मिश्र, जगदीशचन्द्र माथुर, उपेन्द्रनाथ 'अश्क', लक्ष्मीनारायण लाल, धर्मवीर भारती आदि पहले ही महत्त्वपूर्ण भूमिका अदा कर चुके थे। राकेश ने इस धारा को एक नया मोड़ ही नहीं दिया, वरन् उन्होंने विश्व के समकालीन नाट्य लेखन के समानान्तर ऐसी नाट्य रचना भी की जो आधुनिक भाव-बोध के नए आयाम प्रस्तुत करती है।

आधुनिक भाव-बोध पश्चिम में उन्नीसवीं शती के मध्य से बड़ी तीव्रता से जागा और देखते-ही-देखते चारों ओर फैल गया जिससे अनेक कला-आन्दोलन और चिन्तन प्रस्फुटित हुए। फलतः यथार्थवाद, प्रकृतवाद, प्रतीकवाद, अभिव्यक्तिवाद, एपिक थिएटर, अतियथार्थवाद, असंगतवाद आदि अनेक मतवादों ने समय-समय पर नाट्य लेखन और मंचन की अन्यान्य प्रयोगधर्मिताओं और विविधताओं को जन्म दिया। इनके कारण नाटक की क्षेत्रीय और देशीय सीमाएँ टूटीं और विश्व-भर में वे समानान्तर प्रवृत्तियाँ उभरीं, जिनमें नाटक की समस्त देशीयता के बावजूद उसकी एक सामान्य पृष्ठभूमि सामने आई। वस्तुतः समान जीवनानुभूतियों और मानवीय स्थितियों के बीच नाटक ने एक ऐसा सामान्य स्वरूप ग्रहण किया जिसने सारे पाश्चात्य जगत् को आकर्षित किया और कालान्तर में जिसका प्रभाव भारत पर भी पड़े बिना न रहा।

पारस्परिक निर्भरता और एकता का यह माहौल जिन साहित्यिक और कला-आन्दोलनों के माध्यम से बना था उसके पीछे सामाजिक परिवर्तन और वैज्ञानिक चिन्तन का विशेष हाथ रहा है। उन्नीसवीं शती ज्ञान-विज्ञान, सामाजिक परिवर्तन और वैचारिक क्रान्ति की शती रही है। डार्विन, फ्रायड और मार्क्स के सिद्धान्तों के प्रसार के साथ नए वैचारिक वातावरण का निर्माण हुआ; ईश्वर के सम्बन्ध में विचार बदले; आस्थाएँ समाप्त हुईं तथा मानव की नियति और सहज प्रवृत्तियों के घेरे में उसकी दुर्बलताओं-विशेषताओं की चर्चा आम हुई। यही नहीं, मध्यम वर्ग के उदय के साथ समानता का भाव जागा और औद्योगिक क्रान्ति के साथ पुरातन जीवन-पद्धति का खाका ही नहीं बदला, बल्कि नई समस्याएँ भी उभरीं। इस प्रकार, अस्तित्व की जागरूकता को लेकर आधुनिक नाटक की जो शुरुआत हुई उसमें इब्सेन (1828-1906) ने महत्त्वपूर्ण भूमिका निभाई। इब्सेन ने अपने नाटकों में नए प्रश्न उठाए और अपने सारे कृतित्व

को प्राचीन नाटक की रूढ़ियों के अस्वीकार में प्रतिफलित किया। उसकी महानता इस बात में है कि उसने नाटक को स्वच्छन्दतावादी प्रवृत्तियों के बीच से उठाकर यथार्थ के धरातल पर खड़ा किया और अन्ततः उसे यथार्थ जीवन और जगत् से, उसके त्रासद अनुभवों तथा जीवन्त समस्याओं से संयुक्त किया। उसने अपने नाटकों को कई अर्थमयी भंगिमाएँ दीं और पात्रों, नाट्य-स्थितियों और संवादों के बीच से बहु-आयामी संकेत उभारने के प्रयास किए। इसीलिए उसके नाटक कई स्तरों पर भावना, क्रिया-व्यापार और बौद्धिक ऊहापोह के सामंजस्य को ही नही वरन् पक्षधरता के आग्रह के साथ-साथ काव्यमयी दृष्टि के सौन्दर्य को भी उजागर करते हैं। उसकी नाट्य-प्रतिभा ने विश्व-भर के नाटककारों को प्रभावित किया जिससे वह नाटककारों का एक पूरा खेमा तैयार करने में सफल हुआ।

आगे चलकर इब्सेन की यथार्थवादी नाट्यकला का अनेक देशों में बहुमुखी विकास हुआ। नाटककारों ने उसे वाद नहीं, धर्म की तरह अपनाया। किन्तु यथार्थ को नया आयाम देने में बर्नार्ड शॉ और चेख़व का कोई मुक़ाबला नहीं। बर्नार्ड शॉ (1856-1950) ने नाटक को गहरी पैठ और बौद्धिक सजगता प्रदान की और सच्चे अर्थों में उसे सामाजिक आलोचना का शस्त्र बनाया। रंगमंच की पूरी जानकारी के बावजूद नाटक को साहित्यिक विधा के रूप में प्रतिष्ठित करने का श्रेय शॉ को ही है। सच्चाई यह है कि कवित्वपूर्ण भावना और बौद्धिक सूझबूझ, विदूषक वृत्ति और चिन्तक की वाक्पटुता का जैसा अद्‌भुत सामंजस्य उसके नाटकों में मिलता है, वैसा कहीं और नहीं। दूसरी ओर चेख़व (1860-1904) ने यथार्थ की कारुणिक अभिव्यक्ति की। उसने जीवन को उस मानवीय बिन्दु से देखा जहाँ से सहानुभूति की किरण नज़र आती है। किन्तु उसने किसी एक का एकांगी समर्थन नहीं किया—न सत् का, न असत् का। सर्वत्र एक प्रकार की तटस्थता और सन्तुलित भावना रखी, जो व्यक्ति के भीतर अस्तित्व के दोनों छोरों को खोजती है। उसने यथार्थ और स्वप्न का ऐसा ताना-बाना बुना जिसमें एक का रंग दूसरे को फीका नहीं करता। उसके नाटकों में हास्य और करुणा के रंग घुले-मिले हैं जो एक विलक्षण भावनात्मक वातावरण की सृष्टि करते हैं। अपने समय से वह बहुत आगे था। देर से स्वीकृत हुआ; किन्तु बाद में कई नाटककारों का अगुआ बनकर रहा जिनमें ऐलमर राइस (1889-1967), यीट्स (1865-1939), सिंज (1871-1909), ओ' नील (1888-1953), इलियट (1888-1965) आदि का नाम किसी-न-किसी रूप में लिया जा सकता है। उन्नीसवीं शती के तीसरे दशक तक यथार्थ और स्वप्न के संघर्ष से उपजी चेख़व की जीवन-दृष्टि नाटककारों के बीच लोकप्रिय रही जिसने इतालवी नाटककार पिरांडेलो (1867-1936) के लिए उर्वर भूमि तैयार की।

हिन्दी में भी युगीन परिस्थितियों के गर्भ से एक प्रकार की यथार्थवादी चेतना पनपी जिसके दर्शन भारतेन्दु युग में ही होने लगते हैं; किन्तु छायावाद के उदय के साथ उसकी सारी तीव्रता मिट-सी गई। प्रसाद तक इब्सेन का नाम पहुँचा तो सही, पर 'इब्सेन का भूत' उन्हें त्रस्त किए रहा। एक समय ऐसा ज़रूर आया जब उसी के प्रभाव में

ध्रुवस्वामिनी लिखी गई। फिर भी प्रसाद न सही, कुछ लोग थे जिन्होंने आगे बढ़कर धारा के अनुकूल चलने का सजग प्रयत्न किया। भुवनेश्वर, लक्ष्मीनारायण मिश्र, उदयशंकर भट्ट, अश्क, गोविन्ददास आदि ने यथार्थवादी नाटक लिखे। खेद इतना ही है कि इब्सेन और शॉ की परम्परा से जो यथार्थ उन्होंने ग्रहण किया उसे उसका बाहरी स्वरूप मात्र कहा जा सकता है। समस्या-नाटकों के नाम पर लक्ष्मीनारायण मिश्र का नाम ख़ूब उछला। पर आग्रही आदमी की पैठ कृतित्व को विचारणा दे सकती है, कला नहीं। लक्ष्मीनारायण मिश्र के साथ ऐसा ही हुआ। इनमें 'अश्क' एकमात्र ऐसा नाटककार रहा जिसने रंगमंच के जीवन्त मुहावरे के बीच से अपने कृतित्त्व को उभारा; किन्तु कुल मिलाकर हिन्दी में यथार्थवाद पैनी दृष्टि और गहरी अनुभूति को लेकर नहीं आया। उसने हिन्दी को समर्थ नाटककार नहीं दिया—इब्सेन और चेख़व को तो जाने दीजिए—यूरोप के दूसरी श्रेणी के नाटककार जैसा भी नहीं। फिर भी यह मानना होगा कि हिन्दी का यथार्थवादी नाट्य लेखन एक तरह से व्यर्थ नहीं गया। जगदीशचन्द्र माथुर ने स्वातंत्र्योत्तर नाट्य साहित्य को वह आधार-बिन्दु प्रदान किया जिसकी परम्परा और प्रतिक्रिया को लेकर आधुनिक हिन्दी नाटक के विकास में मोड़ आए।

यह नई दिशा हिन्दी को जैसे-तैसे मिल गई—कुछ लोगों की ग्राहिका वृत्ति या प्रतिभा, या फिर प्रयोग की चाह के कारण; किन्तु पश्चिम का नाट्य लेखन एक लम्बे अर्से तक यथार्थ के बाहरी खोल से बड़ी मुस्तैदी से जूझता रहा। यथार्थ का वास्तविक स्वरूप क्या है ? यह प्रश्न इब्सेन के परवर्ती नाटककारों को बहुत समय तक उलझाता रहा। यथार्थवाद के आगे स्वयं यथार्थ का प्रश्न उठ खड़ा हुआ ! इब्सेन ने यथार्थवादी दृष्टि दी, चेख़व ने स्वप्न और यथार्थ के बीच आन्तरिक यथार्थ की काव्यमयी व्यंजना की; किन्तु परवर्ती नाटककारों ने तो दुविधाओं और शंकाओं की शृंखला खड़ी कर यथार्थ की सत्ता के सामने ही प्रश्न-चिह्न का संकट खड़ा कर दिया। इसका फल यह हुआ कि स्वयं यथार्थवाद के भीतर ही शंका और चुनौती का स्वर मुखर हो उठा। इसकी पहली-पहली अनुगूँज पिरांडेलो (1867-1936) तथा स्ट्रिंडबर्ग (1849-1912) के नाटकों में उभरी और फिर उन्नीसवीं शती के उदय के साथ प्रतीकवाद, अभिव्यक्तिवाद, अतिप्रकृतवाद आदि विचार-आन्दोलन उठ खड़े हुए जो सब यथार्थवाद के ख़िलाफ़ एकजुट हो उठे।

इन सबकी तीव्र प्रतिक्रिया थी कि जिसे 'यथार्थ' कहा जाता रहा है वह मिथ्या भ्रम मात्र है। महत्तर यथार्थ इस सबसे कही बहुत आगे है। फलतः स्थूल और आध्यात्मिक यथार्थ, बाह्य जगत् और अन्तर्जगत के यथार्थ की सही पहचान करने के लिए जो विभिन्न वाद उठ खड़े हुए, उन्होंने अपने-अपने ढंग से निजी यथार्थ की खोज की और उस तलाश में व्यक्ति और उसकी आन्तरिकता को तरजीह दी—आन्तरिकता को इसलिए भी क्योंकि इन्होंने सत्य की सत्ता को वैयक्तिक माना। फलतः सतही जीवन के मिथ्याभास की अपेक्षा आन्तरिक जीवन की गहरी वास्तविकता और जीवन के ऊपरी ढाँचे की अपेक्षा उसकी अन्तर्निहित चेतना नाटकीय कृतित्त्व की आधार शिला बनी।

इस प्रकार यथार्थ के सही स्वरूप की तलाश ने नाटककार को सजग किया। यहाँ तक कि वे नाटककार भी—जैसे पिरांडेलो, स्ट्रिंडबर्ग, ब्रेख़्त आदि—जो पहले यथार्थवादी थे, अब यथार्थ के नए क्षेत्रों को तलाशने में लग गए। यह सब अचानक और अयाचित रूप में नहीं हुआ। वस्तुतः डेकार्ट, लॉक, कान्ट, शॉपेनहावर, नीत्शे आदि चिन्तक यथार्थ के प्रति नई जिज्ञासा पहले ही जगा चुके थे। नाटक के क्षेत्र में इसकी जो प्रतिक्रिया हुई, उससे रचना में स्वानुभूति, आन्तरिकता और इन्द्रिय संवेद्य ज्ञान का महत्त्व बढ़ा; नाटककार को तो जैसे एक सूत्र मिल गया : 'जो है, मैं उसे नहीं देखता। अगर कुछ है तो वही, जो मुझे दिखाई देता है !'

इस आत्मपरक दृष्टि के उन्मेष के साथ फ्रांस में 1880 के आसपास प्रतीकवाद (जिसे नव्य स्वच्छन्दतावाद भी कहा जाता है) का आविर्भाव हुआ जो बहुत जल्दी ही 1900 ई. तक लुप्तप्राय भी हो गया। मूलतः यह काव्य का आन्दोलन था; किन्तु कालान्तर में इसने नाटक को भी समान रूप से आत्मसात् कर लिया फलतः नाट्य रचनाओं में भी तर्क और बुद्धि के स्थान पर इन्द्रिय बोध और अन्तर्बोध को उच्चतम सत्य की अवधारणा का साधन बनाया गया और जीवन की अन्तरतम अनुभूतियों की अभिव्यक्ति के लिए भावना और कल्पना को वहन करनेवाले प्रतीकों का प्रयोग सर्वसामान्य हुआ। इससे रचना की समग्रता, काव्यमयी अनुभूति और भाषा की सर्जनात्मकता को बल मिला। कवि अथवा कवि-स्वभाववाले नाटककार ही प्रायः प्रतीकवादी नाट्य लेखन की ओर अधिक प्रवृत्त हुए जिनमें मॉरिस मेटरलिंक (1862-1949), रोस्ताँ (1868-1918), हॉफमैन्स्थल (1873-1928), हाउप्टमान (1862-1946), क्लॉडेल (1868-1955), यीट्स (1865-1939), आन्द्रेयव (1871-1919), ऑस्कर वाइल्ड (1854-1900), लोर्का (1899-1936) आदि उल्लेखनीय हैं। यद्यपि प्रतीकवाद के कुछ तत्त्व आज भी कई नाटककारों के बीच लोकप्रिय हैं, किन्तु रहस्यात्मकता, भावुकता तथा वैचारिक एकसूत्रता के अभाव में यह आन्दोलन बहुत स्थायी नहीं रहा।

लगभग इसी काल में नाटककारों का एक दूसरा वर्ग अभिव्यक्तिवाद के प्रभाव में नाट्य रचना में संलग्न था। प्रतीकवाद का उद्गम फ्रांस था, अभिव्यक्तिवाद जर्मनी की देन कहा जा सकता है। 1900 के आसपास जर्मनी में अभिव्यक्तिवाद का बिल्ला उन रचनाओं पर लगाया जाने लगा था जो वैन्गॉफ की कलाकृतियों का अनुसरण करती थीं। 1910 से लेकर 1920 तक तथा अभिव्यक्तिवादी नाट्य रचना को अनेक प्रतिभाशाली नाटककारों का प्रबल सहयोग मिला। इसमें पिरांडेलो को एकदम अभिव्यक्तिवादी कहने में संकोच हो सकता है; किन्तु इसमें कोई सन्देह नहीं कि यथार्थवादी प्रवृत्तियों का नाटककार होने के बावजूद वह क्रोचे का समर्थक था और यथार्थ के आन्तरिक और बाह्य स्वरूपों के सम्बन्ध में उसकी मान्यताओं का अनुसरण करते हुए ही उसने अपनी नाट्य कृतियों में आन्तरिक यथार्थ को परिभाषित किया। उसने यथार्थ को वस्तु में नहीं, विचार में खोजने का प्रयास किया और इस प्रकार व्यक्तिपरक और अभिव्यक्तिवादी

नाट्य लेखन के लिए आवश्यक भूमिका तैयार की। पिरांडेलो की भाँति ही हाउप्टमान, बेडेकाइंड, स्ट्रिंडबर्ग, कैसर आदि का दाय भी कम महत्त्वपूर्ण नहीं रहा। अभिव्यक्तिवादी नाट्य लेखन को फ्रांस में कोई सहयोग नहीं मिला, किन्तु पूर्वी यूरोप में कारेल चैपक, कॅरोल रोस्तोवोरस्की, सीन ओ' केसी जैनी अनेक प्रतिभाएँ इस दिशा में कार्यरत रहीं।

इनमें सबसे अधिक गतिशील व्यक्तित्व स्ट्रिंडबर्ग का रहा। प्रारम्भ में वह जोला को आदर्श मानकर चला था। वह इब्सेन से घृणा करता था, पर लेखन में उसकी जागरूक दृष्टि का हामी था। अपने प्रारम्भिक काल में इसीलिए उसने पर्याप्त यथार्थवादी नाटक लिखे; किन्तु जीवन की वैयक्तिक अनुभूतियों की आकुलता ने उसे आत्मपरक नाटक लिखने की ओर प्रवृत्त किया। उसने बाह्य जगत् में अपने को केन्द्र में रखकर देखा। फलतः यथार्थ के प्रति उसकी दृष्टि सदा आत्मपरक रही और यही आत्मपरकता उसे कालान्तर में अवचेतन के लोक में ले गई जहाँ से उसके स्वप्न नाटकों की अवतारणा हुई है। इन नाटकों की अवधारणा के सम्बन्ध में उसने स्वयं ही कहा है : 'कुछ भी हो सकता है, सब-कुछ सम्भव है। दिक् और काल का कोई अस्तित्व नहीं। यथार्थ की महत्त्वहीन पृष्ठभूमि पर कल्पना स्मृतियों निर्बंध मानसिक उड़ानों एवं कामचलाऊ संवादों के सहारे अपना तन्तु बुनती है। चरित्र खंडित हैं और बहुरंगी, जो कभी भाप बनकर हवा में उड़ जाते हैं, कभी ठोस हो जाते हैं; कभी इकट्ठे हो जाते हैं, कभी बिखर जाते हैं। किन्तु सबको एक ही चेतना एक सूत्र में बाँधकर रखती है और वह है स्वप्नदर्शी की चेतना।'[1] जैसे-जैसे वह मानव-नियति से परिचित होता गया, उसकी बाह्य जगत् को नकारने की प्रवृत्ति भी उसी मात्रा में बढ़ती गई और धीरे-धीरे जीवन को एक मृग-मरीचिका, एक स्वप्न-भंग, एक उल्टी लटकी तस्वीर समझने का आग्रह ही उसकी सर्जन शक्ति का परिचायक बन गया। उसने इतने अधिक प्रभाव स्वीकार किए, इतनी अधिक दिशाओं में प्रयोग किए कि उसे सबसे अधिक बेचैन, अस्थिर और प्रयोगवादी नाटककार कहा जा सकता है। उसका कृतित्व आस्था और अनास्था के समग्र आयामों को छूता है जिसके अन्तर्गत प्रत्यक्षवाद, नेरीश्वरवाद, आधिभौतिकवाद, समाजवाद—सभी कुछ आ जाते हैं। यही नहीं स्त्री-पुरुष के सम्बन्धों को उसने एक नई दृष्टि से देखा और परखा। मानवीय त्रासदी को एक नए धरातल पर उजागर करने का श्रेय भी उसी को है। रचनाओं के पीछे सक्रिय आत्मपरक तत्त्व, भोगे हुए यथार्थ, यथार्थ के बीच महत्तर यथार्थ की खोज और अन्ततः उसकी अभिव्यक्ति में उसके अपने आत्मिक ओज ने उसके नाटकीय व्यक्तित्व को अद्भुत गरिमा प्रदान की थी। इसका ही परिणाम था कि बाद के यथार्थवादी, अभिव्यक्तिवादी, अतिप्रकृतवादी, दादावादी, ऐब्सर्डवादी नाटककारों ने उससे प्रेरणा के सूत्र निकाले और आगे चलकर वह लेखन की नई प्रवृत्तियों का मसीहा बना।

अभिव्यक्तिवादी नाटककारों की सबसे बड़ी देन यह रही कि उन्होंने एक ओर प्रकृतवाद की रूढ़ियों को तोड़ा, दूसरी ओर मानव की आवरण-रत अन्तरात्मा को उघाड़कर उसे पारदर्शी रूप में प्रस्तुत करने का प्रयास किया। अभिव्यक्तिवाद

स्वच्छन्दतावाद की भावुकतापूर्ण उड़ान और यथार्थवाद की सतही दृष्टि के विरोध में पैदा हुआ। इस विरोध के बीच से कथ्य, वस्तु और शिल्प के स्तर पर जो नए प्रयोग हुए उनका नई दिशा के निर्धारण में महत्त्वपूर्ण योग रहा। अवचेतन में निहित आत्मनिष्ठ आन्तरिक जीवन की अभिव्यक्ति, कथावस्तु का सरलीकरण, चरित्रों की वर्ग-रूप प्रस्तुति, संवादों की मितव्ययता, छोटे दृश्यों और प्रतीकों का प्रयोग अभिव्यक्तिवाद की कुछ सामान्य विशेषताएँ कही जा सकती हैं। इसी प्रकार रंगमंच के क्षेत्र में भी अभिव्यक्तिवाद ने अतिरंजनापूर्ण आकृतियों, विकर्षक रूप-रेखाओं, विसंगत रंगों, अभिनय की यान्त्रिक पद्धति तथा सांकेतिक स्वाद-योजना के माध्यम से कुछ ऐसे उपकरण जुटाए जो ऐसे लेखन की मूल प्रवृत्ति से मेल खाते हैं।

अभिव्यक्तिवाद की भाँति ही यथार्थ की पुनर्व्याख्या के उद्देश्य से फ़्रांस में अतियथार्थवाद का जन्म हुआ। प्रथम महायुद्ध ने स्पष्टतः जीवन, जगत, बौद्धिक विलास, नैतिकता तथा सामाजिक व्यवस्था का खोखलापन उघाड़कर रख दिया था। सभ्य राष्ट्रों ने मानवता के पोषित महान् आदर्शों का स्वयं ही खंडन करके दिखा दिया था। महायुद्ध के दरमियान विवशता का यह बोध बड़ी तीव्रता से गहराया जिसकी अभिव्यक्ति दादावाद में हुई। दादावाद युद्ध की काली छाया का प्रभाव लेकर आया।

अतियथार्थवादियों का विश्वास था कि वास्तविक यथार्थ बाहर की दुनिया में नहीं, वरन् उस मानव-क्षेत्र में निहित है जिसका अन्वेषण अभी तक हुआ ही नहीं। अवचेतन को मुख्यता देते हुए उन्होंने घोषणा की कि सच्ची कला की परिणति घोर निराशा, अनास्था, मूल्यहीनता और अनस्तित्व की भावना में हुई दिखाई देती है। अतियथार्थवाद के प्रवर्त्तन का सूत्र अपोलोनियर (1880-1918) के सुप्रसिद्ध नाटक **ल मैमील डी टेरेसिया** (टेरेसिया के स्तन) में बताया जाता है (जिसका लेखन 1903 में और मंचन 1917 में हुआ था) जिसे उसने 'द्राम सुर-रियलिस्टे' कहकर प्रस्तुत किया था। इसी के आधार पर बाद में आन्द्रे ब्रेटन ने अपना अतियथार्थवाद (सुर-रियलिज़्म) का घोषणा-पत्र प्रस्तुत किया। इसी नाटक से प्रेरणा लेकर ट्रिस्टन जारा पहले ही दादावाद का श्रीगणेश कर चुका था, किन्तु नाजियों के प्रभाव स्वरूप उसने जल्दी ही जर्मनी में दम तोड़ दिया और उसकी चिता से ही अतियथार्थवाद का स्वर मुखर हो उठा। जो पहले दादावादी थे, अब अतियथार्थवाद के समर्थक बन बैठे। अतियथार्थवादियों की नाट्य दृष्टि का अन्दाज़ अपोलोनियर के पूर्वोक्त नाटक की प्रस्तावना से लगाया जा सकता है : 'नाटक को यथार्थ का अनुकरण नहीं होना चाहिए/यह सही है कि नाटककार को सभी दरकार वास्तविकता के भ्रमों का इस्तेमाल करना चाहिए/और यह भी कि उसको अब/दिक् और काल की सीमा को स्वीकारना अनिवार्य नहीं/उसकी दुनिया उसका नाटक है/जिसके अन्दर वह ईश्वर है, सर्जनकर्त्ता है/जो अपनी इच्छा पर आवाज़ों, भंगिमाओं, चेष्टाओं और रंगों को व्यवस्थित करता है ! इसलिए नहीं कि उसे ज्यों-का-त्यों तस्वीर जैसा दर्शाए/जिसे कहते हैं जीवन का एक खंड/वरन् इसलिए कि वह जीवन को ही स्वयं प्रस्तुत करे/और उसके सत्यों को भी।'

उन्होंने बुद्धिवादी दृष्टि की सर्वथा उपेक्षा की और मानव की असंगत चेतना को ही सर्जन का स्रोत ठहराया। फलतः असंगति, वीभत्स कल्पना आदि का सहारा लेकर अतियथार्थवादियों ने ऐसे नाटक लिखे जिनमें विस्फोटक प्रवृत्तियों, रोमांचकारी दृश्यों, अन्तर्विरोधमूलक मुद्राओं, त्रासद और फार्स की सम्मिलित नाट्य स्थितियों, मुखौटों और अर्थहीन संवादों की भरमार मिलती है। अतियथार्थवाद की देन बहुत स्थायी और महत्त्वपूर्ण तो सिद्ध नहीं हुई, किन्तु इस वर्ग के नाटककारों में ज्याँ कॉक्तू, फर्नैण्ड क्रौमलिंक, रॉजर वित्राक आदि ने पर्याप्त ख्याति अर्जित की। इनके अतिरिक्त ऐसे नाटककारों की संख्या भी कम नहीं है जो सीधे अतियथार्थवाद के ध्वज के नीचे नहीं आते, किन्तु उनकी रचना-प्रक्रिया पर उसका यथेष्ठ प्रभाव पड़ा है। इनमें आरमण्ड सालाक्रे, जूलियन तोरमाँ, रेमण्ड रनेल, लोर्का आदि उल्लेखनीय हैं।

दो विश्वयुद्धों के बीच नाटक इन वादों के दौर से तो गुज़रा ही, कुछ और प्रवृत्तियाँ भी उभरीं। इसी दौरान ऐतिहासिक नाटक को प्रश्रय मिला और नाटककारों में समसामयिक जीवन की अभिव्यक्ति के लिए इतिहास के उपयोग की आकांक्षा जागी। फलतः इस समय रूस, फ़्रांस, ब्रिटेन, अमेरिका में पर्याप्त संख्या में ऐतिहासिक नाटक लिखे गए। रोमाँ रोलाँ (1868-1944), फ्रांज वर्फेल (1890-1945), मैक्स्वेल ऐण्डर्सन (1888-1959) आदि को इसमें ख़ूब सफलता मिली। उसी प्रकार कुछ नाटककार मिथकीय विषयों की ओर भी उन्मुख हुए। इस दिशा में ज्याँ जीरादू (1882-1944) का नाम उल्लेखनीय है। बाद में ज्याँ अनुइ (1910—) ने इस दिशा में महत्त्वपूर्ण प्रयोग किए। लोक कथाओं को लेकर कवि की-सी दृष्टि को माध्यम बनाकर गार्सिया लोर्का (1899-1936) ने अनेक नाटक लिखे। जीरादू और लोर्का की मूल संवेदना कवि की थी और काव्यात्मक रंगमंच के निर्माण में उनकी नाट्य कृतियों ने महत्त्वपूर्ण योग दिया। फलतः इस समय तीव्रता से यह अनुभव किया जाने लगा था कि नाटक और मंच को उच्च स्तर पर प्रतिष्ठित करना है तो एक बार फिर नए संदर्भों में उसे काव्य, कल्पना और भाषा के नए संस्कार से युक्त करना होगा। और यह कार्य ऑर्डन, लोर्का, इलियट आदि ने बख़ूबी कर दिखाया।

द्वितीय महायुद्ध के दौरान एक बार फिर जीवन की निस्सारता का तीखा अहसास हुआ। विज्ञान के विकास के नाम पर संघातक अस्त्र-शस्त्रों और शैतान जैसी विशालकाय मशीनों के द्वारा एक ऐसा अभिशाप मनुष्य के हाथ लगा जिसने उसकी सुरक्षा को ही नहीं, मानवीयता को भी ख़तरे में डाल दिया। यांत्रिक सभ्यता ने उसे जैसे अलगाव और विसंगति के कगार पर ला खड़ा कर दिया; आध्यात्मिक दृष्टि से विहीन भौतिकवाद से हिंसा, रक्तपात, होड़, शोषण और अनैतिकता को बढ़ावा मिला। विश्वयुद्ध ने जिस रूप में मनुष्य को झकझोरकर रख दिया, उससे रही-सही मान्यताएँ भी हिल गईं। महान् मानवीय संकट की इस स्थिति में आस्था और विश्वास भी उड़ गए। जो बच रहा वह मात्र अव्यवस्था का अहसास था जिसकी भँवर में व्यक्ति बुरी तरह घिर चुका था। इसलिए मानव के अस्तित्व के सामने एक और प्रश्न-चिह्न लग गया था।

प्रश्नों की इसी किंकर्त्तव्यविमूढ़ता के बीच से इसी समय एक चिंतन उभरा जिससे निविड़ अन्धकार में राह की तलाश शुरू हुई। इस अनजानी राह का अगुआ बना कीर्केगार्द (1813-1855), जो अपने जीवन काल में उपेक्षित रहा, किन्तु एक शती बाद अचानक लोगों को वह मसीहा दिखाई पड़ा। और फिर चाहे साहित्य हो या दर्शन, कुछ भी उसके प्रभाव से न बच सका। उसी की भाँति नीत्शे (1844-1900) और दोस्तोवस्की (1821-1881) ने उन्नीसवीं शती में आधुनिक भाव-बोध की नींव डाली जिस पर बाद में कार्ल यास्पर्स, हेडेगर, मार्सल, ऐल्बर्ट कामू, फ्रांज काफ्का, सार्त्र, बर्दियेफ़ आदि ने एक नए दर्शन का महल खड़ा किया जो अस्तित्ववाद के नाम से प्रसिद्ध हुआ।

इसी अस्तित्ववाद से विसंगतवादी (ऐब्सर्ड) नाटकों का जन्म हुआ है। वस्तुतः युद्धोत्तर नाटकों के विकास में अस्तित्ववाद का महत्त्वपूर्ण योग रहा है, यद्यपि विसंगत नाटक की जड़ें उन्नीसवीं शती में पिरांडेलो, स्ट्रिंडबर्ग, ऐल्फ्रेड ज़ार्री, अपोलोनियर, अनुई आदि की नाट्य कृतियों में विद्यमान हैं। यहीं नहीं, दादावाद के प्रभाव में गॉल तथा ब्रेख़्त कई ऐसे नाटक पहले ही लिख चुके थे जो विसंगत नाटकों के बहुत निकट पड़ते हैं। किन्तु सही अर्थों में विसंगत का तीव्र बोध अस्तित्ववाद की ही देन है। उसी के फलस्वरूप 1950 के आसपास विसंगत नाट्यधारा बड़े प्रबल वेग से सामने आई।

भारत में वैसी स्थिति अकल्पनीय थी, क्योंकि युद्ध वैसा तीव्र अहसास यहाँ दे ही नहीं गया। फिर भी यह कहना ग़लत होगा कि इस कालखंड में हमारे साहित्य ने कोई मार्ग ही तय नहीं किया। द्वितीय महायुद्ध का काल हिन्दी साहित्य-रचना में भी निर्णायक रहा। जब पश्चिम में विसंगतवाद फूल-फल रहा था, हिन्दी की नई कविता और कहानी एक नए दौर से गुज़रने का उपक्रम कर रही थी और नाटक सामाजिक यथार्थ से जूझ रहा था। अस्तित्ववाद ने उसे बाद में प्रभावित किया; किन्तु विसंगत नाटक अपनी अतिसीमाओं में तब भी हिन्दी ने दूर रहा, अब भी बहुत दूर ही दिखाई देता है।

पश्चिम के विसंगत नाट्य लेखन को युग के वैचारिक संकट का द्योतक कहा जा सकता है। जीवन के आधारभूत ढाँचे के प्रति असंतोष को व्यक्त करते हुए यह अस्तित्व की समस्याओं से जूझता, बौद्धिक स्तर पर उतना नहीं जितना अपनी मौलिक अनुभूति, अपने भोगे हुए क्षणों के साक्ष्य और संवेदना पर निर्भर करता है। विज्ञान अन्धी गली तक पहुँच गया है, युद्धोत्तर काल का यह अहसास इस विचार को पुष्ट करने में सफल हुआ है कि युग की अनेक समस्याओं का समाधान और रुग्ण मानवता का निदान विज्ञान के पास नहीं है। मनुष्य की सहायता न विज्ञान कर पाया है और न परम्परा के चलते आए सड़े-गले जीवन मूल्य। अस्तित्ववादी चिंतक यह मानते हैं कि जगत् में कोई शाश्वत सुनिश्चित नैतिकता नहीं; कुछ भी पूर्व-निर्धारित नहीं—मनुष्य स्वयं जगत् में अपने को परिभाषित करता है किन्तु जो जगत् उसे मिला है वह प्रयोजनहीन है, गोबर का ढेर है, बिलकुल ऊपर से थोपा हुआ—सड़ांध लिए हुए कूड़े जैसा, जिस पर उबकाई आती है। इसमें वह पीड़ा, अपराध, आतंक, त्रास, अलगाव और मृत्यु जैसी विसंगतियों से घिरा है। वरण की क्षमता के बावजूद उसकी स्थिति बेतुकी और

ऊलजलूल है। सबसे बड़ी विडम्बना तो यह है कि अर्थहीन जगत् की विभिन्न स्थितियों और मनुष्य की विपुल आकांक्षाओं के बीच बहुत बड़ी खाई है जिससे उसकी पुंसत्वहीनता, विवशता और अजनबीपन की नियति ही प्रकट होती है। कुल मिलाकर यह बोध अस्तित्व की निरर्थकता और नास्ति भाव को ही सिद्ध करता है। विसंगत नाटक इसी दर्शन और बौद्धिक चिंतन पर टिका है।

1945 से 1965 का कालखंड विसंगत नाटकों के लिए विशेष उर्वर रहा है। कामू, ज्यां पाल सार्त्र, ज्याँ जेने, सैम्युअल बेकेट, यूजीन आयनेस्को, आदामोव, ऐल्बर्ट आल्बी, हेरल्ड पिंटर आदि नाटककारों ने इस धारा को पुष्ट करने में अपनी विलक्षण प्रतिभा का परिचय दिया। ये सभी नाटककार प्रकृति और दृष्टि में भिन्न होते हुए भी वस्तुतः एक ही खेमे में आते हैं, क्योंकि मूलतः सभी परम्परागत जीवन-मूल्यों का बहिष्कार करते हुए जीवन को विसंगति का पर्याय मानते हैं। वे सभी अस्तित्व की विरूपता की स्थिति को स्वीकार कर चलते हैं जिससे भद्दे, भोंडी, बेतुकी, अनर्गल स्थितियों का नियोजन एक सामान्य प्रवृत्ति दिखाई देती है। मानवीय स्थिति की सही विद्रूपता का दर्शन कराने के लिए वे प्रायः दुःस्वप्न तथा अतिकल्पना को माध्यम बनाते हैं और अतिरंजनापूर्ण कथावस्तु या सनसनीखेज दृश्य-योजना का अटपटा उपयोग कर पाठक/प्रेक्षक को झटका देते हैं। इसी प्रकार उनकी पात्र-योजना भी कम विलक्षण नहीं होती। विसंगत नाटकों के चरित्र असामान्य व्यक्ति होते हैं—चोर, उचक्के, आवारा, समाज से दंडित, बहिष्कृत, विक्षिप्त, सनकी, कुंठित, निरर्थक जीवन जीनेवाले; फिर भी मानवीय संवेदना और विक्षोभ की भावना से परिपूर्ण; एकदम हास्यास्पद और करुण। उनकी जीवन-पद्धति के बीच से ही हास्यास्पद स्थिति उभरती है, किन्तु उनकी नियति उनको अतिभावुक और अतिनिराश व्यक्ति के रूप में प्रस्तुत करती है। कुल मिलाकर इन नाटकों में प्रहसन और त्रासदी के सम्मिलित तत्त्वों का सुन्दर सामंजस्य मिलता है। आयनेस्को और बेकेट अत्यधिक भय और त्रास का अनुभव देते हैं। सबसे अधिक चौंकानेवाला तत्त्व होता है उनका कथ्य, जो बिम्बों, प्रतीकों और समासोक्तियों के माध्यम से अपने को उद्घाटित करता है। सूनेपन, कुछ न होने की स्थिति, प्रतीक्षा, अकेलेपन और अलगाव की अनुभूति कराने में वे दिमाग़ पर जमी काई को खुरच डालते हैं, इसमें कोई सन्देह नहीं।

विसंगत नाटक नाट्य संरचना के क्षेत्र में कथ्य की भाँति ही विसंगति को ही आधार लेकर चलते हैं। मूक अभिनय, अतिरंजित आंगिक चेष्टाओं, मुखौटों, विरूप दृश्यों और ऊलजलूल संवादों का सजग प्रयोग कर वे वातावरण और अर्थ की अद्भुत सृष्टि करते हैं। भाषा के सम्बन्ध में भी उन्होंने चौंकानेवाले प्रयोग किए हैं। जहाँ वे एक ओर यह मानते हैं कि भाषा की शक्ति चुक गई है, वहाँ रोज़मर्रा की घिसी-पिटी शब्दावली और निरर्थक-सी लगनेवाली उक्तियों से भी वे अर्थपूर्ण काव्यात्मक स्थितियों की सृष्टि कर दिखाते हैं जिससे भाषा पर उनका विलक्षण अधिकार प्रकट होता है।

इस प्रकार, इब्सेन से लेकर पिंटर तक नाटक एक वृत्त को पूरा करता है। यह

विसंगत नाटक चरम उत्कर्ष का द्योतन करता है। यह बात दूसरी है कि अब धीरे-धीरे परम्परा मृतप्राय होती जा रही है और एक बार फिर नाटककार नव-यथार्थवाद की ओर उन्मुख होते जा रहे हैं, और आशा की जाती है कि भविष्य नव-यथार्थवादियों के हाथ में होगा; किन्तु इससे नाटक ने जो लम्बी यात्रा तय की है उसका महत्त्व किसी प्रकार घटता नहीं।

अब प्रश्न उठता है कि सारी चर्चा का क्या मतलब है ? इन सब बातों का हिन्दी नाटक से क्या सम्बन्ध है ? यहीं नाटक की विश्वजनीनता का प्रश्न उभरता है जिसकी चर्चा हम प्रारम्भ में कर चुके हैं। हमारे यहाँ यह कहा जाता रहा है कि पूर्व का पश्चिम से कोई सरोकार नहीं; हमारे साहित्य और चिंतन का अपना अलग दायरा है जिसे औरों से कुछ लेना-देना नहीं। इधर जो शोध-ग्रन्थ लिखे गए हैं उनमें जहाँ खोल से बाहर निकलने का प्रयत्न हुआ भी है, वहाँ नाटक पर पाश्चात्य प्रभाव की बात बड़ी बेढंगी शैली में की गई है और इस तथ्य को नज़रअन्दाज़ किया गया है कि जब दो धाराएँ एक-दूसरे के सम्पर्क में आती हैं तो नए रंग उभरते हैं। प्रसाद, लक्ष्मीनारायण मिश्र, जगदीशचन्द्र माथुर, धर्मवीर भारती, लाल और राकेश ने हिन्दी नाटक को विश्व की विशद् नाट्य धारा में समाहित करते हुए यही रंग उभारे हैं। आज पूर्व और पश्चिम की विभाजक रेखाएँ मिटती जा रही हैं। अखिल विश्व एक-सी परिस्थितियों और अनुभवों के बीच से गुज़र रहा है। साहित्य चाहे पूर्व का हो या पश्चिम का, सर्वत्र आज मानवीय संकट को बड़ी तीव्रता से अनुभव किया जा रहा है। अन्तर इतना ही है कि चरम भौतिक विकास के कारण पश्चिम में यह संकट भयंकरता से गहरा रहा है; पूर्व में अभी शुरुआत ही है—हमने अभी उसका स्वाद चखना ही शुरू किया है। पश्चिम का चिन्तन और लेखन एक लम्बे अरसे से इस अहसास की प्रक्रिया से गुज़र रहा है और उसने उसे एक तीखी प्रसव-पीड़ा में जन्म दिया है जबकि हम अपने को एक बने-बनाए अनुभव के साथ जोड़ रहे हैं। हमारी आधुनिक साहित्य चेतना इतिहास के गर्भ से नहीं उपजी—समय का बहुत-सा पानी जैसे बिना हमें भिगोए ही पुल के नीचे से निकल गया। यथार्थवाद के बाद, जैसा कि विश्व साहित्य में हुआ है, हमारा लेखन विभिन्न प्रतिक्रियाओं के बीच से गुज़रा ही नहीं। नव-स्वच्छन्दतावाद, अभिव्यक्तिवाद, दादावाद, अतिप्रकृतवाद, विसंगतवाद—सबको छलाँगकर हमारा नाट्य साहित्य आधुनिकता के नाम पर मोहन राकेश पर आकर टिका। मोहन राकेश के **आषाढ़ का एक दिन** से लेकर, बीज नाटकों और पार्श्वनाटक (**शायद हाँ**; तथा **छतरियाँ**) तक विसंगतवाद की छाया मिलती है।

फिर भी यह अचानक ही नहीं हो गया। प्रगतिवाद से आगे की बात करें तो 1943 में **तार सप्तक** के प्रकाशन के साथ एक अन्वेषी दृष्टि का उन्मेष मिलता है। इसमें भी कोई सन्देह नहीं कि नई कविता का आन्दोलन भी एक अन्तर्राष्ट्रीय आन्दोलन है जिसने संसार में पनपते अस्वीकार, विफलता-बोध और नैराश्य को मुखर किया है। कविता की तुलना में नई कहानी ने शायद युग के बदलते तेवरों को जल्दी पहचाना। किन्तु नाटक के क्षेत्र में यह जड़ता बहुत देर से भंग हुई। मोहन राकेश कहानीकार भी थे—नाटककार

के रूप में उन्हें जो विरासत मिली वह नाटक से नहीं, नई कहानी से मिली। नाटक के क्षेत्र में उनकी विरासत शून्य है। (**आषाढ़ का एक दिन** के रोमान तथा ऐतिहासिक कथानक की बात जाने दीजिए) प्रश्न उठता है कि क्या राकेश ने पाश्चात्य परम्परा से कुछ विरासत में प्राप्त किया ? इस प्रश्न के उत्तर में दृष्टि अस्तित्ववाद पर टिक जाती है। अस्तित्ववाद काफ़ी समय तक नए कहानीकारों का धर्म-सा बना रहा। स्वतन्त्रता के बाद हमारे देश में महँगाई, मुद्रास्फीति, काला धन, लूट-खसोट, बेराजगारी व असमाजवादी नारों ने एक ऐसी स्थिति खड़ी कर दी जिससे अस्तित्व के कई प्रश्न उभरकर ऊपर आए। स्वयं राकेश के शब्दों में : 'निर्माण हुआ बड़े-बड़े भवनों का, सरकारी और अर्द्ध सरकारी संस्थाओं, समितियों और आयोगों का, कारख़ानों और मशीनों का, बाँध और विकास योजनाओं का और शासकीय शब्दकोशों का। इस निर्माण की सतह के नीचे इंसान का जो रूप सामने आया, वह बहुत ही विकृत था। लगा कि आस-पास के बड़े-बड़े परिवर्तनों के साये में लोग निरन्तर पहले से छोटे और कमीने होते जा रहे हैं।...ज़िन्दगी का सारा अन्दरूनी ढाँचा भुरभुरी मिट्टी की तरह झड़ता-ढहता जा रहा है।'[2]

पश्चिम में तीस साल पहले महायुद्ध के कारण जो पस्ती, नपुंसकता और निरर्थकता लोगों पर हावी थी, वह मानसिकता हमें स्वतन्त्रता की विरासत के रूप में प्राप्त हुई। इसलिए भारत की भूमि अस्तित्ववादी विचारधारा के अंकुरण के लिए उर्वर सिद्ध हुई। हिन्दी कविता और कथा साहित्य पर उसकी पूरी छाप पड़ी, नाटक पर भी। पर कौन देश ऐसा है जहाँ अस्तित्ववाद के प्रभाव में कभी कोई रचना नहीं हुई हो ! राकेश के नाटकों में विश्व नाटक की गतिविधियों और प्रवृत्तियों के सन्दर्भ में ऐसा ज़रूर देखा जा सकता है; किन्तु यह उन पर किसी प्रकार का दोषारोपण नहीं। एक तो अस्तित्व के संकट को उन्होंने स्वयं अपने जीवन में झेला है जिससे उनका लेखन बाहरी/आयातित चिन्तन पर ही निर्भर नहीं करता। दूसरी बात यह है कि अस्तित्ववादी चिन्तन आज की भारतीय परिस्थितियों के अनुकूल होने के नाते समसामयिकता का सही बोध कराता है। साथ ही वह एकदम अभारतीय भी नहीं है। कई भारतीय और विदेशी विद्वानों ने अस्तित्ववाद को भारतीय और पाश्चात्य दर्शन के बीच का सेतु कहा है।[3] मोहन राकेश ने पूर्व और पश्चिम के उसी सेतु पर अपने लेखन को टिकाया है और विश्वजनीनता की तलाश के बीच ही उसे सच्ची निजता दी है।

तो क्या सचमुच राकेश हिन्दी नाटक को निजता दे पाए ? किन्तु जैसा कि उन्होंने स्वयं कहा है : 'निजता एक बहुत सीमित अर्थ में भी ग्रहण की जा सकती है और बहुत व्यापक अर्थ में भी।'[4] राकेश के नाटकों में व्यापक अर्थ में निजता है और उसी अनुपात में विश्वजनीनता भी। 'अपने समवर्ती लेखन के सम्बन्ध में जानकारी होना एक बात है, किन्तु उसे अपना आदर्श मान लेना बिलकुल अलग बात है'[5]—वे इस तथ्य से अवगत थे। राकेश ने जो लिखा है उसमें उनके विलक्षण व्यक्तित्व, अनुभव, चिन्तन और प्रयोग का भी बहुत बड़ा हाथ है। औरों के साथ राकेश के लेखन की यही निजता है। पाश्चात्य

प्रभाव के कारण प्रगतिवाद के बाद हिन्दी में जो व्यक्तिवादी परम्परा पनपी, राकेश उसी के वाहक बने, पर सामाजिक जागरूकता भी अप्रत्यक्ष रूप से व्यंजित करने में वे समर्थ हुए। वे व्यक्ति और समाज को सम्बन्ध सापेक्ष मानकर चले। इसीलिए व्यक्ति की समाज को झेलने की पीड़ा में समाज भी मुखर हुआ है। उनके नाटकों में समाज का युग सत्य और लेखक का व्यक्ति रूप में युग सत्य दोनों का भरपूर सामंजस्य है। हिन्दी में वे पहले नाटककार हुए जिसने नाटक को आधुनिक परिप्रेक्ष्य दिया।

इसमें राकेश अकेले हों ऐसी बात नहीं। स्वतन्त्रता के बाद सारे देश नें नाटक में जो मोड़ आया और रंग-आन्दोलन को जो प्रेरणा मिली उससे हिन्दी में ही नहीं, बंगला, मराठी, गुजराती, कन्नड़ आदि भारतीय भाषाओं के नाटकों में भी एक नई चेतना जागी। वस्तुतः राकेश की उपलब्धि समस्त भारतीय नाटक की उपलब्धि है। इसीलिए उनके नाटकों के सन्दर्भ में जहाँ स्ट्रिंडबर्ग, पिरांडेलो, कामू, बेकेट, आयनेस्को आदि की चर्चा की जा सकती है, जहाँ विजय तेंदुलकर, बादल सरकार, गिरीश कर्नाड, बसन्त कानेटकर, आद्य रंगाचार्य, चि. त्र्य. खानोलकर, विनायक पुरोहित, मोहित चटर्जी, मधुराय, उत्पल दत्त, शम्भु मित्र आदि के नाम भी भुलाए नहीं जा सकते, जिनकी अनुदित नाट्य कृतियों ने हिन्दी लेखन के क्षेत्र में नए वातावरण की सृष्टि की। इसमें कोई सन्देह नहीं कि हिन्दी नाटक और रंगमंच के इस समकालीन विकास में हिंन्दीतर लेखन और रंगकर्मी का बहुत बड़ा हाथ रहा है। सच्चाई यह है कि स्वत्रन्त्रता के बाद शायद पहली बार विभिन्न क्षेत्रीय भाषाओं के नाटक पूर्व और पश्चिम की समस्त उपलब्धियों को आत्मसात् कर एक समान अखिल भारतीय स्तर पर उपस्थित हुए। आश्चर्य की बात यह है कि उन सबका स्वर एक है और वे एक-से प्रभावों और शिल्पगत प्रयोगों से युक्त हैं। यह कम महत्त्व की बात नहीं है कि आज भारतीय नाटक पारस्परिक प्रेरणा और विनिमय के साथ राष्ट्रीय और अन्तर्राष्ट्रीय दोनों आयामों को स्पर्श करने लगा है।

नाटक को इन आयामों तक पहुँचाने का श्रेय रंगमंच के नव विहान को भी है। इब्राहिम अल्काजी, सत्यदेव दुबे, शम्भु मित्र, श्यामानन्द जालान, कारंत, हबीब तनवीर, ओम शिवपुरी, राजेन्द्र नाथ आदि रंगकर्मियों ने जो रंगमंच साधा है, उससे विश्व रंगमंच पर हुए प्रयोगों की परम्परा का निदर्शन हुआ है और नाटककार रंगमंच के प्रति अधिक जागरूक होकर सामने आया है। संयोग की बात ही समझिए या अन्वेषण की एक प्रक्रिया कि जिस प्रकार आज पश्चिम का चिन्तन परम्परागत भारतीय जीवन-दर्शन के आमने-सामने आकर खड़ा हो गया है, उसी प्रकार समान्तर पाश्चात्य रंगमंच का कला-बोध भी भारतीय रंग-परम्परा के काफ़ी नज़दीक आ गया है। नाटक और रंगमंच दोनों क्षेत्रों में पूर्व और पश्चिम की निकटता का यह अहसास मोहन राकेश जैसे जागरूक नाटककार के लिए प्रस्थान बिन्दु की भाँति रहा है।

और इसी के साथ यह प्रश्न भी उठता है—सब लोगों के बीच हमारा नाट्य लेखन कहाँ पर है ? आज फिर पश्चिम में नाटक के क्षेत्र में गतिरोध आ गया है; और जिस भारतीय नाटक के उन्मेष की बात कर रहे हैं, उसमें भी कुछ आलोचकों को रूपवादिता

की बू आती लगती है। फिर बात पश्चिम पर आकर टिकती है—पश्चिम में जो आज चालू है, क्या हम उसी का साथ दे रहे हैं, या फिर उस स्थिति से आगे बढ़ने की सोचते हैं ? पश्चिम में विसंगत नाटकों ने निरर्थकता, फालतूपन, नपुंसकता, निराशा और पराजय की मानसिकता को उघाड़ा है; और भारत में भी कुछ-कुछ वैसी ही परिस्थितियाँ आज सक्रिय हैं जो हमारे साहित्य को उसी ओर ले जा रही हैं। किन्तु क्या इस नियति को स्वीकारना ही काफ़ी है ? क्या आदमी की घिनौनी और बेबसी की तस्वीर ही सब-कुछ है ? राकेश के नाटक ऐसे प्रश्न जगाते हैं। वे अस्तित्व के संकट में घिरे हैं। अस्तित्ववादी प्रवृत्तियों का उनमें निदर्शन हुआ है। उनके बीज नाटकों और 'छतरियाँ' में कुछ विसंगत नाटक की छाप मिलती भी है। किन्तु शायद जान-बूझकर वे भारतीय परिस्थितियों के अनुरूप विसंगत नाटक की ओर उन्मुख नहीं हुए या फिर पश्चिम के विसंगतवार (ऐब्सर्डिज्म) की अनुगूँज उन तक बहुत देर में पहुँच पाई। वे आरम्भ में अस्तित्ववाद और बाद की कृति **आधे अधूरे** में यथार्थवाद से जूझते रहे और यदि जीते तो उसके आर-पार ही शायद अपना आगे का लेखन भी निर्धारित करते। पर आज जो उनका लेखन हमारे सामने है, वह पाश्चात्य प्रभावों पर निर्भर है और पश्चिम में जबकि हर वाद पाँच साल में बदलकर पुराना हो जाता है तब हिन्दी नाटक की पश्चिम-प्रभावित यात्रा की क्या स्थिति होगी, इनका अनुमान लगाया जा सकता है। एक नया संकेत लेकर राकेश हिन्दी नाटक में आए थे। उस वक़्त बड़ा शोर उठा था। अब वह शोर थम गया है किन्तु उनके नाटक आज भी जीवित हैं। लोगों की उनके प्रति दृष्टि बदली है। जिस प्रभाव में उनका लेखन हुआ था, वह धूमिल होता जा रहा है। ऐसी स्थिति में राकेश के सामने नाटकों की व्याख्या भले ही न बदले, मूल्यांकन बदलता रहेगा।

आत्महंता यायावर

कुछ लोगों की ज़िन्दगी में बिखराव होता है। मैं अपने को ऐसे ही लोगों में बिखरना और बिखरना मेरे लिए जितना स्वाभाविक है, पाता हूँ। सँभालना और समेटना उतना ही अस्वाभाविक।

—मोहन राकेश : 'परिवेश' की भूमिका

निराला के बाद हिन्दी साहित्य में जिस आदमी के चारों ओर सबसे ज़्यादा 'मिथ' बुनी गई, वह था मोहन राकेश। उनका कुछ व्यक्तित्व ही ऐसा था और कुछ उनके हमदम, दोस्त, प्रशंसक और आलोचक भी ऐसे रहे हैं कि उन्होंने उन्हें या तो निहायत घटिया आदमी के रूप में पेश किया या फिर मसीहा बनाकर रख दिया।

राकेश का व्यक्तित्व अन्तर्विरोधों से परिपूर्ण था। अलग-अलग लोगों के सामने उनकी अलग-अलग 'इमेज' थी। इसीलिए शायद उनके बारे में एक साथ कई विरोधी बातें प्रचलित थीं, जैसे—'राकेश अपने दोस्तों के लिए जीते हैं, राकेश बहुत आत्मकेन्द्रित हैं; उनकी दोस्तियाँ निभती हैं तो दूसरों की वज़ह से; वे तो एक क़दम भी न चल सकें; राकेश को एक घर की तलाश है जहाँ सही मायनों में कोई उनका अपना हो, जहाँ उन्हें सकून मिल सके; राकेश तो बोहीमियन प्रवृत्ति का इन्सान है; उन्हें कनाट प्लेस के गलियारे, कॉफी हाउस की मेज़-कुर्सियाँ, रेलवे स्टेशन की बेंचें तथा शिमला-श्रीनगर की पहाड़ियाँ अधिक भाती हैं; वे कहीं टिक नहीं सकते; किसी के हो नहीं सकते; राकेश को घटियापन बर्दाश्त नहीं है, न दूसरों का न अपना; राकेश बहुत घटिया इंसान है, वीजेल की तरह दूसरों के अस्तित्व को चूस लेते हैं, उसे खोखला कर देते हैं; फिर बड़ी बेबाकी से पल्ला छुड़ा जाते हैं, बदले में कुछ देना वे नहीं जानते।'[1] और भी कई बातें, जैसे—'राकेश कहीं भी स्थिर नहीं रहे; दिल्ली, जलंधर, अमृतसर, शिमला और मुम्बई कहीं भी नहीं। बँधे वे कभी भी न रहे—न व्यक्तियों से, न जगहों से, न नौकरियों से। कपड़े बदलने की तरह उन्होंने बीबियाँ बदलीं, घर उन्हें डराता रहा और रेस्तराओं-डाकबंगलों की ज़िन्दगी ही रास आती रही। (या) बाहर से राकेशजी जितने सीधे और सरल दीखते थे, उतने वास्तव में थे नहीं।[2] फिर एक दिन लगता है कि जैसे वह ज़िन्दगी से बहुत उठा हुआ व्यक्ति है—ऐक्सट्रीमली इंटेलेक्चुअल और कभी-कभी ऐक्सट्रीमली

इमोनेशल... ।[3] और फिर उनका ईगो—अहं—जो 'किसी भी स्थिति में कम्प्रोमाइज नहीं करने देता था।' 'वो आदमी बाहर जितना इन्फॉर्मल लगता था, मन से उतना ही फॉर्मल था।'[4] इस तरह की कई बातें।

शायद यह मोहन राकेश एक अजीब मिट्टी से बना था। बेहद दोस्तों से घिरा हुआ; हँसी-ठहाकों में खोया हुआ; ज़िन्दगी का मुस्तैद सिपाही, मुँहफट और ज़िन्दगी को अपनी ही शर्त पर जीनेवाला। राकेश एक था, पर उसी के अन्दर एक दूसरा भी राकेश था जिसकी एक निजी भूख थी जो कभी मिटी नहीं, जिसने असुरक्षा की भावना को झेला था, जो अकेले ज़िन्दगी से जूझा था; जो जीवन में सराबोर रहकर भी ज़िन्दगी से हार गया था। बहुत सारे टूटते सम्बन्धों की परवाह न करनेवाला वह निर्मम व्यक्ति कहीं बहुत गहरे स्वयं इतना टूटा था कि माँ कि मृत्यु को न सह सका। कई मामलों में वह अहंवादी था और कभी न झुकनेवाला इंसान, पर वास्तविक जीवन में वह कमज़ोर था, भावुक था। इनमें असली राकेश कौन था, यह चर्चा का विषय हो सकता है। उसे किसी ने समझा भी या नहीं, यह भी कहना मुश्किल है। 'उन्हें अन्दर से समझना बड़ी तपस्या थी। जिन मित्रों के साथ वह अत्यन्त इनफॉर्मल रहे वे उन्हें इससे अधिक नहीं जान सके। और जिनके साथ नजदीक थे, वे भी उससे अधिक नहीं जान सके।'[5] शायद वास्तविक राकेश ही अपने को अच्छी तरह जानता था। इसीलिए उनका यह कथन ही अपने बारे में ज़्यादा सही था— मैं न यह हूँ, न वह हूँ।'[6] शायद सही बात **लहरों के राजहंस** में नन्द के मुख से निकली है : 'इस रूप में हो या उस रूप में, अब किसी भी रूप में मैं अपने को झुठलाकर नहीं जी सकता। क्योंकि मैं यह भी हूँ और वह भी। इनमें से कोई एक नहीं, जैसा कि तुम सब अलग-अलग विश्वास करना चाहते हो कि मैं हूँ... ।' एक बिन्दु पर होते हुए भी वे दूसरे बिन्दु पर होने के लिए विकल रहते थे।

फिर भी राकेश यह थे, यह नहीं थे, या वे न यह थे न वह—इस तरह की बातें उनके बारे में हमेशा कही जाती रहीं ऐसी बात नहीं कि वे न जानते हों कि वे क्या हैं। राकेश ने अपने 'आइने के सामने' लेख और **परिवेश** में अपना ही सही विश्लेषण किया है। उनके स्वभाव की अस्थिरता और विसंगति में उनकी आकुलता ही उत्तरदायी थी और यह आकुलता आनेवाले क्षण को जल्दी ही पा लेने की आकुलता थी : 'जहाँ हो, वहाँ से उठो; कहीं और चल दो। जो कर रहे हो उसे छोड़ो, कुछ और करने लगो। पढ़ने लगो तो दो-दो मिनट में पन्ने पलटकर देख लो कि अध्याय कहाँ समाप्त हो रहा है, खाना खाने बैठो तो चार मिनट में सब-कुछ निगलकर हाथ धो आओ। काँच का गिलास हाथ से छूट जाए तो उसके ज़मीन छूने से पहले ही मान लो कि वह टूट गया है।'[7] राकेश ने स्वयं माना है कि अस्थिरता में अतिवादी प्रवृत्तियाँ आने लगती हैं जिसका उन्हें अपने में अहसास था : 'चाय पियो तो इतनी गरम कि जबान जल उठे; पानी पियो तो इतना ठंडा कि गला दुखने लग जाए; हँसो तो इतना कि मन उदास हो जाए; चुप रहो तो ऐसे, कि लोग परेशान हो उठें।'[8] उनके ठहाके उनकी आन्तरिक पीड़ा के परिचायक थे। राकेश ने जल्दबाज़ी और अन्तर्विरोधों का जीवन जिया है—दो छोरों की

तरह; किन्तु 'एक सूत्र है जो इन सबको आपस में जोड़ता है। वह है अपने को और अपने परिवेश को लेकर अस्वीकृति और असंतोष '[9] इसी से राकेश कभी-कभी कुछ इस तरह से जीते रहे कि 'जीना कुछ इस तरह लगे कि हर समय ढलान के सिरे पर खड़े हैं।'[10] जीवन जीने की इन पद्धति के पीछे उनका अपना बाल्यकाल का अनुभव रहा है जिसकी उन्होंने **परिवेश** में विस्तार से चर्चा की है।

उनकी डायरी में कई स्मृतियाँ अंकित हैं जिनका उनके जीवन पर सीधा प्रभाव पड़ा। उनके अनुसार 'मेरा जन्म एक तंग शहर की छोटी-सी गली के एक अँधेरे घर में हुआ और जो वातावरण मुझे आरम्भ में मिला उसमें श्लाघनीय कुछ नहीं था।' उन्हें अपने सपनों का घर कभी नहीं मिला। अपनी डायरी (26.6.58) में उन्होंने जलंधर, लाहौर, जोधपुर, मुम्बई, शिमला में बीस घर बदले थे जिनमें दिल्ली की गिनती नहीं। घर की ग़रीबी, भटकन, पिताजी, क़र्ज़, देरी, सुख और आत्मीयता की तलाश तथा बेबसी उनके बाल्यकाल से साथ लगी रही। अभावों ने उनके जीवन में असुरक्षा पैदा की थी। एक बार उन्होंने अपनी पत्नी को भी यह बात कही थी : 'मैंने जीवन में इतनी असुरक्षा भोगी है क्योंकि जीवन में अकेला इतना जूझा और लड़ा हूँ।'[11] वस्तुतः कई बार इसके लिए उन्हें कई भूमिकाएँ धारण करनी पड़ीं यद्यपि 'असली राकेश वह नहीं था।' असली राकेश की जड़ें अतीत के अनुभवों और भविष्य के सपनों में फैली थीं। किसी भी व्यक्ति के लिए ऐसी स्थिति में द्वन्द्व जीना सम्भव था।[12] इन्हीं द्वन्द्वों के बीच उनकी ज़िन्दगी बनती और टूटती रही। वे बार-बार नए सिरे से जीते रहे। यह जीना भी फिर-फिर उसी तरह जीना रहा। राकेश के शब्दों में (डायरी, 15.9.63) 'सच कैसा हूँ मैं ? कहीं बहुत रूखा, कहीं बहुत कोमल। एक क्षण बहुत ख़ुश, दूसरे क्षण बहुत उदास; किसी क्षण अपने पर बहुत विश्वास होता है, किसी क्षण बिलकुल नहीं रहता।...साल दर साल गुज़रते जाते हैं।' ऐसा इसलिए भी था क्योंकि राकेश कभी भावात्मक समानुकूलन नहीं कर पाए। दूसरों की अपेक्षाओं के अनुसार जीने को वे 'आत्मघात की प्रक्रिया' मानते रहे जो 'जीवनभर चलती रही।' (वही, 20.10.67)

वस्तुतः सम्बन्धों और अतीत के अनुभवों के दंश से वे अपने को कभी मुक्त नहीं कर पाए—जीने का कोई क्षण अकेला और स्वतन्त्र न रहकर आगे-पीछे के क्षणों में खोया रहता है। जो बीत गया है मन में अपनी जड़ें फैलाए रखता है। जो अनागत है, उसकी ओर उसकी डालियाँ हिलती रहती हैं। जीवन का हर दिन पिछले दिन के अन्दर से उगकर आता लगता है—आनेवाले कल को जल्दी से अपने अन्दर से उगा लेने को व्याकुल।[13] अतीत की भूमि में भविष्य को उगाने की आकुलता ही राकेश के अपने जीवन में बिखराव और साहित्य में सर्जन के रूप में प्रकट हुई। व्यक्ति के रूप में वे टूटे किन्तु टूटन के इसी अनुभव ने उन्हें जीवन और साहित्य का सच्चा बोध दिया।

साहित्य से जुड़े, जीवन से टूटे व्यक्ति को अकेलेपन का अहसास तीखा होता है, किन्तु क्षतिपूर्ति के रूप में उन्होंने बहुत-से लोगों को अपनाया था। अपने चारों तरफ लोगों को इकट्ठा करना शायद वे जानते थे। 'नई कहानी' और नाटक के अन्दोलन

में एक पूरा गुट उनके साथ था। अपने चारों ओर तारीफ़ और रहस्य का वातावरण खड़ा कर देना उनके लिए मुश्किल नहीं था। उनकी पूर्व-पत्नी के शब्दों में 'राकेशजी एक बहुत बड़े पोलिटिशियन भी थे। नई कहानी आन्दोलन और थियेटर मूवमेंट इसी बात के द्योतक हैं। उन्होंने जब भी तीर छोड़ा तो ठीक निशाना-बाँधकर ही छोड़ा।.. .उन्हें हा-हुल्लड़ मचाकर जनता में भगदड़ मचाने से एक और ही आनन्द मिलता था। इसीलिए उनके छोड़े गए तीरों से आहत हुए योद्धा उनके चले जाने के बाद आज भी पीड़ा से कराह रहे हैं।[14] इस तीरन्दाज़ ने भी उनके साथ कई ऐसी बातें जोड़ दीं जो लेखक को पब्लिसिटी देती हैं। इसके अतिरिक्त उनके चारों ओर ऐसे लोगों की कमी भी न थी जो उन्हें 'अच्छा लेखक और बुरा आदमी' प्रचारित करते रहे। किन्तु शायद उन लोगों की संख्या ही ज़्यादा थी जो उन्हें उतना बड़ा लेखक नहीं मानते थे जितना बड़ा आदमी। यहाँ मैं विश्वप्रकाश दीक्षित की कही बात दुहरा रहा हूँ कि मृत्यु के दिन 'सभी श्रद्धांजलिदाता इतने दीन थे कि उन्होंने अपना एक मित्र, एक हितैषी खोया था, साहित्यकार या कला मर्मज्ञ नहीं।'[15] यहाँ तक तो ख़ैर है। वास्तव में राकेश के तथाकथित दोस्त अब यह 'कमिट' करने में भी डरने लगे हैं कि उनका साहित्य किस कोटि का है और वे सचमुच कैसे आदमी थे !

ख़ैर, वे कैसे भी आदमी रहे हों, 'असम्भव' या 'दुर्द्धर्ष', वे एक शानदार आदमी थे जिन्होंने ज़िन्दगी को पूरी तौर से जिया। बहुत कम लोग होते हैं जो साहित्य को जीवन में जीते हैं और जीवन के दीपक को एक पूरी लौ में जलाने के लिए सारा तेल जला डालते हैं। राकेश में जीवन के लिए एक अजीब आकुलता थी, अपना सब-कुछ दाँव पर लगा देने का फक्कड़पन था। अपने को भरा-भरा महसूस करने के लिए उन्हें अपने को रीता करना पड़ा। हर छोटी-मोटी ख़ुशी के लिए उन्हें मुँह-माँगा मूल्य चुकाना पड़ा। अपनी ही अपेक्षाएँ, अपने ही द्वन्द्व, अपनी ही मानसिकता उन्हें अन्दर-ही-अन्दर खाती रही। राकेश ने **एक और ज़िन्दगी** जी थी जिसकी कड़वाहट बहुत गहरे पैठ गई थी। जीने के लिए उन्होंने ज़िन्दगी को विभाजित कर लिया था। एक ज़िन्दगी थी जो नितान्त उनकी 'अपनी' थी, और यह 'अपनी' ज़िन्दगी ही कहीं बहुत कड़वी, कसकती हुई और चुभनवाली थी जिसे भुलाने के लिए उन्होंने ठहाकों, दोस्तों और कॉफी हाउसोंवाली 'दूसरी' ज़िन्दगी ओढ़ रखी थी। राकेश का अपना 'ईगो', द्वन्द्व और जीवन को दूसरों के प्रभाव और अपनी अच्छाइयों और बुराइयों के साथ अपनी शर्तों पर जीने की आकांक्षा उन्हें विवशताओं और बेचैनियों की तरफ धकेलती गई। लड़कियों के प्रति आकर्षण भी एक कारण बना। दोस्तों को चुनने में भी वे बहुत दिलदार रहे। विवाह कर आत्मीयता से सराबोर होना चाहा, पर सारे प्रकरणों ने 'अस्थिर और असन्तुलित' ही किया और समय के साथ 'हर चीज़ का स्वाद फीका पड़ता गया।' पर प्यार की तृष्णा भटकाती रही, क्षत-विक्षत करती रही। वे पहाड़ी सौन्दर्य स्थलों से **आख़िरी चट्टान** तक अपने लिए फिर-फिर ज़िन्दगी की तलाश में फिरते रहे। कोई खोल ज़्यादा काम नहीं आया। अनीताजी के शब्दों में '...राकेशजी अन्दर से क्षत-विक्षत हो चुके थे। किसी

रिश्ते, किसी दोस्ती या किसी कमिटमेंट में अब उन्हें विश्वास नहीं रह गया था और अगर विश्वास था तो उसे फिर से जोड़ने या मज़बूत करने की हिम्मत बाकी नहीं रह गई थी। सारा-सारा दिन या तो गुमसुम बैठे रहते या फिर पुरवा को गोदी में लेकर ढेर-ढेर-सा प्यार देते रहते। सारी-सारी रात मुझे पास बिठाए रहते, कहते : 'पता नहीं क्यों अकेला होने से दहशत होने लगी है मुझे। मेरे पास बैठो—मुझे अकेला न छोड़ो।'[16] और वह दिन भी आया जब वे आकुलता में स्वयं ही टूट-बिखरकर अकेले चले गए। इतनी छोटी उम्र में वे मरे नहीं, उन्होंने 'अपने को खुद मारा था।' बटुकजी ने लिखा है : 'दाह-क्रिया के अवसर पर हज़ारों की भीड़ में मैं भी था। मैंने घूम-घूमकर लोगों की मुद्राएँ देखी थीं, कुछ सुना भी था।...मोहन राकेश की अबोध पुत्री ममतामयी माँ की गोद में चढ़ी कह रही थी—पपा अगर इतनी ज़्यादा शराब न पीते तो वह न मरते।'[17] पता नहीं बेटी की बात में कितना सच था; पर मारनेवाली चीज़ शराब नहीं थी; शराब तो एक अस्त्र थी। मृत्यु उनके जीवन में एक घटना नहीं, एक बहुत पहले से चला आता अहसास था : 'जाने क्यों एक बेचैनी मन में बसी रहती है कि यहाँ नहीं रहना है, बल्कि जहाँ कहीं भी होऊँ वहाँ नहीं रहना है।'[18] वही 'रोज़ की अनचाही ज़िन्दगी के लम्बे सिलसिले की एक-एक कड़ियाँ एक ही ढंग से रोज़ जोड़ते जाना' उन्हें पसन्द न था। सदा उनके अन्दर ऐसा कुछ घटित होता रहा जिसमें मृत्यु का अहसास कहीं बहुत गहरे छिपा था। (डायरी, 2.5.58) वे एक अर्थ में आत्महंता थे। उन्हीं की इस प्रवृत्ति की छाप कहीं उनके नाटकों के मुख्य पात्रों में भी समाहित दिखाई देती है। हर आनेवाले कल को बिलकुल अलग और नए रूप में देखने की आकांक्षा रखनेवाले पर मन को बार-बार मारनेवाले राकेश ने एक ही क्रम में जीने की मानवीय नियति को झुठलाने का प्रयास किया; किन्तु सारा प्रयत्न उन्हें कितनी विरोधों की ज़िन्दगी दे गया। अन्ततः वे मानने लगे थे कि 'यह सब मेला इन्सान किसलिए लगाए रहता है।' फिर भी जीवन-भर वे जो करते रहे, वह 'जीवित रहने के हित में करते रहे। राकेश कहा करते थे, 'मैं एक असम्भव व्यक्ति हूँ, उसके साथ मैं यह भी कहना चाहता हूँ कि मैं बहुत ईमानदार आदमी भी हूँ।'[19] निजता के साथ असम्भव व्यक्तित्व को एक साथ निभाने के प्रयत्न में वे ज़िन्दगी से हाथ धो बैठे। ऐसा संकट हर सच्चे सर्जक के सामने आता है।

राकेश ने अपने जीवन और समाज में जिस मानवीय संकट को महसूस किया, उसने साहित्य को अपूर्व क्षमताएँ दी हैं। राकेश मसीहा थे या नहीं, यह मैं नही जानता; पर उनके साहित्य का स्वर मसीहाई ज़रूर है। इस अर्थ में नहीं कि उसमें कहीं ज्ञान, प्रवचन पर चिन्तन की बातें हैं, बल्कि इस अर्थ में कि उसमें अस्तित्व सम्बन्धी कुछ मौलिक प्रश्न उठाए गए हैं। राकेश ने हिन्दी नाटक में पहली बार अस्तित्व के प्रश्न उठाए और उनको नई भंगिमाएँ दीं। उनकी रचनाएँ एक विशिष्ट धरातल पर जीवन का अर्थ खोजती हैं। और यह अर्थ वे शून्य में नहीं खोजते। जीवन की विसंगतियों और उनके प्रति आक्रोश के बीच से उनके नाटकों के अर्थ उभरते हैं। यह आक्रोश एकदम भौतिक नहीं, वरन्, आधिभौतिक स्तर को भी छूता है क्योंकि लेखक की खोज भौतिकता

के बीच भी अभौतिकता की, सम्भव के बीच भी असम्भव की खोज है। यह खोज उसे आकुल करती है, चिन्तन के गतियामों में भटकाती है और अन्ततः अकेला छोड़ जाती है। जो दुनिया उसे मिली है, वह उसकी अपनी नहीं है। वह उसके आमने-सामने खड़ा है। इसीलिए वह उसमें अपने को असहाय, निरुपाय और नितान्त अकेला अनुभव करता है। उसे एक सत्य की तलाश है जिसे वह अपना कह सके। जो उसने देखा-परखा है उससे वह अपनी आस्था नहीं जोड़ पाता, क्योंकि उसकी अपनी भी एक दृष्टि है। अपनी आस्था जगाने की कोशिश में नकार और स्वीकार, विद्रोह और वास्तविकता के द्वन्द्व में फँसा वह टूटे सपने सँजोता और परखता है। कहते हैं, जो जितना बड़ा मसीहा होता है, वह उतने ही बड़े सपने सँजोता है।

राकेश की एक और भी विशेषता है; उनमें भावना का लगाव इतना सीधा और सच्चा लगता है कि अपनी रचनाओं में वे पुरोधा लगते हैं जिसने अपना ईश्वर खो दिया है, जिसकी पूजा का पात्र खंडित हो चुका है। फिर भी वे अपने लेखन को मानवीय पीड़ा और तड़प के अनुष्ठान के रूप में लेते रहे। उन्होंने दर्द-भरे सवाल ही नहीं उठाए वरन् सवालों से घिरी अपनी एक दुनिया भी खड़ी की जिसमें पीड़ाएँ और यंत्रणाएँ हैं, संशय और विवशताएँ हैं, किन्तु मोक्ष या मुक्ति नहीं है। फिर भी राकेश अपने लेखन को विचारों के ढाँचे में इस प्रकार सँजोना चाहते थे कि वह एकदम ऊँचे जीवन का सत्य—गौस्पेल—लगे। लोग उन्हें ठीक समझें न समझें, वे अपने को ठीक समझते थे। जो कुछ उन्होंने लिखा, वे उनके उद्गाता थे। उसे उन्होंने जीवन में जिया भी था। इसलिए जो कुछ भी उन्होंने लिखा, आन्तरिक दबाव से उद्भूत था; जीवन की निजी आवश्यकताओं की देन था।

कहते हैं वह सर्जक ही क्या, जिसकी अन्तरात्मा की आवाज़ न हो ! राकेश ने जो कुछ लिखा है उसके केन्द्र में व्यक्ति और उसकी स्थितियाँ हैं। उनके नाटक प्रतीति के नाटक हैं। वे मानवीय स्थिति के भयावह संकट, सम्बन्धों के विघटन तथा व्यक्ति के संत्रास की प्रतीति कराते हैं। उनसे पूर्व हिन्दी में जो नाटक लिखे जाते रहे, उनमें देश, समाज या आदर्श मुख्य रहा, व्यक्ति गौण। राकेश ने अपने नाटकों में व्यक्ति को स्थापित किया और उसके अकेले कंठ की आवाज़ को बुलन्द किया। उन्होंने मानव को उसके परिवेश के बीच सही दृष्टि से देखा और फिर उसकी पीड़ा, अजनबीपन और फालतू होने के अनुभव को वाणी दी। पहले मानव और मानव के बीच केवल सत्-असत् का एक कृत्रिम विभाजन था। राकेश ने जीवन के अन्तर्विरोध को एक भिन्न स्तर पर समझा, और कटु अनुभव ने उनका मोहभंग किया। चिन्तन की एक भिन्न मुद्रा जो उन्होंने अख्तियार की, वह अस्तित्ववादियों से उन्होंने उधार ली हो—ऐसा हो सकता है; लेकिन यह मुद्रा मुखौटे की तरह से ओढ़ी गई हो, ऐसा नहीं लगता। राकेश ने उसे भी अर्जित या पुष्ट किया था। असुरक्षा, आन्तरिक विक्षोभ, निर्वासन, अकेलेपन, अधूरेपन और जीवन की शून्यता के अहसास को उन्होंने कहीं अन्दर से भोगा था। बहुत से अनुभव बाल्यकाल के थे। इसी प्रकार स्त्री-पुरुष की सम्बन्धहीनता को उन्होंने किसी

समय वैवाहिक जीवन में झेला था। आत्मीयता की भूख उनमें प्रबल थी, निजी अहं उन्हें सर्वथा भटकाता रहा। तीन-तीन ब्याह कर भी वे रमे नहीं न उन्हें 'घर' दे सके। 'घर' उनकी आवश्यकता थी—वह सुरक्षा का प्रतीक था। उसमें उन्हें जिस सांत्वना और सुख की तलाश थी, वह उन्हें कभी न मिली। माँ आधार स्तम्भ थीं। उनके मरते ही राकेश का भावात्मक जगत् जैसे बिखरकर टूट गया था। अन्ततः कालिदास, नन्द या महेन्द्रनाथ जैसे उनके अन्दर घिर आए थे ! यह परिणति एक दिन में पैदा होनेवाले अनुभव की न थी—स्थितियों और चिंतन की मिली-जुली संचित अभिव्यक्ति थी। जीवन को प्रारम्भ में जो जितनी शक्ति से जीता है, वह उतना ही थका-हारा भी महसूस करता है। हिन्दी में कुछ को छोड़कर जीवन को इतनी समग्रता से शायद ही किसी ने भोगा हो !

राकेश के खंडित व्यक्तित्व ने उन्हें जिन अन्तर्विरोधों में फँसा दिया था, वे उनके जीवन से चलकर उनके साहित्य में प्रतिबिम्बित हुए हैं। उनका साहित्य एक तरह से उनकी भीतरी और बाहरी आकुलता और अभीप्सा की देन है। कालिदास राकेश का ही प्रतिनिधि है। भोगे हुए यथार्थ के कारण राकेश का साहित्य प्रामाणिक अनुभव देता है। उनका साहित्य उनके जीवन के अनुभवों पर खड़ा किया गया है। इसीलिए व्यक्ति राकेश को उनके साहित्य में खोज लेना कठिन काम नहीं है। उनकी कहानियों और उपन्यासों में—विशेषतः **अन्तराल** और **अँधेरे बन्द कमरे** में—उन्हें अलग से पहचाना जा सकता है। **आषाढ़ का एक दिन** में कालिदास मल्लिका से कहता है : 'मैंने जब-जब लिखने का प्रयत्न किया, तुम्हारे और अपने जीवन के इतिहास को फिर-फिर दुहराया।' यह बात राकेश पर भी अक्षरशः लागू होती है। वे कालिदास में भी हैं, नन्द और महेन्द्रनाथ में भी। उन्होंने मल्लिका को भी देखा था, सुन्दरी और सावित्री को भी, शायद बिन्नी, किन्नी और अशोक को भी। कालिदास की तरह आश्रित लेखन को उन्होंने झेला था। सर्जन की क्षमता के लिए उन्होंने एक के बाद एक नौकरियाँ छोड़ी थीं। अपनी आन्तरिक स्वतन्त्रता के वे इतने हामी थे कि दूसरे कलेवर में जीने के बजाए अपने वास्तविक रूप में जीना उन्हें प्रिय था।

आषाढ़ का एक दिन के अन्य पात्र उनके कितने निकट थे, यह कहना मुश्किल है; किन्तु राकेश की डायरी में मल्लिका को खोजा जा सकता है। इसी प्रकार अनीता के संस्मरण और राजेन्द्रपाल का लेख पढ़कर अम्बिका और राकेश की अपनी माँ के बिम्ब परस्पर घुल-मिल जाते हैं। वे ममतामयी नारी थीं जिन्होंने असमय वैधव्य को झेला था। मौन उनकी अदम्य शक्ति थी और अन्तःसलिला के समान प्रच्छन्न स्नेह उनकी निधि। अम्बिका की मूल अवधारणा यही है। उसके मरते ही मल्लिका टूट जाती है; माँ के मरने पर राकेश भी कहीं बहुत गहरे में आहत हुए थे। राकेश ने अपनी डायरी (1.6.57) में माँ की ममतामयी मूर्ति का रूप उभारा है; उसमें इसका प्रमाण मिलता है। माँ के 'अनात्मरत और निःस्वार्थ जीवन' की जो बात उन्होंने कही उसमें स्थिरता, 'डेडिकेशन' तथा योगस्थ प्रकृति की ऐसी विशेषताएँ गिनाई हैं जो अम्बिका में साफ़ झलकती हैं। कई बार ऐसा लगता है कि नारी के इस मातृरूप के प्रति उनका

स्थायीकरण हो गया था जिससे अन्य रूपों से वे समझौता नहीं कर सके।

नन्द का द्वन्द्व भी शायद कभी उन्होंने भोगा हो तो कोई आश्चर्य की बात नहीं। वे घर, परिवार, पत्नी सब-कुछ चाहते थे, पर पूरी तरह से शायद कुछ भी नहीं चाहते थे। इसीलिए अपनी चाहत को भी उन्होंने एक क्रम में बाँध डाला था। पत्नी से उन्होंने कहा था कि उनकी ज़िन्दगी लेखन है, दोस्तों में बँटी है और तुम तीसरे नम्बर पर हो। फिर भी यह ज्ञात तथ्य था कि 'तीनों ही मेरे लिए आवश्यक हैं।' राजेन्द्रपाल से कहे उनके ये शब्द दूसरों को अपने लिए आवश्यक समझने, किन्तु अपने को अपने लिए और भी आवश्यक समझने की प्रवृत्ति को और भी स्पष्ट करते हैं : 'तुम ज़िन्दगी में अच्छे बेटे बन सकते हो, अच्छे पति बन सकते हो या फिर कुछ कर सकते हो। मैं अच्छा बाप, अच्छा पति बनने और ज़िन्दगी में कुछ कर लेने के बीच त्रिशंकु की तरह लटक रहा हूँ।'[19] नन्द भी ऐसा ही त्रिशंकु है। जब राकेश **लहरों के राजहंस** लिख रहे थे, उन्हीं दिनों एक नए प्रणय की लहरों पर तैर रहे थे। एक समर्पण का भाव था जो होंठों पर आने को आकुल था; किन्तु राकेश सुन्दर थे, समझते थे, 'मन की बात कहकर आदमी छोटा हो जाता है।' उनकी प्रेम-पात्रा (बाद में पत्नी) के सामने भी यही उलझन थी।[20] उलझन तो देर-सबेर सुलझ गई। राकेश ने वह नाटक उसी 'एक को' समर्पित भी किया; पर सुन्दरी को अन्ततः यह मूलमंत्र थमा दिया—मन की बात कहकर आदमी छोटा हो जाता है। नाटक के अन्त तक पहुँचते-पहुँचते छोटे होने की बात के डर से वह अन्तर्मुख हो जाती है। वस्तुतः कामना और प्राप्ति में राकेश कहीं एक नहीं हो सके उन्होंने विवाह तो किए : 'कुछ अपनी स्टार्वेशन के कारण और कुछ इंसानी भलमनसाहत के कारण। किया तो पर मन भाँवरें पड़ते समय भी उदास ही था।...विवाह के साथ ही मैंने अपने को कुछ रुके हुए पाया, रुके हुए ही नहीं, जड़ और स्तम्भित। मैं जहाँ था वहाँ अपने को पाने के लिए या मानने के लिए कि मैं वहाँ हूँ, तैयार नहीं था।'[21] नन्द में मोहन राकेश की अपनी छवि भी है। उन्होंने अपनी डायरी (7.6.57) में कहा है कि वे शीघ्र ही दूसरों के प्रभाव में आ जाते थे। नन्द में भी यह प्रवृत्ति दिखाई देती है।

आधे अधूरे के केन्द्र में जो घर है वह राकेश का अपना घर नहीं था, पर घर वह भी नहीं था।[22] बड़े होकर जितना ही उन्होंने अपने घर के सपने सँजोए, उतनी ही 'घर नाम की चीज़ उनसे रुसवा रही।' पहले दो विवाह-सम्बन्धों ने भी उन्हें सुखी पारिवारिक जीवन का अहसास नहीं दिया। पर **आधे अधूरे** का पारिवारिक यथार्थ राकेश से कहीं अधिक उनकी पत्नी अनीता का भोगा यथार्थ लगता है। **चन्द सतरें और** के प्रारम्भिक पृष्ठ इसके प्रमाण हैं। और इसके साथ ही ब्रजमोहन शाह का यह संकेत भी : 'बेहतर होता यदि इस नाटक में मोहन राकेश के साथ अनीताजी का भी नाम आता, क्योंकि यह नाटक उन्हीं की एक कहानी पर आधारित है।'[23] अनीता ने राकेश की यारबासी और मित्रों की महफ़िल लगाने के जो विवरण दिए हैं, वे **आधे अधूरे** में महेन्द्रनाथ के मित्रों के सम्बन्ध में सावित्री की बातों से मेल खाते हैं। दोस्तों के लिए अपने को ढालने की आत्मघाती प्रवृत्ति महेन्द्रनाथ और मोहन राकेश में समान दिखाई देती है। वस्तुतः

आधे अधूरे का सारा फलक **चन्द सतरें और** के विवरण से साफ़ झलकता है।[24]

और उधर मोहन राकेश को देखिए : 'घर के बाहर राकेशजी का व्यवहार कुछ और होता था और घर में एकदम और। एक अरसा लगा मुझे इस ताल-मेल को बिठाते, यद्यपि बाद में आकर राकेशजी ने मेरे साथ जबरदस्ती करनी छोड़ दी। लगता था मेरा घर और मेरा राकेश मुझसे छिन गया है।'[25] 'छः महीने में ही हम दोनों एक-दूसरे के लिए और भी बेगाने हो गए ' 'राकेश महफ़िलबाज़ थे। इसलिए पत्नी को राकेश से कुछ क्षण भी सिर्फ़ अपनी मर्ज़ी से काटने तक को न मिलते ।'[26] राकेश अपनी इस कमज़ोरी को पहचानते थे। 20 अक्टूबर '67 को अपनी डायरी में उन्होंने अपनी इस आत्महंता प्रवृत्ति का संकेत भी दिया था : 'दूसरों की अपेक्षाओं के अनुसार अपने को ढालना—यह केवल आत्मघात की प्रक्रिया है जो जीवन-भर चलती रह सकती है। परन्तु कुछ ऐसा क्रम है रोज़ की ज़िन्दगी का, कि यह सब अनजाने में होता चलता है। राकेश ने अपनी डायरी (20.9.57) में लिखा है कि वे अपने वर्तमान से मुक्ति चाहते हैं पर उसकी सम्भावना नहीं लगती वे कभी बँधे नहीं, इसलिए अपनी आत्मघाती प्रवृत्तियों से कभी मुक्त नहीं हो सके। उनके ही (डायरी 10.1.58 के) शब्दों में 'मैं वह अभिशप्त हूँ जिसे बन्धन बाँधता नहीं, मुक्ति मुक्ति नहीं देती।' राकेश के मुख्य पात्र इस नियति का निर्वाह करते मिलते हैं।

वस्तुतः राकेश ने अपने घर की कामना को परवशता के बीच गढ़ा था। 16 वर्ष की उम्र में उन्होंने क़र्ज़ के बोझ के नीचे दबे पिता की मौत देखी थी जिस पर मकान मालिक ने अपने वहशीपन की छाप लगा दी थी : 'मैं तब तक मुर्दा नहीं उठने दूँगा, जब तक किराया नहीं अदा किया जाएगा।' यह जीवन का पहला तल्ख़ अनुभव था जिसने उसे 'सलाखों के पास से आकाश की गहराइयों में' जीवन को दिखा दिया था। यहीं से उन्होंने अस्तित्व की तलाश शुरू की और अहसास की उस कटुता को जिया जिसका 'वास्तविक स्रोत अन्दर का रिसता हुआ बिन्दु है।' यही उन्हें अथ से इति और इति से अथ के बीच भटकाता रहता जिसमें एक अजीब-सा अस्तित्व का तर्क उन्हें घेरे रहा : 'कुछ है जो चेतन पर कुंडली मारे रहता है और मुझे अपने से मुक्त नहीं रहने देता।' नन्द की यह मसीहाई पीड़ा राकेश की अपनी आन्तरिक कुढ़न थी जो उन्हें एक सलीब से दूसरे सलीब तक भटकाती रही। ऐसी व्याकुलता राकेश के नाटकीय पात्रों—महेन्द्रनाथ, कालिदास और नन्द—में भी दिखाई देती है जिन्हें ज़िन्दगीभर घर नहीं मिला था। उनके शब्दों में, रेगिस्तान में घर कैसे बनाया जा सकता है ? इस कारण वे अधूरे घरों के वासी बने रहे। उनके सब नाटक इस स्थिति के परिचायक हैं। राकेश का घर जितनी बार टूटा उसकी गवाही उनका साहित्य देता रहा। इन्द्रनाथ मदान तो **चन्द सतरें और** की भूमिका में यहाँ तक लिखते हैं कि राकेश ने 'वास्तव में एक ही पूरा नाटक लिखा है जो अधूरा है।' यह पूरा सत्य नहीं है। रचनाकार रचना में अपने को दफ़नाता नहीं। वह चाहकर भी अपने जीवन के सब पर्दे नहीं उठा सकता। राकेश के कथन से (डायरी, 18.1.59) मदान की बात अतिशयोक्ति लग सकती है। कभी मुक्ति नहीं मिली,

पर अपनी मुक्ति को खोजते-खोजते मानव की मुक्ति की जैसी वकालत राकेश ने की, वैसी आज तक कौन नाटककार कर पाया है ?

ये कुछ उदाहरण हैं जो राकेश के वैयक्तिक जीवन और उनके लेखन के बीच तन्तु स्थापित करते हैं, उसे प्रामाणिकता प्रदान करते हैं। वह और कुछ नहीं, भोगा हुआ यथार्थ है जो उनकी रचनाओं में बोलता है। पर यह कहना ग़लत होगा कि वह किसी एक व्यक्ति का यथार्थ है या उसमें सार्वजनीनता नहीं है। वस्तुतः राकेश ने जिस यथार्थ को देखा और भोगा है, वह स्थूल परिवेश से ही नहीं, अन्तर्दृष्टि से भी गर्भित है। आज प्रताड़ित मानव अस्तित्व के जिस संकट को झेल रहा है, उससे वे पूर्णतः परिचित थे। इसीलिए उसे समग्र आत्मपरकता के साथ व्यक्त करने में वे सफल हुए। उन्हीं के शब्दों में 'लेखक का वास्तविक कमिटमेंट किसी विचारधारा से न होकर अपने से, अपने समय से और समय के जीवन से होता है। यदि वह सचमुच अन्दर से कमिटेड है, तो वह अन्धे की तरह लकड़ी लेकर अँधेरे में अपने अकेले के लिए रास्ता नहीं टटोलता, बल्कि अँधेरे और आतंक को पैदा करनेवाली शक्तियों के साथ अपने समूचे अस्तित्व से लड़ जाना चाहता है।'[25] व्यक्ति और रचनाकार के रूप में कुछ लोग अलग-अलग होते हैं, पर राकेश सम्भवतः ऐसे नहीं थे। व्यक्ति राकेश में अन्तर्विरोध था, पर रचनाकार राकेश और व्यक्ति राकेश में कोई अन्तर्विरोध नहीं है।

राकेश का लेखन उनके निजी अनुभवों की देन है, किन्तु उनका लेखन वैयक्तिक नहीं है। उन्होंने अपने लेखन को वास्तविक स्थितियों का पर्याय नहीं माना वरन् उसे विशिष्ट अर्थ में ढालने का प्रयास किया। उनके नाटकों में उनके भोगे हुए जीवन और परिचितों के चेहरों को देखा जा सकता है। किन्तु सब-कुछ उन्होंने प्रतिबिम्ब रूप में ग्रहण नहीं किया। उसे अपना चिन्तन, अर्थ और व्याख्या प्रदान की और समसामयिक जीवन और भाव-बोध से जोड़ने का प्रयास किया। पर इस जुड़ाव में स्वतन्त्रता के बाद की मोहभंग की स्थितियों और जन-जन की पीड़ा को व्यक्त करने की क्षमता राकेश यदि अपने वैयक्तिक अन्तर्द्वन्द्वों के चित्रणों का अतिक्रमण कर जुटा पाते तो दीन-दुर्बल जनों के संघर्ष में वे अपनी आत्महंता प्रवृत्तियों पर भी नियन्त्रण पा सकते थे। पर स्वयं अपने से जूझते व्यक्ति के लिए समाज की लड़ाई में सम्मिलित होना सम्भव न था। यह युग की सीमा थी और राकेश की भी।

एक आन्तरिक संघर्ष को उजागर करना भी शायद राकेश और उनके युग की माँग रही हो। हो सकता है इसलिए भी वे सुविधापूर्वक अपने लेखन में व्यक्ति के परिमोचक बने—एक नई मानवता के; दुर्बलों, असहायों और लघु मानव की दुनिया के; आधे-अधूरे और टूटे सपनों के सौदागरों के। आदर्शवादी लेखन की आड़ में आदमी की सच्ची तस्वीर ओझल हो गई थी। राकेश ने अस्तित्व में निहित करुणा, विवशता, विफलता, पीड़ा से घिरे आदमी की पहचान कराई। उसके ऊपर लदे सारे मुखौटे हटाए और नंगा करके चौराहे पर खड़ा कर दिया जहाँ 'अपने को ढकने के लिए उसके पास आवरण नहीं है।' अस्तित्व के संकट के बीच उन्होंने मानव की निजता की तलाश की। आन्तरिक जीवन

की तलाश तथा स्व के अनुभव का लक्ष्य लेकर उन्होंने उस आदमी को देखने-परखने की कोशिश की जो अपने लिए स्वयं एक प्रश्न बन चुका है। इस आदमी की जो पहचान उन्होंने की है, वह बहुत आशावादी नहीं, परन्तु यथार्थ ज़रूर है जो कहीं-न-कहीं अपनी मार्मिकता और अनुभव की सच्चाई में हृदय को छूती है। एक युग था जब आदर्शवाद का झूठा आवरण आदमी की असलियत को प्रकट नहीं होने देता था; जब देव और दानव कोटि के पात्र होते थे; जब जीवन सत् और असत् की रेखा से बँटा था। राकेश ने पुरानी मान्यताएँ नकारी हैं। आदमी की जो नई तस्वीर उन्होंने पेश की, उसमें प्रभा-मंडल ग़ायब है; पर वह इतनी घिनौनी भी नहीं कि उबकाई आए। यहीं वे पश्चिम के बहुत-से अस्तित्ववादी लेखकों से कुछ हटकर दिखाई देते हैं। मानवीय सीमाओं, समस्त दुर्बलताओं के बावजूद आदमी को दुलारने-पुचकारने की क्षमता उनमें सर्वत्र दिखाई देती है। अपनी अभिशप्त नियति से घिरे उनके पात्र इसलिए हमसे सहानुभूति छीन लेते हैं।

वे व्यक्ति की मुक्ति के सन्देशवाहक थे। मोहन राकेश के पास उस अर्थ में मुक्ति का कोई मार्ग नहीं था, पर अहसास वे सर्वत्र समेटे दिखाई देते हैं। हिन्दी नाटक को कथ्य और शिल्प के स्तर पर उन्होंने परम्परा और रूढ़ि से मुक्ति दिलाई। ख़ास बात यह है कि घिसी-पिटी परम्पराओं से विच्छिन्न होकर उन्होंने हिन्दी नाटक को आधुनिकता का बोध प्रदान किया। आधुनिकता का यह बोध अपने में कितना पूर्ण है, इस पर मतभेद हो सकता है। राकेश नाटककार से भी पहले कहानीकार रहे हैं इसलिए कुछ लोगों ने कहा है कि उनके नाटककार स्वरूप पर कहानीकार हावी है। हमारी दृष्टि में शायद अन्तर्प्रज्ञा से ही उनको नाटक की सही पहचान थी। वे नाटक की दुनिया के असली स्रष्टा थे। प्रसाद के बाद वे अकेले व्यक्ति थे जिन्होंने नाटक के क्षेत्र में गतिरोध को तोड़ा और उसे आधुनिकता का बोध दिया। फिर भी इस बात में सच्चाई है कि आधुनिकता का यह बोध 'नई कहानियों' के द्वार से नाटक में प्रविष्ट हुआ है। हम डॉ. इन्द्रनाथ मदान के कथन से सहनत हैं कि **आषाढ़ का एक दिन** में कालिदास का टूटा-हारा घर लौटना और परिचित होकर भी अपरिचित हो जाने की नियति 'नई कहानी' के दौर की है। इसी प्रकार **लहरों के राजहंस** की पूर्व कल्पना एक कहानी के रूप में की गई थी। इस नाटक में बाद में राकेश ने जिन प्रतीकों और अर्थ-संकेतों का प्रयोग किया है, वे भी नई कहानी के दौर के ही हैं। आज 'नई कहानी' की प्रासंगिकता समाप्त होती जा रही है और उसका स्थान सचेतन कहानी, अकहानी तथा समानान्तर कहानी लेती जा रही है। तब कथ्य की दृष्टि से राकेश के नाटकों की सार्थकता का प्रश्न एक बार ज़रूर उठेगा। किन्तु वे अचानक अप्रासंगिक सिद्ध नही होंगे। राकेश के आधुनिकता बोध ने इब्सेन से लेकर अनुइ तक की उपलब्धियों के बीच से अपना रास्ता बनाया है। पाश्चात्य जगत् में जो प्रयोग हुए उनसे उन्होंने बहुत कुछ सीखा—कुछ अंगीकार किया तो बहुत कुछ परखकर छोड़ा भी। इसका परिणाम यह हुआ कि वे प्रयोग करने में समर्थ हुए, पर भटकने से बच गए।

वे इस बात के लिए सदा याद किए जाएँगे कि वे हिन्दी नाटक की एक नई दिशा के अग्रदूत थे। उन्होंने एकदम कथ्य का कोण बदला, शिल्प और भाषा के बलाघात को बदला। उनके नाटक न कवित्व से दबे हैं, न दर्शन से; न अनुभूति से हीन हैं और न सामयिक अनुभव से रिक्त। वे न यथार्थवादी हैं और न यथार्थवाद के विरुद्ध। उन्होंने यथार्थ और अयथार्थ का ऐसा समन्वय किया है कि उनके रंग परस्पर घुल-मिल जाते हैं। **लहरों के राजहंस** में हर बारीकी यथार्थपरक है; पर परिणति फैंटास्टिक लगती है। **आषाढ़ का एक दिन** में यथार्थ के साथ रोमान भी जुड़ा है। वे यथार्थ से चलकर यथार्थ से परे जाते हैं। इसीलिए वादों की क्रिया-प्रतिक्रिया का अन्धानुकरण उन्होंने नहीं किया; अभिव्यक्तिवाद, प्रतीकवाद, विसंगतवाद को वहीं तक अपनाया जहाँ तक वे उनकी रचनाओं में खप सकते थे। उनके आत्यंतिक प्रयोगों को उन्होंने नहीं अपनाया। उन्होंने अयथार्थवादी प्रवृत्तियों को स्वीकार तो किया है, पर किसी सीमा तक। विसंगत नाटकों की विद्रूपता को भी स्वीकारा, पर उस सीमा तक जहाँ तक वह अयथार्थ और असत्य न लगे। उनका कथ्य अनास्था और अनस्तित्व का अस्त्र लेकर चलता है, किन्तु तलाश उसमें आस्था और अस्तित्व की ही होती है।

मोहन राकेश ने परम्परागत कथ्य को ही नहीं शिल्प को भी रूढ़ियों से मुक्ति दिलाई है। उनके नाटक सुगठित और सुरचित नाटक हैं। शिल्प पर इतना श्रम शायद ही किसी ने किया हो और इतनी जागरूकता शायद ही किसी में रही हो। उनमें बहुत बारीक पच्चीकारी है जो नाटक की बुनावट में घुल-मिल जाती है। प्रतीकों, बिम्बों, संवादों और स्थितियों के आयोजन में वे एक जागरूक कलात्मक दृष्टि का परिचय देते हैं। इसका मुख्य कारण अपने लेखन के प्रति ईमानदारी थी। अनीता राकेश के शब्दों में, 'राकेशजी सिर्फ़ व्यवहार में ही नहीं बल्कि अपने लेखन में भी उतने ही ईमानदार थे। उनके कम लिखने के कई कारण थे...और सबसे महत्त्वपूर्ण कारण सेल्फ रिजेक्शन था। **स्याह सफ़ेद, काँपता हुआ दरिया, अँधेरे बन्द कमरे में, लहरों के राजहंस, आधे अधूरे, पैर तले की ज़मीन** के छः-छः, सात-सात ड्राफ़्ट मौजूद हैं। अपनी ही कृति को लेकर इतनी ज़बर्दस्त पेश्यंस थी उनमें। अपने लेखन का मानदंड उनके लिए यह नहीं था कि लोगों ने उसे कहाँ तक पसन्द किया या स्वीकारा। उनका मानदंड उनकी निजी ईमानदारी ही थी, बस।'[27]

यह ईमानदारी कथ्य और शिल्प दोनों स्तर पर थी। अनुभूति की तीव्रता, भोगे हुए जीवन-क्षणों की प्रामाणिकता और सृजन में उभरी उसकी शिल्प-क्षमता इसी की देन कही जा सकती है। भाषा के स्तर पर उन्होंने जो स्वतन्त्र प्रयोग किए उनके पीछे यही बात ख़ास थी। वे निरन्तर **आषाढ़ का एक दिन** से लेकर बीज नाटकों तक भाषा की तलाश करते रहे। यह भाषा उस शानदार अतीत की न थी जिसको लेकर प्रसाद नाटक लिख चुके थे; यह भाषा सामान्य बोलचाल की भी न थी जिसमें आजकल कई साधारण नाटक लिखे जा रहे हैं। यह भाषा सर्जनात्मक स्तर तक उठी भाषा है जो अपने कथ्य से पूरी तरह जुड़ी है। पाश्चात्य नाटककार जब भाषा के चुक जाने और भाषा-विहीन रंगमंच

की बात कर रहे थे, तब राकेश उस हवा में बहने को तैयार न हुए। उन्होंने नाटक में शब्द की असाधारण भूमिका को घोषणा कर यह दिखा दिया कि वे अपने चिन्तन में कितने ईमानदार हैं। यद्यपि लोग उनकी बात से चौंके, पर उनकी बात में सत्य का अंश ज़रूर था। इसी प्रकार वे एकदम रंगमंच की पाश्चात्य परम्परा के हामी नहीं थे। भारतीय रंगमंच का रूप-विधान सही अर्थों में 'नाटकीय प्रयोगों के अभ्यन्तर से जन्म लेगा' वे ऐसा विश्वास लेकर नाट्य-लेखन के क्षेत्र में उतरे थे।

ये सब राकेश की महानता के उपादान थे, किन्तु राकेश की लोकप्रियता का एक कारण उनके नाटकों में निहित विचार-संकेत भी है। उनके नाटकों का समसामयिक अनुभव पाठक/प्रेक्षक को विलक्षण आत्मीयता दे जाता है। हाँ, एक प्रश्न ज़रूर उठता है : राकेश मानव की लघुता, उसकी सम्बन्धहीनता, विवशता, शून्य और नास्तिभाव का जो चित्र प्रस्तुत करते हैं, वह हमको कहाँ ले जाता है ? यह ठीक है कि उनके पात्र अस्तित्व का संकट झेलते हैं, पर अन्त में जाकर संघर्षविमुख हो जाते हैं। यह जीवन का यथार्थ हो सकता है, पर जीवन जीने का सहारा नहीं। राकेश में स्थितियों के ह्रास, विघटन और पराजय का अहसास ही मुख्य है, मानव की भविष्य की ओर उन्मुख रचनात्मक दृष्टि का अभाव है। एक अभिशप्त जीवन उनके नाटकों में अन्ततः मुखर हो उठता है। उनके पात्रों में जीवन और जगत् के प्रति एक विराट् और गहन संपृक्ति नहीं रह जाती। इससे अस्तित्व की समस्या हल नहीं होती। उसमें विवशता की स्वीकृति है, विक्षोभ कहाँ है ? वह अनुभव और युगबोध कहाँ है जो युग के मर्म को भेद सके ? यह मान भी लें कि 'यदि कोई कलाकार कहे कि उसके पास ऐसी कोई विशिष्ट दृष्टि है जो दूसरों का मार्ग-दर्शन कर सकती है, तो उसका कोरा दम्भ है...। वह तो ज़्यादा-से-ज़्यादा इतना ही कर सकता है कि अपने पाठक या दर्शक के मन में बैठे द्रष्टा को जगा दे और उसे एक ऐसे बिन्दु पर लाकर खड़ा कर दे जहाँ से वह यदि चाहे तो स्वयं अपना पथ आलोकित कर सके।'[28] फिर भी सवाल दृष्टि जगाने या बिन्दु-विशेष पर पहुँचने का रह जाता है। राकेश ऐसे संकेत देते हैं; यह भी ठीक है कि उनके नाटक अन्ततः एक 'मूड' या भावस्थिति में ले जाते हैं जो सोचने को बाध्य करती है; किन्तु उनमें वह पैनापन नहीं जो मस्तिष्क को मथ डाले। आदमी को अस्तित्व के संकट के बीच प्रस्तुत करना नाटक के लिए एक नई दिशा हो सकती है, किन्तु मनुष्य की क्षमताओं, संभावनाओं को भी न आँक सकना, एक बहुत बड़ा अभाव है। आदमी हर हालत में एक परकीय तत्त्व नहीं है; वह जगत् की विकास-प्रक्रिया में एक विवशता का अनुभव मात्र नहीं है। यह ठीक है कि अन्तिम समाधान सम्भव नहीं, किन्तु प्रश्नों को बड़ी तीव्रता से उठाया तो जा सकता है। इसलिए कुछ समय तक राकेश की नाट्य-कृतियाँ भले ही बहुत लोकप्रिय रहें, परिस्थितियों के बदल जाने के बाद उनकी प्रासंगिकता पर कभी भी प्रश्न-चिह्न लग सकता है !

जहाँ तक आनेवाले कुछ दशकों का सवाल है, राकेश के नाटकों का महत्त्व अक्षुण्ण है। वे पहले नाटककार हैं जो हिन्दी की क्षेत्रीय सीमाओं से उठकर राष्ट्रीय स्तर के

नाटककारों में गिने गए और अपने समय के प्रसिद्ध भारतीय नाटककारों की पाँत में बिठा दिए गए। इब्राहीम अल्काजी के शब्दों में 'भारतीय रंगमंच के इतिहास में ऐसा पहली बार हुआ जब एक नाटककार ने अपनी मातृभाषा में नाट्य-रचना के कुछ महीनों के भीतर ही राष्ट्रीय स्तर पर एक दर्शक-समुदाय पा लिया; अपने नाटकों की कम-से-कम एक दर्जन प्रमुख निर्देशकों द्वारा व्याख्याएँ देख लीं और उच्चतम राष्ट्रीय पुरस्कार भी प्राप्त कर लिए। ऐसी स्थिति में नाटककार के मनोबल और उसकी प्रतिष्ठा की वृद्धि का अनुमान सहज ही लगाया जा सकता है।'[29]

दूसरी ओर उतनी ही तीव्रता से कटु आलोचनाएँ भी हुईं। हिन्दीवालों ने उनके नाटकों के प्रदर्शनों में निर्देशकों को सराहा, नाटकों को सामान्य कहा। कई तरह की बातें उनके बारे में की गईं, जैसे—उनमें रोमानी बोध है; उनके नाटक अपने को दुहराते हैं; बीज नाटक से लेकर वृक्ष-नाटक तक उनका 'ब्लू प्रिंट' एक जैसा है; वे सीधे और सपाट हैं; उनमें घिसी-पिटी बाते हैं; उनमें शिल्प-विधान की नवीनता नहीं है; उनके लक्ष्य और लेखन में संगति नहीं; उनके नाटकों में जो दार्शनिक मुद्रा दिखाई देती है वह उधार ली हुई है; आदि। फिर भी राकेश का महत्त्व ऐतिहासिक है। वे भारतेन्दु और प्रसाद के समान आधुनिक हिन्दी नाटक की एक कड़ी हैं। राकेश एक बड़ी 'उपलब्धि' न सही, एक बड़ी 'सम्भावना' तो रहे ही हैं।

नाट्य कृतियाँ

खोलो द्वार

धूल लगी है, पद काँटों से बिंधा हुआ, दुःख अपार,
किसी तरह से भूला भटका आ पहुँचा हूँ तेरे द्वार।
पैरों से ही लेपटा-लिपटाकर लूँगा निज पद निर्धार,
अब तो छोड़ नहीं सकता हूँ पाकर प्राप्य तुम्हारा द्वार।
सुप्रभात मेरा भी होवे, इस रजनी का दुःख अपार,
मिट जावे जो तुमको देखूँ, खोलो प्रियतम खोलो द्वार !

—जयशंकर 'प्रसाद'

कहते हैं, 'उष्ट्र' शब्द का सही उच्चारण न कर पाने पर कालिदास को विद्योत्तमा ने घर से बाहर निकाल दिया था और जब वह पढ़-लिखकर लौटा तो उसे अपने घर के दरवाज़े बन्द मिले। उसने दस्तक दी, कहा 'कपाटं देहि !' विद्योत्तमा ने वाग्वैष्ट्यि की पहचान की, तब जाकर अन्दर आने दिया। **आषाढ़ का एक दिन** (1958) में भी कोई एक कालिदास है। उसके जीवन में उसे घर से बाहर कर देनेवाली कोई विद्योत्तमा नहीं आई; किन्तु एक मल्लिका ज़रूर है जिसने अपने दरवाज़े से महत्त्वाकांक्षी कालिदास को यश की चाह में स्वयं उज्जयिनी भिजवा दिया था। मल्लिका ने कभी उनके लिए हृदय के कपाट बन्द नहीं किए। टूटा-हारा जब वह ज़िन्दगी के उत्तरार्द्ध में उसके द्वार आया तो उसने उसका पहले की तरह स्वागत किया; पर सहसा जब नई स्थिति का उसे परिचय मिला तो जिस दरवाज़े से घुसा था, उसी से बाहर भी निकल आया। इसी नाटक में एक पात्र और है, विलोम। वह इस घर में कभी अयाचित था, अब अयाचित नहीं आता; पर जब भी आता है, वह पाता है दरवाज़े बन्द, जब देखो दरवाज़े बन्द !

समस्या बन्द या खुले दरवाज़े की नहीं, मानव नियति की है जिसे लगाव और अलगाव अनजाने ही रच डालते हैं। वास्तव में, घर की खोज उस आत्मीयता की खोज है जो सबसे बड़ी विडम्बना बन जाती है। कालिदास में अपने परिवेश के लिए एक ऐसी भूख दिखाई देती है जो उसे विगत अनुभवों से जोड़ती है। निर्वासित व्यक्ति का गृह-विरह (नॉस्टालजिया) सर्वथा मानवीय है और अस्तित्व की मूल प्रवृत्ति है। यह आत्मा की, अन्तःप्रज्ञा की नैतिक परिणति है। मानव अपने जीवन में भोगे हुए पूर्ण क्षणों को

पकड़ने की कोशिश में रहता है जिससे वह अपनी निजता, अस्मिता और अस्तित्व की पूर्णता अनुभव कर सके। और जैसा कि दोस्तोवस्की ने **द पजेस्ड** में कहा है कि 'कुछ क्षण ऐसे भी होते हैं जब समय रुक जाता है और शाश्वत हो जाता है।' कालिदास भी थका, टूटा-हारा किसी ऐसे ही क्षण में मल्लिका के द्वार पर पहुँचता है जब समय स्वयं रुक जाता है या उसके पाँव स्वयं रुक जाते हैं। काल के इसी ठहराव के क्षण में उसे यह अनुभव होता है जैसे कि वह पीछे रह गया, समय आगे भाग गया—वह 'किसी की प्रतीक्षा नहीं करता !'

ऐसे क्षणों में ही कभी अन्तःप्रज्ञा को ऐसी बिजली कौंधती है कि वे क्षण अस्तित्व के उच्चतम बोध के साथ जुड़ जाते हैं। आत्म-साक्षात्कार के ऐसे क्षणों में समय अपने को बलवान सिद्ध करता है। वह तब और भी केन्द्रीय महत्त्व का विषय बन जाता है जब व्यक्ति वर्तमान में अस्तित्व के संकट से ग्रस्त होता है। यदि कालिदास के जीवन में यह संकट न आता तो सम्भव था उसे मल्लिका के द्वार पर जाने की आवश्यकता ही न पड़ती। पर नीत्शे ने ठीक कहा है—जीवन जैसा कि तुमने जिया है या जी रहे हो, उसी को तुम्हें एक बार और अनेक बार जीना पड़ता है;[1] सुख-दुख उसी प्रकार लौट आते हैं। यह ठीक है कि कालिदास भविष्योन्मुख हैं; मल्लिका उसे एक दिशा की ओर अग्रसर कर देती है। और तब तक अतीत एक अनुपयोगी तत्त्व के रूप में दबा रहता है। किन्तु जैसे ही काश्मीर पर आपत्ति आती है, सब-कुछ खो जाने पर कालिदास के मन में अतीत की रोमानी याद और 'विरह-भावना' लौट आती है। वह अतीत की ओर लपकता है, किन्तु वर्तमान का यथार्थ उसे ध्वस्त कर देता है। मल्लिका के द्वार पर पहुँचकर उसे लगता है जैसे 'सब-कुछ बदल गया है।' दृष्टि और दृश्य दोनों बदले हैं जिसके कारण कालिदास को लगता है कि उसकी पहचान ही मिट चुकी है। वस्तुतः काल का सन्दर्भ पाकर व्यक्ति, परिवेश और मूल्य का बदलना स्वाभाविक है। प्रत्येक क्षण अपने को फिर से सर्जित करता है; दूसरे में परिणत होता है और अपने लिए मूल्यों का निर्धारण करता है। इसी प्रक्रिया में जब कालिदास को यह मालूम होता है कि मल्लिका विलोम की 'भार्या' बन चुकी है और एक बच्ची की माँ है, तो पिछला सम्बन्ध-सूत्र कालगत क्रिया के सन्दर्भ में मूल्य खो बैठता है।

कालिदास 'अथ' से प्रारम्भ करना चाहता है—वह समय के चक्र को मोड़ना चाहता है जो अव्यावहारिक सिद्ध होता है। राकेश इच्छा के समय के साथ घटित होनेवाले द्वन्द्वों से परिचित थे। अपनी सुप्रसिद्ध कहानी **एक और ज़िन्दगी** में उन्होंने इसी सन्दर्भ में कुछ ज्वलन्त प्रश्न उठाए थे; 'मन में आशंका उठती है—क्या सचमुच पहले की ज़िन्दगी को मिटाकर नए सिरे से ज़िन्दगी शुरू कर सकता है ? ज़िन्दगी के कुछ वर्षों को वह एक दुःस्वप्न की तरह भूलने का प्रयत्न कर सकता है ? कितने इन्सान हैं जिनकी ज़िन्दगी कहीं-न-कहीं किसी-न-किसी दोराहे पर ग़लत दिशा की तरफ़ भटक जाती है। क्या यह उचित नहीं है कि इंसान उस रास्ते को बदलकर अपने को सही दिशा में ले आए ? आख़िर आदमी के पास एक ही ज़िन्दगी तो होती है—प्रयोग के लिए भी और जीने के

लिए भी। तो क्यों आदमी एक प्रयोग की असफलता को ज़िन्दगी की असफलता मान ले ?...(किन्तु) जब उसका एक प्रयोग सफल नहीं हुआ तो कैसे कहा जा सकता है कि दूसरा प्रयोग सफल होगा ?' जाहिर है कि जो राह गुज़र चुकी है उसके पद-चिह्न आगे की राह के लिए सहायक सिद्ध नहीं होते और जीवन की शाम को जीवन की सुबह की तरह नहीं किया जा सकता। किसी एक समय का तथ्य दूसरे समय के सन्दर्भ में विसंगत हो जाता है। वे आलोचक धन्य हैं जो कालिदास के अन्ततः मल्लिका के द्वार से बाहर निकल आने को अनैतिक पलायन करार देते हैं। वस्तुस्थिति को देखते हुए कोई भी विवेकशील व्यक्ति वही निर्णय लेता जो राकेश के कालिदास ने लिया है। राज्य की अपेक्षाओं में अपने को ढालकर वह एक अनुभव अर्जित कर चुका था। अब मल्लिका की अपेक्षा के अनुसार अपने को ढाल लेना उसके लिए दूसरा आत्मघात होता। इसी लिए वह बन्धन का नहीं, स्वाधीनता का मार्ग चुनता है। यह भी एक प्रकार का, एक नए जीवन का 'अथ' ही है।

एक नए जीवन का अथ इस रूप में कि कालिदास उस अनुभव में अपनी आत्मा का साक्षात्कार करता है। नाटक के अन्त में आत्मकथन इसका प्रमाण है। स्वार्थ, महत्त्वाकांक्षा, सुविधा का मोह—सब-कुछ उसमें है; पर वही उसकी मानवीयता है और उन्हीं के बीच से गुज़रकर वह अपनी निजी सत्ता की तलाश करता है। वह अपना सब-कुछ गँवाकर वहाँ तक पहुँचता है। उसकी सबसे बड़ी त्रासदी यही है कि वह अनजाने एक दोहरा जीवन जीता है और अनचीन्ही दिशाओं में बहकता है। दरिद्रता उसे राज्याश्रय की ओर ले जाती है; असुरक्षा का भाव उज्जयिनी से काश्मीर तक भटकाता है और फिर एक दिन आत्मीयता की खोज में मल्लिका के द्वार पर ला पटकता है। किन्तु अपने ही अन्तर से उद्भूत अंध-शक्तियों की प्रेरणा से वह जिनसे जुड़ा था, कट जाता है और जो कभी उसके अस्तित्व के साथी थे ही नहीं, उनसे जुड़ा पाता है। फलतः वह किसी से भी न जुड़ पाया—न अपनों से, न परायों से। यह विसंगति अपनी अस्मिता को सही रूप में स्वीकार न कर पाने के कारण घर कर जाती है। अपने स्वभाव की तरलता से ग्रस्त वह किसी भी रेखा से घिरता गया—अन्दर से ठोस बनने के बजाय वह आत्मरक्षा के कवच ओढ़ता गया, और अन्ततः क्षत-विक्षत हुए बिना न रहा। उसे कोई भी ज़िन्दगी रास नहीं आई—न पशुपाल की, न सम्मानित कवि और राजपुरुष की। किन्तु यह स्थिति पलायन की नहीं, 'असत् आस्था' की है। वैसे अन्त में वह अपने को उस असत् आस्था से उबार लेने का संकेत देता है। मल्लिका के द्वार तक ही उसके जीवन की राह समाप्त नहीं हो जाती। वह अपने को विश्लेषित करता है; अपने किए के लिए, कैफियत पेश करता है और अपनी आन्तरिक आवश्यकताओं के सन्दर्भ में अपने को 'जस्टीफाई' भी करता है। वस्तुस्थिति के जान लेने के बाद मारक द्वन्द्वों की राह उसके लिए समाप्त हो जाती है। एक प्रकार का वरण उसे करना ही था।

इस वरण में उसे सम्भावनाएँ दीखती हैं। इन सम्भावनाओं का आधार उसका अपना लेखन है। वह कहता है : 'मल्लिका मुझे वर्षों पहले यहाँ लौट आना चाहिए था,

ताकि यहाँ वर्षा से भीगता, भीगकर लिखता—वह जो मैं तब तक नहीं लिख पाया और जो आषाढ़ के मेघों की तरह वर्षों से मेरे अन्दर घुमड़ रहा है।' जो जीवन और परिवेश उसका अपना नहीं था, उससे मुक्ति सुखमय अस्तित्व के लिए एक अनिवार्यता थी। उसने अपने लिए एक अप्रामाणिक जीवन चुना, जिधर उसकी प्रवृत्ति न थी। वस्तुतः कालिदास का पहला और आखिरी वरण उसका लेखन ही हो सकता था। सार्त्र का रॉक्वेन्टिन और बेकेट का मेलोय भी इसी प्रकार लेखन का वरण करते हैं। लेखन निराश व्यक्ति के लिए शरण्य है जो अस्तित्व की समरूप अनुभूति देता है—घनानन्द की तरह 'लोगहिं कवित्त बनावत मोहि तो मेरे कवित्त बनावत' जैसा अनुभव। लेखन निरन्तर तनाव और टकराव की स्थिति में रखता है और अन्दर से भरा-भरा महसूस कराता है। इसीलिए कामू ने कहा है : 'सृजन को दोहरे जीना है।'[2] मुक्ति भी सृजन का महत्त्वपूर्ण आयाम है। साथ ही रचनाकार अपनी रचना में अपने स्वत्व को भी प्राप्त करता है। यही नहीं, जैसा कि बर्दयेव ने कहा है : 'कला के क्षेत्र में सर्जनात्मकता विजय की आकांक्षिणी होती है।'[3] कालिदास में भी सर्जक की यह आकांक्षा सर्वत्र दिखाई देती है। वह एक जगत् में रहकर दूसरे की सृष्टि करता है और उसका व्यतिक्रम कर आगे बढ़ता है। लभ्य से अलभ्य की ओर बढ़ते हुए उसके चरण कहीं बँधकर नहीं रहते ! इसीलिए यह कवि-प्रवृत्ति मानसिक रुग्णता से पलायन का अवसर प्रदान करने की अपेक्षा स्वयं रुग्णता का लक्षण बन जाती है।'[4] फिर भी कवि के लिए अपने लेखन की तुलना में कोई और वस्तु अधिक वरेण्य हो, ऐसा नहीं लगता।

आज के सन्दर्भ में कालिदास इसीलिए सार्थक लगता है। उसे हम अपने युग की विडम्बनाओं के बीच खड़ा पाते हैं। समस्या मूलतः अलगाव और विच्छिन्नता की है। अपने मूल और परिवेश से उखाड़कर वह अकेलेपन, संत्रास और सम्बन्धहीनता के जिस वातावरण में धकेल दिया जाता है, वह आधुनिक मानव की नियति बन गया है। कालिदास 'सृजनात्मक शक्तियों का प्रतीक' हो सकता है, पर इतना मान लेना ही काफ़ी नहीं है—वह आधुनिक जीवन में व्याप्त मानव के 'गृह-विरह' की चेतना का अंग भी है। वस्तुतः उसके माध्यम से मूल परिवेश से हटकर अर्जित या आरोपित जीवन-पद्धति की पृष्ठभूमि में मानव की आन्तरिकता और उसकी नियति की बड़ी ही विडम्बनापूर्ण अभिव्यक्ति हुई है। विडम्बना यह है कि आदमी को एक ऐसे वातावरण में जीना होता है जिसमें वह अनचाहे जीने के लिए छोड़ दिया जाता है। सारा अस्तित्व इन दो आयामों में निहित है—एक यह है कि व्यक्ति को स्वयं जीना होता है; उसकी जगह कोई और नहीं ले सकता। और दूसरा यह कि यह स्थिति अन्य स्थितियों को जन्म देती है—परिवेश भी उनमें से एक है। आदमी अपने परिवेश को जीता है और यही परिवेश विचारों और समस्याओं को उत्पन्न करता है और प्रतिक्रियाओं को भी। यह द्वैत जैसे अस्तित्व में ही निहित है। आदमी मात्र ज़िन्दगी नहीं जीता—ज़िन्दगी तो वह क्रिया है जो आदमी और उसके परिवेश के बीच घटित होती है।

कालिदास प्रारम्भ में गाँव की सामुदायिक ज़िन्दगी के परिवेश में पलता और बड़ा

होता है। वहाँ गाय चराना बड़ी उपलब्धि है और कविता करना दायित्वहीनता। इसलिए मातुल, अम्बिका सभी उसके कवि-व्यक्तित्व पर आक्षेप करते हैं। अपने में कवि और पशुपालन की विसंगति को जीता हुआ उस परिवेश से भी वह निरीह प्राणी की तरह जूझता है। बाद में वह राजकवि बनकर उज्जयिनी जाता है—वहाँ कविता को सराहना के लिए उपयुक्त भूमि मिलती है पर वह आन्तरिकता नहीं दिखाई देती जो पिछले परिवेश की देन थी। यहाँ भी परिवेश उसके विरुद्ध जाता है। नया परिवेश नई अपेक्षाओं को जन्म देता है। फलतः कहीं भी आन्तरिक अपेक्षाओं और परिवेश की आवश्यकताओं का समन्वय नहीं हो पाता। यह कामू के इस कथन को चरितार्थ करता है कि जीवित रहना विसंगति को जीवित रखना है। आश्रय और सत्ता के मोह में वह अपनी सर्जनात्मक क्षमता को ही बाधित नहीं करता, शासन का अस्त्र भी बनता है वह मुक्ति और शक्ति के सन्दर्भों से इस प्रकार जुड़ जाता है कि उनके बीच उसका सारा अस्तित्व विसंगत और त्रासद होने को बाध्य है।

कालिदास अन्दर-ही-अन्दर जो भोगता है, उस विसंगति के साथ कई औरों की विसंगतियाँ भी जुड़ जाती हैं—अम्बिका, मल्लिका, मातुल, विलोम—सबके-सब उसके भागीदार बनते हैं। कालिदास काल के साथ इच्छा के द्वन्द्व की जैसी विनाशकारी परिणति भोगता है, वह उसे झकझोर देती है। यह उसके मन को हल्के आक्रामक भाव से भर देती है। यह आक्रामक प्रवृत्ति बहुत कुछ निराशा की देन है। पारिवारिक सूत्रों और स्नेह-सौहार्द के स्पर्शों से वंचित, मातुल की गाय चराने में जीवन को सार्थक करनेवाले कालिदास के लिए निराशा की क्या कमी हो सकती थी ? फलतः वह गाँव-भर में मल्लिका के अतिरिक्त सबसे अलगाव महसूस करता है। उज्जयिनी में जाकर यह अलगाव और बढ़ता है और जब सब जगह से मुँह मोड़कर वह मल्लिका के पास गाँव में लौटता है तो उसकी कोई सीमा नहीं रह जाती। तब किसी समय जो चीज़ें संगत थीं, असंगत लगने लगती हैं—मल्लिका, उसका घर, उसका प्रेम और स्वयं अपना अस्तित्व। जो कुछ उसके आसपास (विशेषतः मल्लिका में) घटता है उसमें भी वह अपने को संपृक्त पाता है। फलतः लज्जा और अपराध-भावना की अनुभूति उसे आत्मस्वीकार की पीड़ा देती है।[5] ये भावनाएँ आदमी में तब आती हैं जब वह अपने को उस रूप में देखने लगता है जिस रूप में दूसरे उसे देखते हैं। अतः कालिदास कह उठता है : 'मैंने बहुत बार अपने सम्बन्ध में सोचा है मल्लिका, और हर बार इस निष्कर्ष पर पहुँचता हूँ कि अम्बिका ठीक कहती थी।' इस आत्मस्वीकार में वह कितना नंगा, कितना अकेला हो जाता है ! यह सब उसे निस्सारता का अनुभव दे जाता है। यह अनुभव मातुल के अनुभव से जुड़कर और गहरा हो जाता है। राजप्रासाद में रहकर गाँव तुड़वाकर अब वह यथार्थ की पहचान में इतना कर्त्तव्य-मूढ़ हो जाता है कि कहता है : 'समझता हूँ कि जो कुछ मैं समझ पाता हूँ, सत्य सदा उसके विपरीत होता है।' और फिर विलोम की विसंगति भी कालिदास से कम नहीं—वह मल्लिका के साथ घर जरूर बसा लेता है, पर जब देखो द्वार बन्द ! कालिदास के लिए ये द्वार खुले होने पर भी हमेशा के लिए

बन्द हो जाते हैं और विलोम के लिए बन्द रहने पर भी खुले हैं। सारा नाटक इसी विडम्बना पर टिका है—

किवाड़ बंद मिले शहर के मकानों के,
कयाम भी मुयस्सर न हुआ, सफ़र भी गया।[6]

इस विसंगति का आधार है—मनुष्य की इच्छा का समय के साथ द्वन्द्व, अपनी भूमि से उखड़ने और एक भिन्न कर्म और भावभूमि पर अपनी महत्त्वाकांक्षा के कारण 'दूसरा' व्यक्ति हो जाने की नियति। इसी के कारण मल्लिका कालिदास को बाँध नहीं पाती और प्रियंगु उसे राजसत्ता का भूखा बना डालती है। समय के प्रवाह में यह परायापन दुधारी तलवार की तरह काम करता है और विच्छिन्न कर एक त्रासदी को जन्म देता है। विसंगतियों के बीच उलझा यह सारा कथ्य एक रोमानी, भावपूर्ण कथानक के बीच से उभरता है। कथ्य और कथानक का यह वैषम्य ज़रूर एक सुखद आश्चर्य दे जाता है। **आषाढ़ का एक दिन** की उठान ही काव्यमयी है। आषाढ़ के मेघों का घिरना, बरसना और फिर मल्लिका का भीगना—सब एक विलक्षण अनुभव के रूप में अंकित हुआ है : 'वह बहुत अद्भुत अनुभव था माँ, बहुत अद्भुत। नीलकमल की तरह कोमल और आर्द्र, वायु की तरह हलका और स्वप्न की तरह चित्रमय।...मेरा तो शरीर निचुड़ रहा है माँ ! कितना पानी इन वस्त्रों ने लिया है।' यह वर्षण मेघ का ही नहीं, तरल युवा मन का भी है। यह भीगना शरीर या वस्त्रों का ही नहीं, मन और यौवन का भी है। यहीं मल्लिका के रोमानी, अल्हड़, मुखर और समर्पणशील चरित्र की भूमिका तैयार हो जाती है। वस्तुतः मल्लिका के हृदय की गहराई में आषाढ़ के ये पहले बादल कहीं कालिदास से जुड़ जाते हैं। इसीलिए जीवन-भर साथ रहते हैं और कभी उष्ण तो कभी शीत स्पर्श दे जाते हैं।

इस चरित्र की भावमयी आत्मा को एक अलग रंग देने के लिए राकेश ने उसकी माँ अम्बिका के विरोधी चरित्र को साधन रूप में प्रयुक्त किया है। जीवन की कविता के विरोधी रंग लिए हुए अम्बिका उस मौन ज़िन्दगी का प्रतिनिधित्व करती है जहाँ गहरे कुएँ में पड़े लोटे की तरह शब्द मुँह से बहुत नीचे उतर गए हैं। उसकी अनुभूति-प्रवणता के नुकीले छोर बढ़ी उम्र और जागतिक स्थितियों के संघातों से खुंडे हो चुके हैं। माँ के शब्द-कार्पण्य के बीच मल्लिका जीवन की कविता को मुक्त छन्द की तरह जीती है। वे दोनों प्रतिकूल बिम्ब उभारते हैं, पर सहसा बहुत भिन्न भी नहीं। मल्लिका युवती अम्बिका है और अम्बिका जीवन के थपेड़ों से ढली मल्लिका—एक आदि और दूसरा अन्त। इसीलिए जिन बरसते मेघों में भीगते हुए मल्लिका उमंगित होती है, उन्हीं की कल्पित विडम्बना पर अम्बिका रो उठती है !

यहीं से आषाढ़ के प्रथम मेघों के रंग काले होने लगते हैं। राजपुरुष गाँव में पहुँचते हैं और उनके साथ ही उनकी काली छाया भी उन उल्लास-भरे मेघों में घुलमिल जाती है : 'कभी वर्षों में ये आकृतियाँ यहाँ दिखाई देती हैं। और जब दिखाई देती हैं तो कोई-न-कोई अनिष्ट होता है। कभी युद्ध की सूचना आती है, कभी महामारी की।'

पिछली महामारी में ये आकृतियाँ मल्लिका के पिता की मृत्यु से जुड़ गई थीं। अबकी बार मौत के शिकार ये मेघ होते हैं जो मल्लिका के मन की पुलक को 'मेघदूत की सजल कल्पना' में बदल देते हैं। और फिर इसी चेतना से आ जुड़ता है आहत हरिण-शावक का प्रसंग। उसे लेकर कालिदास के पास उपस्थित होना, अम्बिका का मन-ही-मन रुष्ट होना, कालिदास का उधर ध्यान न देकर हरिण-शावक से बतियाना—सब कालिदास की पीड़ा को उभारते हैं। वास्तव में हरिण-शावक ही नहीं, कालिदास भी आहत है। उसे आहत करनेवाले बाण मातुल के व्यंग्यों के हैं, अम्बिका के हैं, अपवादों के हैं और वे बाण फिर राजसत्ता के हो जाते हैं। राजसत्ता हरिण-शावक के लिए जितनी नृशंस सिद्ध होती हैं, उतनी ही बाद में कालिदास के लिए भी।

इस प्रकार हृदय को सिक्त कर देनेवाला मेघ का उन्मुक्त वर्षण और हरिण-शावक की आहत निरीह कोमलता, मल्लिका का यौवनागम का मधुर अनुभव और अम्बिका की नपी-तुली, कटी-छँटी मानवीयता—नाटक में ये विरोधी रंग जीवन की कविता को रचते दिखाई देते हैं। इस कविता के बीच से नाटक शुरू होकर उन स्थितियों तक पहुँचता है जहाँ व्यावहारिक जीवन के स्वप्न, यथार्थ और आशा-आकांक्षा का द्वन्द्व मुखर हो उठता है। यह द्वन्द्व नाटकीय है, कवित्वपूर्ण नहीं। पर द्वन्द्व न सही, द्वन्द्व की भावात्मक परिणति अवश्य कवित्वपूर्ण है। इस द्वन्द्व के बीच अम्बिका की निष्ठुर मुद्रा यथार्थ के कटु अहसास से रोमानियत का रंग गाढ़ा करती है। वह यथार्थ के भीतर इतनी पैनी आँखों से देखती है कि सत्य अपने वीभत्स रूप में महसूस होने लगता है। यथार्थ और आकांक्षा का टकराव मल्लिका और अम्बिका, कालिदास और विलोम, अभाव और कल्पना को आमने-सामने खड़ा कर देता है। इस टकराव में मल्लिका घटना-प्रवाह को एक दिशा देती है और अपनी ही दी गई दिशा के भँवर में स्वयं ही घिर भी जाती है। सदाशयता ही जैसे उसकी अपनी त्रासदी की ज़िम्मेदार बन जाती है।

परिस्थिति, क्रिया और काल की प्रक्रिया से मनुष्य अपने को गढ़ता है। इसमें वह हर क्षण अपने को बनाता और मिटाता है। जीवन का एक-एक क्षण जब अपनी सक्रियता में व्यक्ति से जुड़ता है तो व्यक्ति वही नहीं रहता जो वह कभी था। सारे नाटकीय संघर्ष के दौरान कालिदास, मातुल, मल्लिका आदि पात्र इसी अहसास को दिलाते हैं। विलोम इसी अहसास से प्रेरित उज्जयिनी को प्रस्थान करते कालिदास पर टिप्पणी करता है : 'तुम अभी तक वही व्यक्ति नहीं हो जो कल तक थे।' और कालिदास आगे अपने व्यवहार से इस आशंका को सार्थक करता है। उज्जयिनी जाकर वह दूसरा व्यक्ति हो जाता है। यह स्थिति ही मल्लिका और स्वयं उसकी कारुण्कि परिणति के लिए उत्तरदायी बनती है। किन्तु दूसरा आदमी बन जाना मानव की नियति भी है। कालिदास जानता है कि नए जीवन की नई अपेक्षाएँ होती हैं। मल्लिका और निक्षेप कालिदास को इस सन्दर्भ में समझने की कोशिश भी करते हैं। जीवन जब साधारण से असाधारण स्थिति में पहुँचता है, तो उसे असाधारण की ही अपेक्षा होती है। इसीलिए मल्लिका बहुत

पीछे छूट जाती है। राजकवि और राजपुरुष की अर्जित प्रतिष्ठा कालिदास से गुप्त कुल की राजदुहिता का वरण करवाती है। कभी अभाव मल्लिका से विवाह न हो सकने का कारण था, अब ऐश्वर्य और असाधारणता उससे भी बड़ा कारण बन जाती है। कालिदास काश्मीर जाते हुए गाँव आता है, पर मल्लिका को नहीं मिलता। यहीं मोहभंग होता है और राजशक्ति के साथ जुड़ी विसंगत क्रूरता मुखर होती है। अनुस्वार, अनुनासिक, दंतुल, रंगिणी, संगिनी, प्रियंगु मंजरी सब उसके प्रतीक हैं। इनके माध्यम से मल्लिका की चेतना में कोमल वक्ष पर रखे पत्थर की भाँति सत्ता का क्रूर यथार्थ उभरता है। गाँव में आकर वातावरण इकट्ठा करने के नाम पर पत्थर, वनस्पति, जन्तु आदि एकत्र करना, मातुल के परिवार को साथ ले जाना, मल्लिका के सामने घर का परिष्कार करने का प्रस्ताव रखना और फिर 'जिसे तुम अपने योग्य समझो विवाह कर लो' की बात करना उसी का अंग है। उस ऐश्वर्यभोगी वर्ग के लिए ग्रामीण प्रकृति और जन, यहाँ तक कि मल्लिका भी, कुतूहल और उपयोग के उपादान मात्र हैं। इसीलिए प्रियंगु मल्लिका की आन्तरिक भावनाओं को कहीं पकड़ ही नहीं पाती। वह जिस प्रकार कृत्रिम साधनों से कालिदास के अभावों को भरना चाहती है, उसी प्रकार की कृत्रिम क्षतिपूर्ति वह मल्लिका के जीवन में भी करना चाहती है। स्थिति की यह मार्मिकता पाठक/प्रेक्षक में तीव्र भावना जगाती है।

यहीं आकर पूर्वस्थिति का/मल्लिका की भावना में भावना के वरण का, क्रूर व्यंग्य उभरता है। विलोम और अम्बिका की वाणी और घटनाओं की परिणति में नाटक की 'आयरनी' जैसे अपने सत्य को प्रतिष्ठित करने लगती है। और ऐसी ही एक स्थिति में जब प्रियंगु मल्लिका से प्रश्न कर बैठती है : 'क्यों, तुम्हारे मन में कल्पना नहीं कि तुम्हारा अपना घर-परिवार हो ?' तो अम्बिका सहसा फूट पड़ती है : 'इसके मन में यह कल्पना नहीं है क्योंकि यह भावना के स्तर पर जीती है !' कितना बड़ा व्यंग्य वह अनायास ही उस पर उगल देती है जिससे मल्लिका तड़प उठती है; किन्तु उससे भी बड़ी तड़प उसकी अपनी है जिसके शब्द गले में अटक जाते हैं। यहीं कालिदास का दोहरा व्यक्तित्व साफ़ नज़र आने लगता है। इससे बड़ी विडम्बना और क्या हो सकती है कि प्रियंगु जिस प्रदेश के सौन्दर्य पर मुग्ध होकर कह उठती है : 'इस सौन्दर्य के सामने सब सुविधाएँ हेय हैं' उसी में जन्म लेकर कालिदास उस सौन्दर्य को छोड़कर सुविधा के पीछे दौड़ पड़ता है। मल्लिका के लिए मेघ अब भी बरसते हैं, किन्तु कालिदास का आकाश जैसे निरभ्र हो जाता है—दया और करुणा से विहीन। अन्ततः कालिदास की उपेक्षा उस अयाचित स्थिति को पैदा करती है जिसकी मल्लिका ने कभी कल्पना भी न की थी। मल्लिका उस विलोम की 'पत्नी' बनने को बाध्य होती है जिसे कभी वह 'अयाचित अतिथि' मानती आई थी और अब जिसके साथ अपने सम्बन्ध को लेकर वह अपने को वारांगना कहती है। मल्लिका बाहर से टूटकर भी अन्दर से कालिदास से जुड़ी रहती है। मन की कविता के छन्द टूटने नहीं देती। पर तन और मन का यह विभाजन उसे व्यक्ति से विशेषण बना डालता है। और फिर एक दिन वही कालिदास जब मल्लिका के द्वार पर

लौट आता है तो नियति ही जैसे कविता रच डालती है। उसे पुरानी मल्लिका बदली हुई लगती है जिस पर वह अपनत्व और अधिकार की बात सोच नहीं सकता। और जब विलोम अपनी लँगड़ी टाँग लेकर आता है तो उस 'बदली हुई' मल्लिका के बन्द दरवाज़ों के भीतर कहीं अधिकार और अपनत्व महसूस होता है : 'अब तो वह अधिकार से आता है। विलोम अब तो इस घर में अयाचित अतिथि नहीं।' यहीं विलोम और कालिदास आमने-सामने खड़े बेतुके और दयनीय लगते हैं। कालिदास समय की निष्ठुरता में अपने व्रण छिपाने का प्रयास करता है; किन्तु विलोम के व्रण तो उसके कटाक्ष ही सहला पाते हैं : 'क्योंकि तुम यहाँ लौट आए हो ?...क्योंकि वर्षों से छोड़ी हुई भूमि आज फिर तुम्हें अपनी प्रतीत होने लगी है ?...जैसे तुमसे बाहर जीवन की गति ही नहीं है। तुम्हीं तुम हो और कोई नहीं ? परन्तु समय निर्दय नहीं है। उसने औरों को भी सत्ता दी है; अधिकार दिए हैं। वह धूप और नैवेद्य लिए घर की देहली पर रुका नहीं रहता। उसने औरों को अवसर दिया है।...तुम्हें उसके निर्माण से वितृष्णा होती है ? क्योंकि **तुम जहाँ अपने को देखना चाहते हो, नहीं देख पा रहे** ?' विलोम के तरकश के सारे बाण कालिदास पर टूट पड़ते हैं। वह तो अन्त में सदय हो भी जाता है; किन्तु परिस्थितियों द्वारा निर्मित यथार्थ बहुत क्रूर होकर सामने आता है और मल्लिका की नई भूमिका में कालिदास का स्थान अपरिभाषित, अनिर्धारित रह जाता है। फलतः ज्यों ही वह बच्ची को चुप कराने के लिए अन्दर जाती है, वह बाहर निकल आता है ! और मेघ कविता बरसाते जाते हैं।

अस्तित्ववादी चिन्तन-संकेतों तथा अम्बिका और विलोम की यथार्थवादी गिद्ध दृष्टि के बावजूद **आषाढ़ का एक दिन** की अवधारणा काव्यात्मक है। उसमें चरित्र और स्थितियाँ मिलकर विडम्बनापूर्ण कविता रचते हैं। इसी आधार पर कुछ लोगों ने इस नाटक का मूल स्वर रोमानी माना है। वैसे भी यह प्रसाद और नए नाट्य लेखन के बीच की कड़ी प्रस्तुत करता है। इनमें एक ओर मेघ, हरिण-शावक और ग्रामीण प्रकृति से सम्बद्ध भाव-बोध है, दूसरी ओर ऐश्वर्य के प्रति विमुखता, अभाव में भाव की प्रतिष्ठा, पूर्वस्थिति की ओर प्रत्यावर्तन तथा अतीत विरह की टीसती भावना मुख्य है जो छायावाद की भाव-प्रवणता का संकेत देती है। मल्लिका का 'भावना में भावना का वरण' छायावाद के अशरीरी प्रेम का ही द्योतक है। और वह जो मन से कालिदास की और तन से किसी और की बनकर रहती है, यह विभाजित व्यक्तित्व भी उसी का प्रतिफल है। 'मैं यद्यपि तुम्हारे जीवन में नहीं रही, परन्तु तुम मेरे जीवन में सदा बने रहे हो'—आस्था का यह स्वर प्रसाद की नायिकाओं का धर्म रहा है। और कालिदास का द्वन्द्व भी कम रोमानी नहीं है। इसलिए भी कि मोहन राकेश उसे न किसी सामाजिक परिवेश से जोड़ पाए और न जीवन-मूल्यों से। यही कारण है कि लेखक इस नाटक में जो समस्या उठाना चाहता है, वह कथानक के बीच से नहीं उभरती और कथावस्तु तथा अर्थ-संकेत में सही ताल-मेल नहीं बिठा पाती।

राकेश ने एक स्थान पर लिखा है : 'एक साहित्यकार के जीवन की मूल समस्या है साहित्यकार के रूप में अपने व्यक्तित्व को बनाए रखने की और शेष समस्याएँ इस एक समस्या के साथ जुड़ी हैं।...लेखक का व्यक्तित्व निस्सन्देह उन सब सुविधाओं की अपेक्षा अधिक महत्त्वपूर्ण है जो राज्य उसे दे सकता है।[7] राकेश का यह भी कहना है कि कालिदास सृजनात्मक शक्तियों का प्रतीक है और **आषाढ़ का एक दिन** में वह 'उस अंतर्द्वंद्व को संकेतित करता है जो किसी भी काल में सृजनशील प्रतिभा को आन्दोलित करता है।'[8] इसी आधार पर यह कहा जाता है कि यह नाटक राज्याश्रय की समस्या को उठाता है और उसके परिणामों को रेखांकित करता है। इसी समस्या को लेकर इसे आधुनिक और समसामयिक भाव-बोध से भी जोड़ा जाता रहा है।

सिद्धान्ततः आज के सन्दर्भ में राज्याश्रय की समस्या बड़ी ज्वलन्त है। राज्यसभा की सदस्यता, बड़े-बड़े पदों और सरकारी कमेटियों में नामजदगी, पद्मश्री और पुरस्कार-प्राप्ति आदि कई प्रकार के प्रलोभनों में आज का लेखक आता रहा है। अभावग्रस्त मध्यवर्गीय भावना के प्रतिनिधि होने के नाते वह उन पर बुरी तरह झपटा है और पंगुता स्वीकार कर उसने अपनी धुरीहीनता का प्रमाण भी दिया है। इस प्रकार प्रतिबद्धता से जी चुरानेवाला सुविधाभोगी लेखक आत्मपरायापन और आत्मनिर्वासन का शिकार हुआ है, इसमें कोई सन्देह नहीं। इसीलिए कहा जा सकता है कि कालिदास के माध्यम से राकेश ने समसामयिक साहित्यकार की मानसिकता का चित्रण किया है।

लेखन और राज्याश्रय कालिदास के लिए दो नौकाओं में संतरण करना जैसा सिद्ध होता है। राज्याश्रय व्यक्ति की स्वतन्त्रता को बाँधता है—कालिदास इससे भी आगे जाकर राजसत्ता का फन्दा अपने लिए स्वीकार कर लेता है। सत्ता आदमी के आन्तरिक अहं की पूर्ति तो कर सकती है, पर उसके सपनों और अभिरुचियों को पूरा नहीं होने देती। विशेषतः उस व्यक्ति की जो अपने क्षेत्र से अलग दख़ल देता है। कालिदास ग़लत चुनाव करता है। इसीलिए राजपुरुष कालिदास और कवि कालिदास टकराहट में आते हैं। व्यक्ति कालिदास संभावनाओं के होते हुए अगर राज्याधिकारी बनता तो संभवतः बुरा नहीं होता। कवि कालिदास राज्याधिकारी/शासक बनता है, यह बुराई पैदा करता है। दोनों के लिए अलग-अलग अर्हताएँ अभिप्रेत हैं। और सबसे बड़ी विडम्बना तो यह है कि राजसत्ता का प्रलोभन उसे अपने परिवेश से काट डालता है। वह वैभवपूर्ण जीवन और उच्च सम्पर्क के सोपानों पर चढ़ता जाता है, पर उसके पैर तले की ज़मीन पीछे छूटती जाती है और जो ज़मीन वह अपने पैरों के नीचे पाता है, वह भी उसे आत्मीयता नहीं दे पाती। फलतः वह अभिजात वर्ग की थोथी संस्कृति का सड़ा अंश बन जाता है और अपने मूल स्रोत से विलग हो जाता है। इससे जो अलगाव पैदा होता है वह उसके लिए घातक सिद्ध होता है। कवि-जीवन की सहज आन्तरिक आवश्यकता को भुलाकर वह अपनी निजता की उपेक्षा करता है। कीर्केगार्द ने ठीक ही कहा है कि मनुष्य चाहे कितना ही महान क्यों न हो, उसकी महानता तब तक संदिग्ध ही रह जाती है जब तक वह अपने चुनाव को अपनी अन्तरात्मा के सामने स्पष्ट नहीं कर लेता।[9]

कालिदास के चुनाव में अधिकार का मोह था, आन्तरिकता का स्वीकार नहीं। इसीलिए वह अपने अस्तित्व में अविद्यमान रहने की नियति भोगता है।

नाटक के प्रारम्भ में ही राकेश ने राजसत्ता की नृशंसता के संकेत दिए हैं। हरिण-शावक का शिकार उनमें से एक है। अम्बिका अपने एक संवाद में राजसत्ता का क्रूर बिम्ब उभारती है। कालिदास घोषणा करता है : 'मैं राजकीय मुद्राओं से क्रीत होने के लिए नहीं हूँ।' रंगिनी-संगिनी, अनुस्वार-आनुनासिक तथा प्रियंगु की उक्तियाँ राजसत्ता के बिम्ब को और गाढ़ा करती हैं। मातुल की वाणी भी उसे मुखर करती है और अन्ततः कालिदास का यह आत्म-स्वीकार सुनने को मिलता है : 'अधिकार मिला, सम्मान मिला...परन्तु मैं सुखी नहीं हुआ। किसी और के लिए वह वातावरण और जीवन स्वाभाविक हो सकता था, मेरे लिए नहीं था। **एक राज्याधिकारी का कार्यक्षेत्र मेरे कार्यक्षेत्र से भिन्न था।** मुझे बार-बार अनुभव होता है कि मैंने प्रभुता और सुविधा के मोह में पड़कर उस क्षेत्र में **अनधिकार प्रवेश** किया है और जिस विशाल में मुझे रहना चाहिए था, उससे दूर हट आया हूँ।...और एक दिन मैंने पाया कि मैं सर्वथा टूट गया हूँ...लोग सोचते हैं कि मैंने उस जीवन और वातावरण में रहकर बहुत कुछ लिखा है। परन्तु मैं जानता हूँ मैंने वहाँ रहकर कुछ नहीं लिखा। जो कुछ लिखा वह यहाँ के जीवन का **संचय** था।'

ये सब कथन राजसत्ता और लेखन को एक-दूसरे के विरोध में खड़ा करते हैं। इसे स्वीकारने में कोई सैद्धान्तिक विरोध भी नहीं है। किन्तु प्रश्न राज्याश्रय के औचित्य-अनौचित्य का ही नहीं, यह भी है कि कालिदास के जीवन में राज्याश्रय कहाँ और किस रूप में बाधक होता है ? नाटककार ने कालिदास के मुँह से कहलवाया भर है, नाटक में उन उपस्थितियों का, उस प्रक्रिया का अहसास कहाँ पर होता है ? कालिदास लेखन और राज्याश्रय के द्वन्द्व को कहाँ पर झेलता है ? सब-कुछ जैसे नेपथ्य में घटता है। मल्लिका के प्रति कालिदास की प्रवंचना आगे आ जाती है। इसीलिए कालिदास और उसकी समस्याएँ मंच पर नहीं उभरतीं, केवल ध्वनित होती हैं। समस्या को उठाने का यह तरीक़ा ख़तरे से ख़ाली नहीं कहा जा सकता। ऐसा प्रतीत होता है जैसे समस्या नाटक के बीच स्वतः स्फूर्त नहीं हो रही है, नाटककार उसे आरोपित कर रहा है। इस सब पर ध्यान न भी दें तो कुछ प्रश्न और भी उभरते हैं : क्या राज्याश्रय सदा लेखन में बाधक रहा है ? यदि ऐसा ही था तो कालिदास ने स्वयं राज्याश्रय में महान् ग्रन्थों की रचना कैसे की थी ? क्या शासनतंत्र से बाहर रहकर ही लेखक स्वतंत्र रह सकता है ? किन्तु यह बात भी नहीं कही जा सकती क्योंकि शासन के बाहर रहकर भी आत्मप्रवंचक रचनाकार अपने लेखन और स्वातन्त्र्य का अपहरण करा लेता है।

राकेश एक ओर लेखक की आर्थिक स्वतन्त्रता को अनिवार्य मानते हैं, दूसरी ओर यह भी कहते हैं कि 'राज्याश्रय के प्रसंग में इतना ही पर्याप्त होगा कि यदि राज्य द्वारा ऐसी परिस्थितियाँ उत्पन्न की जा सकती हैं कि एक लेखक राज्य के द्वारा दी हुई सुविधाओं का भोग करते हुए भी अपने व्यक्तित्व और विचारों की स्वतन्त्रता को बनाए

रख सके...तो उसे स्वीकार करने में कोई बाधा नहीं होनी चाहिए।'[10] कालिदास के सामने कोई बाधा नहीं थी, ऐसा स्पष्ट है। तो फिर वह टूटता क्यों है ? क्या राज्याश्रय उसे तोड़ता है ? क्या यह सत्य नहीं कि कालिदास जिस कारण टूटता है वह वस्तुतः राज्याश्रय नहीं, अपनी भूमि से उखड़ना, दूसरा व्यक्ति हो जाना अथवा राजसत्ता का प्रलोभन है ? राज्याश्रय से वह राज्याधिकार/राजसत्ता की ओर बढ़ता है। यह लिप्सा उसके कार्यक्षेत्र को बदलकर उसे भ्रष्ट करती है। मल्लिका ठीक कहती है : 'मैंने इसलिए भी नहीं कहा था कि तुम जाकर कहीं का शासन-भार सँभालो।' राज्याश्रय उसके जीवन का एक चरण था, शासन-भार सँभालना दूसरा। यद्यपि राज्याश्रय और राजसत्ता की प्राप्ति दोनों परस्पर सम्बद्ध हैं, पर विध्वंसक प्रभाव लेखन पर राजसत्ता का जितना पड़ सकता है, राज्याश्रय का उतना नहीं। सत्ता का प्रलोभन हावी हो जाने के कारण ही सृजन की संभावनाएँ खत्म होती हैं। उसके लिए कालिदास ही दोषी है। इस बिन्दु पर नाटक की समस्या ही बदल जाती है

और यह फिर लेखक-मात्र की समस्या नहीं है। कोई भी व्यक्ति जब सुविधाभोगी हो जाता है तो वह कालिदास की नियति भोगने के लिए बाध्य है। पर इसके साथ ही यह प्रश्न उठता है कि यदि वह राज्याश्रय न स्वीकार करता तो क्या उसकी सृजनशीलता बनी रहती ? क्या वह उस हालत में टूटता नहीं ? मल्लिका सब-कुछ ठुकराकर जिस प्रकार खंडित होकर जीती है, वह भी जी सकता था। पर विसंगति हर दशा में उसे घेरती ! तो क्या कालिदास को इस बात के लिए दोषी ठहराया जा सकता है कि वह राजसत्ता की ओर उन्मुख होता है ? शायद हाँ, इसलिए कि आदमी को अपनी सीमाओं से बाहर न जाना चाहिए। किन्तु नहीं नए लक्ष्यों की खोज भी तो मानव की ईहा का अंग है। कालिदास का द्रष्टा व्यक्तित्व भोक्ता की ओर अग्रसर होता है। ललक मनोवैज्ञानिक है : 'अभावपूर्ण जीवन की एक स्वाभाविक प्रतिक्रिया।' 'मात्र आजीविका का संघर्ष एक लेखक का संघर्ष नहीं होता ?'[11] उसकी आन्तरिक आवश्यकताएँ भी उसे भटकाती हैं। मानव के लिए यथास्थिति से बँधना मुश्किल है और उससे आगे बढ़कर पूर्व-स्थिति की कामना तो और भी मुश्किल है। दोनों स्थितियों में उसके हाथ विसंगति ही आती है !

यहीं, यह प्रश्न उठता है, क्या कालिदास सचमुच 'सृजनात्मक शक्तियों का प्रतीक' है ! नाटक में कालिदास आत्मकेन्द्रित, स्वार्थी, पलायनवादी प्रवंचक के रूप में चित्रित है। क्या कोई कह सकता है कि सृजनात्मक शक्ति इन प्रवृत्तियों में निहित है ? अथवा इस सब को अनदेखा कर यदि द्वन्द्व को ही मुख्य मानें 'जो किसी भी काल में सृजनात्मक प्रतिभा को आन्दोलित करता है' तो ऐसा द्वन्द्व केवल सृजनात्मक प्रतिभा को ही भोगना पड़ता हो, ऐसी बात भी नहीं। हर महत्त्वाकांक्षी व्यक्ति, जो एक भिन्न परिवेश से ऊपर उठता है, इस द्वन्द्व का शिकार होता है। यदि उसे प्रतीक मान भी लें तो वह रोमानी दृष्टि से सर्जित और अपने सन्दर्भों में बहुत व्यापक नहीं है। सवाल यह उठता है कि सृजन के विरुद्ध कौन-सी बाधाएँ उठ खड़ी होती हैं और कालिदास

उनसे कैसे जूझता है।

आषाढ़ का एक दिन के कथ्य का एक और आयाम भी है। वह प्रेम का एक त्रिकोण लेकर चलता है। भावना में जीनेवाली मल्लिका कालिदास के साथ अपने प्रेम-सम्बन्ध को सब सम्बन्धों से बड़ा मानती है। इसीलिए जब उसे राजकवि का सम्मान मिलता है तो वह उसे उज्जयिनी जाने के लिए प्रेरित करती है—रेखा की तरह घेरने की बजाय वह उसके लिए विस्तृत क्षितिज के द्वार उन्मुक्त कर देती है। वह अभाव झेलती है—वारांगना जैसा अस्तित्व भोगती है, किन्तु उसका आस्था-भाव बना रहता है : 'मैं टूटकर भी अनुभव करती हूँ कि तुम बने रहो। क्योंकि मैं स्वयं को अपने में न देखकर तुममें देखती रही।' यह निश्छल भाव उस कालिदास के प्रति है जो बेहद आत्मकेन्द्रित 'कैरियरिस्ट' व्यक्ति है। उसे मल्लिका की याद तब आती है जब परिस्थितियाँ उसे उसके लिए बाध्य करती हैं। और तब वह उस पूर्व-प्रेम की कड़ियों को फिर से जोड़ने का प्रयत्न करता है : '**कुमारसम्भव** की पृष्ठभूमि हिमालय है और तपस्विनी उमा तुम हो। **मेघदूत** के यक्ष की पीड़ा मेरी पीड़ा है और विरह-विमर्दिता यक्षिणी तुम हो। **अभिज्ञान शाकुन्तल** में शकुन्तला के रूप में तुम्हीं मेरे सामने थीं।' तो क्या कालिदास दोहरा जीवन जीता रहा ? दोहरा जीवन तो मल्लिका भी जीती है। तन और मन के इस दोहरेपन को छायावादी/स्वच्छन्दतावादी कहना ज़्यादा उपयुक्त होगा। पर इसे भी एक जीवन-दृष्टि कहा जाएगा, रोमानी ही सही। वस्तुतः कालिदास और मल्लिका का यह प्रेम-भाव उन धारणाओं को ही रेखांकित करता है जो नीत्शे, बाल्ज़ाक, बायरन आदि व्यक्त कर चुके हैं कि प्रेम स्त्री का सम्पूर्ण अस्तित्व है और पुरुष के लिए एक आवश्यकता मात्र है। पुरुष की ज़िन्दगी प्रसिद्धि है और स्त्री की प्रेम। पुरुष का प्रेम और उसका जीवन दो अलग वस्तुएँ हैं; स्त्री के लिए प्रेम एक आस्था है। हिन्दी में भी इस तरह की विचारधारा प्रसाद में सामान्यतः यत्र-तत्र मिलती है।

मल्लिका, कालिदास और विलोम के पारस्परिक सम्बन्ध कुछ प्रश्नों को जगाते हैं : क्या प्रेम बिना विवाह-सम्बन्ध सम्भव है ? या विवाह के बिना प्रेम का निर्वाह हो सकता है ? क्या प्रेम एकांगी हो सकता है ? क्या व्यक्ति तन और मन को विभाजित कर एक (तन) को अनेक और दूसरे (मन) को मात्र एक से जोड़कर अपने को सार्थक कर सकता है ? **आषाढ़ का एक दिन** के त्रिकोणात्मक सम्बन्ध इस सन्दर्भ में कुछ संकेत देते हैं, किन्तु नाटक इन प्रश्नों का उत्तर पाने के लिए लिखा गया प्रतीत नहीं होता।

तो फिर कालिदास और मल्लिका के सम्बन्धों की चर्चा किस रूप में सम्भव है ? वस्तुतः कालिदास का प्रेम एक विरोधाभास है। वह मल्लिका को स्मृति में याद रखता है, वास्तविक जीवन में भूल जाता है। वह उसे अस्तित्व से वंचित कर देता है, पर अपने अतीत से वंचित नहीं करता। **कुमारसम्भव** या **मेघदूत** लिखते हुए वह उसे जिस उमा या यक्षिणी के रूप में याद करता है (या सर्जित करता है), वह प्रेम नहीं, एक पूर्व स्थापित स्मृति का आख्यान, एक संचित अनुभव की अभिव्यक्ति मात्र है। वह जीवन्त मल्लिका को याद नहीं कर रहा—याद कर रहा है अतीत की एक ऐसी मल्लिका को,

जो कालबद्ध है। इसीलिए वह उसमें और उसकी स्थितियों में हुए परिवर्तनों को नहीं देख पाता। तभी तो मिलने पर उसे सब-कुछ बदला-बदला-सा लगता है। दरअसल उसकी भावना की मल्लिका और वास्तविक मल्लिका में कोई समानता नहीं है। इसीलिए वास्तविक मल्लिका का सामना करना उसके लिए संघातक सिद्ध होता है।

सार्त्र के उपन्यास **ला नॉसी** में ऐनी और रॉक्वेन्टिल की यही स्थिति है जिसमें सम्बन्धों की, जीवन और उसके क्षणों की पूर्णता की आशा दुराशा मात्र है। सार्त्र मानता है कि 'क्षण की पूर्णता' एक बुर्जुआ तथ्य है। क्षण की पूर्णता की प्राप्ति का प्रश्न मल्लिका के लिए नहीं है क्योंकि वह दूसरा कालिदास है जो उसके द्वार आया है और स्वयं कालिदास के लिए वह सम्भव नहीं, क्योंकि वह मल्लिका दूसरी है जिसे वह मिलने आता है। समस्त प्रेमाभिव्यक्ति के बावजूद, कालिदास के चले जाने पर एक बार 'उसके पैर बाहर ज़रूर बढ़ने लगते हैं परन्तु बच्ची को बाँहों में देखकर जैसे वहीं जकड़कर रह जाती है।' यह क्या नई स्थिति की स्वीकृति नहीं है, जिससे अथ से प्रारम्भ करने का प्रतिवाद ध्वनित है ?

कालजन्य विसंगति के बावजूद मल्लिका कालिदास की चेतना के केन्द्र में स्थित है। हर आदमी के अवचेतन में नारी का एक बिम्ब होता है। मल्लिका भी कालिदास की नारी-भावना का 'प्रोजेक्शन' (प्रक्षेपण) है; दूसरी ओर मल्लिका स्वयं अपने को कालिदास में 'प्रोजेक्ट' करती है। इसीलिए अपनी सत्ता का बोध वह उसी में पाती है—अपने को अपने में न देखकर उसी में देखती है। और दूसरी ओर कालिदास अपनी रचनाओं में उसे ही पाता है, अंकित करता है। मल्लिका कालिदास की आन्तरिक प्रेरणा है जो उसके अन्दर हृदय की अतल गहराई में राह बनाती है और साथ ही उसके बाहर भी एक राह पर उसे रखती है जिसके माध्यम से वह बाह्य जगत् के सम्पर्क में आता है। मल्लिका उसके अपने अस्तित्व की भूख को पूरा करती है जो उसे सही रूप में समझती है और अपनापन देती है। इसलिए कालिदास के जीवन में वह उसकी अपनी आन्तरिक आवश्यकता के कारण भी महत्त्वपूर्ण हो जाती है। स्थान और स्थिति के बदलते ही कालिदास की अपेक्षाएँ बदलती हैं और उसी अनुपात में 'अन्य' की, उसकी खोज भी—प्रियंगु मंजरी से लेकर वारांगनाओं तक। व्यक्ति होने के लिए 'अन्य' की खोज़ अस्तित्व की अनिवार्य शर्त है। इस खोज में वह या तो अपने को दूसरे से स्वीकृत करवाता है या स्वयं स्वीकृत व्यक्ति को समर्पित हो जाता है। अस्तित्व की यह 'भूख' स्वाभाविक है। मल्लिका उस भूख की प्रस्थान बिन्दु है, विराम स्थल नहीं। किन्तु जब यह भूख एक वृत्त पूरा कर लेती है तो फिर लौटकर कालिदास उसी प्रस्थान बिन्दु पर आ पहुँचता है। वस्तुतः आदमी को चक्कर कटवानेवाली हर भूख अतीत-विरह (नॉस्टालजिया) से युक्त होती है—जहाज़ के पंछी की तरह। किन्तु कालिदास की नियति जहाज़ के पंछी जैसी भी नहीं रह पाती—मल्लिका के अस्तित्व की नई परिणति को एक दंड की तरह स्वीकार कर वह उस बिन्दु से बाहर छिटक जाता है।

कालिदास और मल्लिका दोनों के प्रेम-मूल्यों में अन्तर है। कालिदास के लिए

मल्लिका अनेक मूल्यों में से एक मूल्य है। वह उसके लिए अपना अस्तित्व न्यौछावर नहीं करता, किन्तु मल्लिका के लिए वह सम्पूर्ण अस्तित्व है, एक बिना शर्त समर्पण और आत्मदान। वह भावना में भावना का वरण करती है—वह अमूर्त चाह को भावना का रूप देती, किन्तु क्रिया देने की शक्ति उसमें नहीं है क्योंकि उसके पास मार्ग नहीं है। वह कालिदास को उपादान नहीं मानती। इसलिए अप्राप्ति का भी उसे खेद नहीं। यहाँ तक वह भारतीय पद्धति की आदर्श नारी के रूप में सामने आती है; किन्तु विलोम से ब्याह कर वह एक नया व्यक्तित्व अर्जित करती है। वहाँ उसका रोमान यथार्थ से जुड़ जाता है। कुछ आलोचकों को उसका यह नया रूप खलता रहता है। डॉ. जगदीश शर्मा के शब्दों में, 'भाव के कोष्ठ को रिक्त न होने देकर अभाव के कोष्ठ में न जाने कितनी-कितनी दूसरी आकृतियों को प्रवेश करने देना और उन्हों में से किसी एक की सन्तान को अभाव की सन्तान के रूप मे जन्म देना उस भव्य चरित्र की संगति में नहीं आता जिसका साक्षात्कार नाटक के प्रारम्भ में होता है।[12] हमारी दृष्टि में मल्लिका विलोम से ब्याह कर अपने जीवन को प्रामाणिकता—ऑथेण्टिसिटी—प्रदान करती है। वह अपनी स्वतन्त्रता का उपयोग करती है। यह उस असत् आस्था का विरोध है जो आत्म-बलिदान में ही नारी की परम सिद्धि मानता है। वस्तुतः व्यक्ति को तब अपनी निजता/आइडेंटिटी स्थिर करनी ही होती है जब जीवन कहीं पर पहुँचकर रुक जाता है या बहना बन्द कर देता है। 'होने' और 'है' में बड़ी खाई है। काल आदमी को वही नहीं रहने देता जो वह है और अनागत की भी अपनी अपेक्षाएँ होती हैं जिसका प्रभाव उस पर पड़ता है। अम्बिका, कालिदास, विलोम और स्वयं मल्लिका का अपना अन्तर्जगत तथा बाहर की दुनिया सब मिलाकर उसे वहाँ पहुँचाते हैं जहाँ वह है। वह सुरक्षा के लिए विलोम का हाथ पकड़ती है। वह स्वयं इस आचरण को वारांगना जैसा मानती है। इसलिए कि जो उसने किया है, वह मुखौटा चढ़ाने जैसा है, ऊपर से आरोपित और मिथ्या। बाह्य जगत् और आन्तरिक स्वरूप के बीच उसका आन्तरिक स्वत्व इसलिए उसके आरोपित व्यक्तित्व से घृणा करता है। किन्तु इस 'आरोपित व्यक्तित्व' में भी एक प्रकार की सच्चाई है। वह सब-कुछ अर्पित कर सकती है या इस तरह कुछ भी नहीं, क्योंकि जीवन की धुरी बदल गई है।

मल्लिका के स्वीकार और अस्वीकार में सम और विषम दोनों प्रकार की अनुभूतियों का योग है। मल्लिका के एक अंश पर कालिदास का अधिकार है दूसरे पर विलोम का। वह दोनों के बीच जीती है। वह प्रेयसी भी है और पत्नी भी। पर उसके पूरे अंश पर न कालिदास का अधिकार है और न विलोम का। वह कालिदास की ही होकर रह सकती थी; किन्तु न रह सकने का कारण भी कालिदास ही है। और उससे भी बड़ा कारण शायद अम्बिका है। अम्बिका जब तक जीती रही, कालिदास और उसके प्रेम-सम्बन्ध को आत्मप्रवंचना मानती रही। सम्भवतः मल्लिका के अवचेतन में उसका यह भाव बस जाता है जिसके कारण वह बाद में कालिदास से मुक्त होने के लिए अनजाने विलोम में बँध जाती है। कालिदास से प्रवंचित होकर मल्लिका स्वयं कालान्तर में अम्बिका बन

जाती है। उसके अवचेतन में बसा माँ का बिम्ब चेतन स्थिति को अस्वीकार की ओर ले जाता है। इसीलिए प्रवंचित होने पर कालिदास और स्वयं अपने को दंडित करने के लिए वह विलोम से ब्याह कर लेती है इस व्यवहार में आत्मपीड़क और परपीड़क दोनों प्रवृत्तियों का सम्मिश्रण है। हर हालत में कालिदास और विलोम की 'अन्यता' बनी रहती है। प्रारम्भ में कालिदास के प्रेम में जिस प्रकार वह अपनी आत्मपरकता (सब्जेक्टिविटी) समाप्त कर वस्तुगत अस्तित्व धारण कर लेती है, उसी प्रकार अन्त में अवचेतन में बदला लेने के लिए उसे फिर से धारण कर लेती है और साधन-रूप वस्तु बनना अस्वीकार करती है। इस प्रकार दो पुरुषों के बीच मल्लिका जिन सम्बन्धों को जीती है उनमें प्रेम और घृणा, आत्मदान और आत्म-आग्रह, समर्पण और विद्रोह का अद्भुत समन्वय है। किन्तु इन ग्रन्थियों के बावजूद उसका प्रेम भरा-पूरा लगता है और विसंगति के बावजूद उपलब्धि और उदात्तता का अपूर्व अहसास दिलाता है।

जो भी हो, सम्बन्धों की यह विसंगति काल्पनिक नहीं है। कालिदास की नियति आधुनिक मानव की नियति है। विलोम समसामयिक जीवन की अनास्था का स्वर है तो मल्लिका 'आस्था का विस्तारित रूप'। इस प्रकार यह नाटक युगीन भाव-बोध को कई स्तरों पर छूता है और नए सन्दर्भों को उभारता है। 'बात क्योंकि सन्दर्भों में जुड़ने की है, इसीलिए वहाँ न तो हमें वह कालिदास मिलता है जो **मेघदूत** के अभिशप्त यक्ष की भाँति विरहाकुल और काम-पीड़ित होकर भी अपनी आस्था से विचलित नहीं होता, और न मल्लिका ही जो **कुमारसम्भव** की उमा की तरह सब-कुछ खोकर शिव की प्रतीक्षा में तिल-तिल गलकर संकल्प-शक्ति मिटने नहीं देती। यह इसलिए कि दोनों के चरित्र इतिहास की मिट्टी से गढ़े होने पर भी वर्तमान की घुटन में साँस लेते हैं।'[13] इस स्थिति को लेकर एक विवाद उठ खड़ा हुआ है : क्या **आषाढ़ का एक दिन** को ऐतिहासिक नाटक कहा जा सकता है ? इस सम्बन्ध में नाटककार ने स्पष्ट कहा है कि इस नाटक की रचना एक समसामयिक परिस्थिति को उसकी अपनी नाटकीयता में अभिव्यक्त करने के लिए हुई है, इसी से इसे इतिहासगत सन्दर्भ से अलग रखकर इसके साथ न्याय किया जा सकता है।[14] कहा जा सकता है कि **आषाढ़ का एक दिन** ऐतिहासिक नाटक नहीं है—वह इतिहास का भ्रम खड़ा करने के लिए केवल कुछ ऐतिहासिक सन्दर्भों का उपयोग मात्र करता है। कालिदास के सन्दर्भ में कुछ तथ्य लोक में भली-भाँति ज्ञात हैं, जैसे—उसका काली का भक्त और वेश्यागामी होना, प्रारम्भ में तिरस्कृत-उपेक्षित रहना, राजदुहिता से ब्याह तथा हिमालय का वासी होना। **राजतरंगिणी** में मातृ गुप्त के रूप में काश्मीर का शासक बनने की बात भी प्रमाणित होती है। जयशंकर प्रसाद के **स्कंदगुप्त** में निहित मातृगुप्त की अवधारणा भी, सम्भव है, कहीं सहायक रही हो। और सबसे महत्त्वपूर्ण रहे हैं राकेश के अपने वे निष्कर्ष जो उन्होंने कालिदास के ग्रन्थों से निकालकर अपनी रचना में ढाले हैं : '**मेघदूत** को पढ़ते हुए मुझे लगा करता था कि

वह कहानी निर्वासित यक्ष की उतनी नहीं जितनी स्वयं अपनी आत्मा से निर्वासित की।' फलतः नाटक में सारी भौतिक स्थिति को एक दूसरे बिन्दु से देखने का प्रयास हुआ है।

जिन तथ्यों से कथानक का ताना-बाना बुना गया है, वह इतिहास कम और मिथक अधिक है। मिथक एक ठोस यथार्थ को ग्रहण करता है और उसकी सृष्टि एक मानसिक-आध्यात्मिक प्रक्रिया के रूप में होती है। इसलिए वह जन-मानस में आसानी से स्वीकृत होता है और प्रभाव में मंत्र जैसी शक्ति और विश्वास रखता है। एक विशेष मनःस्थिति में इतिहास भी मिथक बन जाता है और उससे भी अधिक प्रामाणिकता ग्रहण कर लेता है। कालिदास के जीवन से सम्बद्ध तथ्य भी मिथक से जुड़े हुए हैं; नाटककार का उद्देश्य आधुनिक सन्दर्भ में उसी का उपयोग करना रहा है। ऐसी स्थिति में विश्वसनीयता मुख्य हो जाती है, ऐतिहासिकता गौण पड़ जाती है। इसीलिए राकेश का कथन ठीक जान पड़ता है कि हो सकता है व्यक्ति कालिदास का यह नाम भी वास्तविक न हो, पर हमारी आज तक की सृजनात्मक प्रतिभा के लिए इससे अच्छा दूसरा संकेत मुझे नहीं मिला।'[15] स्पष्ट है कि नाटक में इतिहास को निर्मित किया गया है और उसे एक खोल की तरह ओढ़ा गया है। मुख्य वस्तु यह खोल नहीं है वरन् वह अन्तर्द्वन्द्व है जिसे सृजन के सन्दर्भ में नाटककार चित्रित करना चाहता है। व्यक्ति कालिदास को उस अन्तर्द्वन्द्व से गुज़रना पड़ा है या नहीं, वह बात गौण है। मुख्य बात यह है कि बहुतों को उसमें से गुज़रना पड़ता है।[16]

यह शाश्वत आन्तरिक यथार्थ ही सब यथार्थों का यथार्थ है। मोहन राकेश ने इतिहास अथवा मिथक के गर्भ से उसी यथार्थ को पकड़ने की कोशिश की है। यह बात दूसरी है कि यह यथार्थ उतना पुराना नहीं जितनी पुरानी नाटक की कथावस्तु है। असल में आज का मानव उस यथार्थ को भोग रहा है, जो पुरातन काल में अज्ञात था, अजन्मा था। राकेश ने युग और दृष्टि के भेद के साथ इतिहास और उसमें निहित मिथक को फिर से युगीन सन्दर्भ में विश्लेषित और परिभाषित किया है। यह दोहन सहसा अनुचित भी नहीं है; क्योंकि वर्तमान अतीत का ही एक आयाम है। प्रायः समसामयिक जीवन की अभिव्यक्ति के लिए नाटककारों की आत्मपरक दृष्टि अतीत की ओर जाती रही है। राकेश की दृष्टि भी आत्मपरक रही है। फलतः उनके नाटकों में कथावस्तु गौण हो जाती है, एक माध्यम बन जाती है—मानव का अस्तित्व बोध, उसकी आन्तरिकता तथा मानसिकता मुख्य हो जाती है। **आषाढ़ का एक दिन** में भी ऐसा ही हुआ है। उसमें राकेश ने 'अपने ही एक नए और अलग रूप में इतिहास का निर्माण' किया है। यह निर्माण इतिहास से दूर इसलिए लगता है क्योंकि नवीन कथ्य के आरोप से उसकी संस्कारगत प्राचीनता अविशिष्ट नहीं रह पाई। फिर भी उसमें इतिहास का भ्रम या सत्याभास अवश्य पैदा होता है। और यही नाटककार का असली उद्देश्य भी है।

राकेश का उद्देश्य कालिदास के जीवन से सम्बद्ध घटनाक्रम को प्रस्तुत करना नहीं है। वे कुछ आधारों को लेकर कालिदास के चारों ओर कथावस्तु की पुनर्रचना करते हैं। फलतः कथ्य के सामने ऐतिहासिकता निमित्त मात्र बनकर रह जाती है और मुख्यता

उस अन्तर्द्वन्द्व अथवा अनुभव को प्राप्त होती है जिसके बीच से प्रमुख पात्र गुज़रते हैं। किन्तु जब नाटककार ख्यात पात्रों को नए सन्दर्भों में चित्रित करता है तो एक ख़तरा बराबर बना रहता है और वह है उनके ऐतिहासिक बिम्ब की अवहेलना का ! अतीत के साथ महानता का भाव स्वभावतः जुड़ा है। ऐसी स्थिति में कुछ आलोचकों को कालिदास इतिहास के प्रतिकूल (दुर्बल, स्वार्थी और पलायनवादी) लगे या मल्लिका और विलोम का वैवाहिक सम्बन्ध अनुचित जँचे, तो कोई आश्चर्य की बात नहीं। यह मान भी लें कि कालिदास के रचनाकार व्यक्तित्व को सृजनशील प्रतिभा के अन्तर्द्वन्द्व को उद्‌घाटित करने के लिए प्रयुक्त किया गया है, फिर भी यह सवाल जायज है कि कालिदास का जो मिथकीय या इतिहास-सम्मत बिम्ब युगों से बना है, क्या **आषाढ़ का एक दिन** में उसका निर्वाह अनिवार्य नहीं था ? यह ठीक है कि कालिदास का चरित्र एकदम परम्परागत बिम्ब से भिन्न नहीं है। किन्तु इतना तय है कि नाटक में 'अभिज्ञान शाकुन्तल', 'कुमारसम्भव', 'रघुवंश' आदि ग्रन्थों के स्रष्टा कवि का व्यक्तित्व नहीं उभरता। यह भी ठीक है कि राकेश ने आरम्भ में कालिदास को कवि-सम्वेदना दी है, पर उसमें भी वह मल्लिका से पिछड़ जाता है। कालिदास का कवि रूप बार-बार मुखर होने से रह जाता है और उसका कवित्व निर्वासित-सा लगता है। इसका मुख्य कारण यह है कि लेखक नायक विषयक परम्परावादी अवधारणा को खण्डित करना चाहता है। नया नाटक आदमी की तलाश महानता में करने के बजाय लघुता में करता है। इस लघुता में कालिदास और उसका प्रतिद्वन्द्वी विलोम एक ही भूमि पर खड़े दिखाई देते हैं।

आषाढ़ का एक दिन की चरित्र-सृष्टि में सम और विषम तत्त्वों का समावेश है। पात्र परस्पर संघर्ष में ही आते हों, ऐसी बात नहीं है। वस्तुतः चरित्र इकहरे नहीं हैं, इसलिए अंशतः विरोधमूलक होने पर भी पूर्णतः विरोधमूलक नहीं हैं। वे कहीं-न-कहीं एक-दूसरे के पूरक और एक-दूसरे में बँटे हुए दिखाई देते हैं। इसीलिए उनमें आधुनिक मानव का पर्याप्त अन्तर्विरोध है। पर इस अन्तर्विरोध में भी वे आपस में जुड़े हैं। उदाहरण के लिए अम्बिका और मल्लिका, विलोम और कालिदास, कालिदास और मातुल ऊपर से एक-दूसरे के विरोधी चरित्र लगते हैं किन्तु इसके बावजूद कहीं-न-कहीं उनमें समीकरण है। कालिदास एक सकारात्मक चरित्र है, विलोम उसी का नकारात्मक रूप है। एक करुण और भावुक है, दूसरा यथार्थवादी, आक्रामक और अन्तर्द्वन्द्वों को पचाए हुए। एक भोक्ता है और दूसरा द्रष्टा। दोनों में संघर्ष है, पर 'अंडरस्टैंडिंग' भी। दोनों एक-दूसरे के पूरक हैं। विलोम खलनायक तभी तक लगता है जब तक वह कालिदास की खल-वृत्ति को उभारता है। मल्लिका के मन में समाए कालिदास के 'सन्त चरित्र' की धज्जियाँ उड़ाने में वह विदूषक-जैसा बाना पहनता लगता है, पर उस विदूषकत्व में भी उसकी अपनी त्रासदी छिपी है। कालिदास उसकी तुलना में दुर्बल, थका-हारा लगता है, पर इस सबके बावजूद कालिदास उसकी अपेक्षा सम्भावनाओं से परिपूर्ण है। विलोम आक्रामकता के बावजूद विवश लगता है जिसके लिए कोई सम्भावना नज़र नहीं आती। इसलिए वह

कालिदास से कहीं अधिक त्रास्द है। राकेश ने ठीक ही कहा है कि **आषाढ़ का एक दिन** में 'पराजित टूटा हुआ व्यक्ति कालिदास नहीं, अपने में संयोजित विलोम है।'[17] वस्तुतः त्रासद और कॉमिक तत्त्वों के समन्वय से उन्होंने इन चरित्रों को आधिभौतिक आयाम प्रदान किया है।

चरित्रों की परस्पर घुल-मिल जानेवाली यह तरलता मातुल और कालिदास के चरित्रों में भी दिखाई देती है। दोनों का टकराव अन्ततः एक ही परिणति में समा जाता है। जो मातुल कभी कालिदास पर इस बात पर रुष्ट था कि वह राजसम्मान की उपेक्षा करने जा रहा है वही राजप्रासाद में अपनी टाँग तुड़वा बैठता है और उस पंगुता का शिकार बनता है जिसका काल्दिास स्वयं सहभागी है। अन्ततः दोनों की विरोधी चेतना समरूप हो जाती है। इसी प्रकार मल्लिका और अम्बिका भी विरोधी भूमिकाएँ अदा करती हैं, किन्तु अन्ततः दोनों चरित्र अपनी मानसिकता में एक हो जाते हैं। अम्बिका इतनी आत्मकेन्द्रित है कि कोई भी लहर उसे छूती नहीं। वह अपने को सिकोड़कर रखती है। उसकी दृष्टि में छिद्रान्वेषी पैनापन है। यहीं वह विलोम से जुड़ती है। स्वयं उसी के शब्दों में, 'जो अम्बिका के मन में रहती है, मैं मुँह से कह देता हूँ।' कालिदास के सन्दर्भ में विलोम एक तरह से अम्बिका का ही मुखर रूप है। फिर भी अम्बिका और विलोम एक हैं, ऐसा नहीं कहा जा सकता। अम्बिका कालिदास के प्रति कठोर ज़रूर है, किन्तु मल्लिका के कारण अन्दर से कहीं उसका चुपचाप लिहाज़ भी करती है—इसीलिए वह सीधे उसके सामने कोई बात नहीं कहती और न विलोम को ही कहने देने का मौक़ा देती है। मल्लिका और अम्बिका में एक वैषम्य के बावजूद दोनों में एक अदृश्य प्रेम का तन्तु भी है। लगता है, दोनों में कहीं सम्प्रेषण नहीं, किन्तु सच्चाई यह है कि यह सम्प्रेषण इतना गहरा है कि शब्द आँसुओं में बदल जाते हैं। और मल्लिका भी उससे कहीं ऐसी जुड़ी है कि उसके मरने पर बुरी तरह टूट जाती है। फिर भी दोनों बाहर से कितनी अलग लगती हैं : मल्लिका भावुक, उदार, स्नेहिल और क्षण में जीनेवाली; अम्बिका गुमसुम, अनुदार, कटु और त्रिकाल के अखंड अहसास को संचित किए बैठी। फिर भी दोनों एक हो जाती हैं। युवती मल्लिका जीवन के मार्ग पर ढलती उम्र में स्वयं अम्बिका का प्रतिरूप बन जाती है। अम्बिका अपनी पीड़ा तो झेलती ही है, साथ ही मल्लिका की तात्कालिक और भावी पीड़ा को भी अपने भीतर लिए हुए है। वह अपने शरीर में ही नहीं, अपने शरीर के बाहर मल्लिका में भी जीती है। अम्बिका और मल्लिका यथार्थ और भावना को एकदम आमने-सामने खड़ा कर देते हैं।

इस प्रकार अपनी चेतना में कालिदास-विलोम, विलोम-अम्बिका, कालिदास-मल्लिका, विलोम-मल्लिका, मातुल-कालिदास, मातुल-मल्लिका—सब आन्तरिक सम्बन्धों से जुड़े हैं। निपेक्ष जैसा सूत्रवाही पात्र भी दूसरे पात्रों की चेतना से जुड़ा है। वह प्रारम्भिक कालिदास और परवर्ती मल्लिका के बीच का बहुत ही संवेदनशील तन्तु है। एक ही चेतना के विविध रंग और रेखाओं से निर्मित इन पात्रों के कहीं इकहरे विषम रंग उभरते हैं और कहीं वे एक-दूसरे से मिलकर और गहरे हो जाते हैं।

उनके जीवन में कुछ घटता ही नहीं। उसके सामने कोई चुनाव की गुंजाइश नहीं, रास्ता भी नहीं। फिर भी औरों की तुलना में वह कितनी दृढ़ है, कितनी सक्षम ! उसकी ज़िन्दगी इसलिए उतनी त्रासद नहीं जितनी संतप्त है। और यदि त्रासद है भी तो उसकी त्रासदी औरों में बँट जाती है। इसीलिए मल्लिका, कालिदास और विलोम तीनों एक ही बिन्दु से खींची तीन रेखाएँ हैं—किन्तु अपनी शक्ति में मल्लिका कालिदास की 'आस्था का विस्तारित रूप' बन जाती है।

मातुल, दन्तुल, निक्षेप, रंगिणी, संगिनी जैसे पात्र भी कथावस्तु की आवश्यकताओं को ही पूरा नहीं करते, नाटक की संरचना और अर्थवत्ता में भी सहायक होते हैं। वे भी अन्य पात्रों के साथ कथा की गठन, क्रिया व्यापार के संकेत, प्रतीकार्थ तथा सांकेतिक अभिव्यक्ति में महत्त्वपूर्ण भूमिका निभाते हैं। सम्पूर्ण नाटक का कथानक और कथ्य दोनों मुख्य और सूत्रवाही पात्रों की विविध भंगिमाओं द्वारा प्रभावशाली ढंग से अभिव्यक्त होते हैं। और समग्र प्रभाव के स्तर पर जो बात सारे नाटक में हृदय को छू जाती है, वह है एक 'मूड', एक मनःस्थिति जिसमें पाठक/प्रेक्षक अनायास ही पहुँच जाता है। इसका श्रेय नाटक के बिम्ब-विधान और वातावरण-सृष्टि को जाता है। इस दृष्टि से विभिन्न कथा-स्थितियों के बीच मेघ का बिम्ब बहुत सार्थक सिद्ध होता है। नाटक का प्रारम्भ ही उससे होता है जब मेघ मल्लिका के तन-मन को भिगा देता है। भीगने का यह अनुभव एक ओर मेघ और दूसरी ओर कालिदास से जुड़ जाता है। दोनों जैसे एक ही हो जाते हैं। दोनों अनुभवों को एक साथ वह आँखों में भर लेती है। मल्लिका का वर्षा से भीगना एक आन्तरिक अनुभव है। इसीलिए यहाँ मेघ एक ऐसा वस्तुगत बिम्ब बन जाता है जो उनकी आन्तरिक भावनाओं को प्रकट करता है। दूसरे अंक में ये मेघ उज्जयिनी चले जाते हैं। मल्लिका के पास केवल सूखने के लिए फैलाए गीले वस्त्र मात्र रह जाते हैं जो अभाव और उपेक्षापूर्ण दुखद स्थिति में अस्तित्व का गीलापन व्यंजित करते हैं। तीसरे अंक में कालिदास के साथ ही ये मेघ वर्षा और गर्जन को लेकर फिर लौट आते हैं। पर पूर्वमेघ और उत्तरमेघ में अन्तर है। पूर्वमेघ यौवनागमन और प्रेमार्पण का है, उत्तरमेघ उमड़ती-घुमड़ती स्मृतियों का पुंज है; बदली स्थिति का दंश और नियति का व्यंग्य है। इस प्रकार अलग-अलग समय पर मेघ अलग-अलग भावों का उद्दीपन बन जाता है। नाटक के उत्तरार्द्ध में कालिदास, मातुल और विलोम भी वर्षा में भीगते हैं। यह भीगना सबके लिए अलग-अलग अर्थ रखता है।

मेघ के इन बिम्बों के द्वारा राकेश ने एक ही प्राकृतिक परिवेश, किन्तु भिन्न परिस्थितियों और मनःस्थितियों को एकान्वित कर अद्भुत प्रभाव पैदा किया है जिससे नाटकीय घटना-क्रम के बीच काल का अन्तराल भर जाता है; किन्तु कथावस्तु के संयोजन का वास्तविक कौशल एक और समान्तर बिम्ब के सृजन में दिखाई देता है। और वह है राजसत्ता का बिम्ब। प्रियंगु, अनुस्वार, अनुनासिक, रंगिणी, संगिनी, सैनिक आदि सब मिलकर उसकी सृष्टि करते हैं। यह बिम्ब राजसत्ता की शुष्कता और

हृदयहीनता को व्यक्त करता है। मेघ और राजसत्ता के ये दो विरोधी बिम्ब आमने-सामने वैषम्य प्रस्तुत करते हैं। एक भावना का पोषक है और दूसरा भावनापूर्ण मानव को मृग-शावक के रूप में आहत करनेवाला है। सारा नाटक, वास्तव में दो विरोधी बिम्बों का नाटक है। कामू ने कहा है : 'उपन्यास और कुछ नहीं—बिम्बों में निहित एक दर्शन है।'[18] उपन्यास के सन्दर्भ में यह बात सही हो न हो, राकेश के इस नाटक के सन्दर्भ में सही मानी जा सकती है। इसमें विरोधी बिम्बों की टकराहट में समान नाटकीय स्थिति में विषमता का बोध बड़ी तीव्रता से उभरता है। सब-कुछ वही है—वही मेघ है, वही घर है, वे ही लोग; पर बहुत कुछ है जो अन्ततः अनचीन्हा, अनजाना रह जाता है। और ऐसी ही स्थिति में कथावस्तु में नाटकीय व्यंग्य की योजना में नियति का रंग गाढ़ा हो जाता है। मल्लिका का 'भावना में भावना का वरण', विलोम का मल्लिका के घर में अयाचित अतिथि न रहना, अधिकार से आने पर भी द्वार बन्द पाना और राजमहिषी की मल्लिका के प्रति सारी संवेदनाशून्य उदारता इसी का उदाहरण है।

सारा नाटक मल्लिका के घर में घटित होता है। यह घर एक के बाद दूसरे अंक में उजड़ता जाता है और क्रमशः कमल और स्वस्तिक के भित्ति-चित्र मिटते जाते हैं। गन्दे कपड़े, औंधे अन्न-पात्र समय की गति और दुरवस्था का चित्र प्रस्तुत करते हैं। सारा दृश्य-विधान मल्लिका के अन्तर का प्रतीक बन जाता है। इस प्रकार घर का बदलाव बिम्ब कालिदास के ऐश्वर्य के सामने कंट्रास्ट प्रस्तुत करता है। और रंगिणी-संगिनी-प्रियंगु के संवादों और क्रिया-व्यापार में नाटकीय व्यंग्य की सर्जना करता है। इस उजड़ते घर की पृष्ठभूमि में नाटक स्वयं उजड़ती मल्लिका की कथा बन जाता है। किन्तु उजड़ता घर उसी की नहीं, कालिदास, मातुल, अम्बिका सभी के उजड़ने की कहानी कहता है। यह बात दूसरी है कि कालिदास नेपथ्य से ही कथा-सूत्र के साथ अधिक जुड़ा है। मंच पर वह प्रारम्भ और अन्त में ही आता है। मध्य के कथा-सूत्र से वह परोक्ष रूप से सम्बद्ध है। पर मल्लिका आदि से अन्त तक विद्यमान है। जैसा कि कहा जा चुका है, नाटक की कथावस्तु का संयोजन संघर्ष और द्वन्द्व के आधार पर हुआ है। प्रारम्भ में यह संघर्ष रोमान और यथार्थ के बीच है; बाद में निरीह प्राकृतिक जीवन की सरलता पर राजसत्ता का ऐश्वर्य एक व्यंग्यमयी छाप छोड़ जाता है। और अन्त में 'इच्छा के साथ समय का द्वन्द्व' मुख्य हो उठता है। पहला संघर्ष अन्ततः दूसरे में समाहित हो जाता है और फिर समन्वित रूप में मानव-नियति का संघर्ष मात्र बनकर प्रकट होता है। सच कहें तो यही नाटक का मूल संघर्ष है भी। संघर्ष मल्लिका-अम्बिका, अम्बिका-कालिदास, कालिदास-विलोम, विलोम-मल्लिका, मातुल-कालिदास, मल्लिका-प्रियंगु में कई स्तरों पर होता है। पर सभी संघर्ष एक सूत्र में बन्धन में बँध जाते हैं। अन्त में कालिदास के पलायन में बाह्य संघर्ष जैसे लुप्त हो जाता है और उनका अन्तर्मुख होना ही एक त्रासद आत्मिक संघर्ष का अनुभव दे जाता है।

नाट्य-स्थितियों के विकास की दृष्टि से **आषाढ़ का एक दिन** सीधा-सपाट नाटक नहीं

है। उसमें मन्थर गति ज़रूर है। कहीं-कहीं क्रिया का स्थान भावपूर्ण वातावरण ले लेता है। किन्तु स्थितियाँ तनाव से परिपूर्ण हैं। एक स्थिति दूसरी स्थिति में जुड़कर समाविष्ट हो जाती है। एक विशिष्ट क्रम में सम-विषम तत्त्वों का प्रभावशाली उपयोग किया गया है। पहले अंक में अम्बिका और मल्लिका के बीच तनाव की क्षीण स्थिति का, दूसरे में उसकी सघनता और तीसरे में करुणा का अहसास होता है। नाटक पुरातन पद्धति से कुछ भिन्न स्तर पर आरम्भ, विकास और चरम की सीढ़ियों को चढ़ता है। संघर्ष प्रारम्भिक बिन्दु से ही सक्रिय हो जाता है; पर वह तीव्र नहीं है, तीव्रता उसमें आगे जाकर आती है। और चरम तब आता है जब संघर्ष पराजय में बदल जाता है। यों भी कहा जा सकता है कि प्रथम अंक में प्रेम की अनुभूति जगती है और कवि राजसत्ता के सम्पर्क में आता है; दूसरे में उसका मोहभंग होता है और राजसत्ता अपने वास्तविक रूप में उद्घाटित होती है और तीसरे में कालिदास उससे मुक्त होता है। यहीं मातुल के संवाद कालिदास पर राजसत्ता के अलगाव के लिए यथेष्ट वातावरण बनाते हैं। मातुल कालिदास के पलायन और संन्यासी हो जाने की बात कहकर एक बार फिर मल्लिका के हृदय में टीस जगा देता है। तभी कालिदास की आकस्मिक रूप से उपस्थिति और यथार्थ के उद्घाटन की प्रतिक्रिया नाटक को एक और टीस से भर देती है। इस प्रकार पूरा नाटक कालिदास के जीवन के एक पूरे विस्तार को रूपायित करता है जिसे लेखक ने तीन अंकों में उसके तीन खंडों के रूप में प्रस्तुत किया है। ये तीन खंड समय के व्यवधान से युक्त हैं पर मेघ और आषाढ़ के एक दिन की स्मृति उसे भर देती है और इस प्रकार चेतना का एक अन्तर्निहित प्रवाह सबका सामंजस्य उपस्थित कर देता है। इसमें सारा नाटक एक भावपूर्ण काव्यमय वातावरण में घटित होता है, किन्तु अंकों के बीच की सायास आयोजित सूच्य सामग्री अनिवार्य होने पर भी खलती है। फिर भी मनःस्थितियों का नाटक होने के नाते **आषाढ़ का एक दिन** के कथा-संयोजन में एक प्रवाह और सहजता है। कुल मिलाकर यह नाटक शिल्प की दृष्टि से गढ़ा हुआ लगता है। इसमें कथावस्तु की एकाग्रता ही नहीं, स्थिति और भावों का व्यंजनापूर्ण वातावरण, अनुभूति की तीव्रता तथा अर्थतत्त्व का समन्वय अपना पूरा प्रभाव पैदा करता है।

शिल्प के स्तर पर यह नाटक भारतीय और पाश्चात्य दोनों नाट्य-तत्त्वों का उपयोग करता है। उसमें द्वन्द्व भी है और सम की स्थिति भी जिसका एक आयाम कालिदास प्रस्तुत करता है और दूसरा मल्लिका। रंगिणी-संगिनी, अनुस्वार और अनुनासिक का प्रसंग एक ओर नई पद्धति के अनुरूप विसंगति को उभारता है; दूसरी ओर कुछ-कुछ वैसी ही राहत देता है जैसी परम्परागत नाटकों में विदूषक। इसी प्रकार विलोम खलनायकत्व और नायकत्व के बीच स्थित है। कालिदास नायक है पर मंच के पीछे रहता है। कभी विलोम उसके समकक्ष आ जाता है। मल्लिका की चारित्रिक अवधारणा एक ओर छायावादी आग्रह से परिपूर्ण है, दूसरी ओर उसकी जीवन-दृष्टि पर अस्तित्ववादी विचारधारा का आरोप भी है।

आषाढ़ का एक दिन की रंग-सृष्टि भी नाटकीय अवधारणा के समान ही काव्य तत्त्व पर आधारित है। यहीं राकेश 'रंगमंच का कवि' (पोयट-इन-थियेटर) लगता है। आषाढ़ के झरते मेघ, सुखाने के लिए फैलाए हुए गीले कपड़े, अम्बिका की ठोस और मल्लिका की तरल भाव-मुद्राएँ, अनुस्वार और अनुनासिक की बनावटी भंगिमाएँ, ग्रामवासियों का भोला देहातीपन और राजपुरुषों के दर्पपूर्ण चेहरे और अन्ततः दीपक की लौ में डूबा सब चेहरों में मल्लिका का एक चेहरा—ऐसे दृश्य बिम्ब प्रस्तुत करता है जो रंगमंच की कविता रचते हैं।

राकेश ने रंगमंच की यह कविता सामान्य क्रिया-व्यापारों और भावों के द्वारा भी रचकर दिखाई है। उदाहरण के लिए अम्बिका का सूप से अनाज फटकना एक ऐसा ही सहज व्यापार है जो उसकी मनोदशा को व्यक्त करता है। इसी प्रकार मल्लिका कालिदास की पांडुलिपियाँ एकत्र करती है, और इस आकांक्षा से कि वह उन पर महाकाव्य लिखे, वह कई कोरे पन्नों को सँजोकर रखती है; किन्तु कालिदास देखता है कि उन कोरे पन्नों पर आँसू, स्वेद कणों, नखों, सीलन ने समय की कविता लिख डाली है। तब वह कहता है : 'इन पृष्ठों पर अब क्या नया कुछ लिखा जा सकता है ?' इस प्रकार के प्रसंगों और संवादों से राकेश ने प्रतीकार्थ को उजागर किया है। वस्तुतः कोरे पृष्ठ यहाँ मल्लिका के कोरे जीवन और उन पर लिखी अव्यक्त व्यथा के प्रतीक बन जाते हैं। उसकी ज़िन्दगी भी महाकाव्य की भाँति रची जा सकती थी किन्तु जीवन की विसंगति उस पर सीलन, आँसू, आक्रोश और विवशता के नख-क्षत छोड़ जाती है। जीवन उसे महाकाव्य बनने के बजाय विलोम का पत्नीत्व और एक बच्ची दे जाता है। इस बच्ची का नाटक के अन्त में रोकर अपने अस्तित्व का अहसास दिलाना भी अपने में एक प्रतीकार्थ है; एक कठोर यथार्थ है जिसे वह वक्ष से लगाए है और जो एक बार फिर कालिदास को उससे विलग कर देता है। इस प्रकार इस नाटक की सबसे बड़ी शक्ति इसके प्रतीक और बिम्ब-विधान में आँकी जा सकती है। शब्देतर माध्यम का तो यह नाटक सफल प्रयोग करता ही है, साथ ही नाटकीय शब्द का भी इसमें बड़ा सर्जनात्मक प्रयोग हुआ है। इसकी भाषा संवाद की रंगमंचीय भाषा है जिसमें पात्र और स्थिति के अनुरूप त्वरा, ठहराव, गति, भावुकता, रोमान, कठोरता, कोमलता सब-कुछ है। अम्बिका की भाषिक और संवाद रचना में उसके हृदय की पीड़ा, अवरोध, चिन्ता है तो विलोम अपने शब्दों में आक्रामक है। प्रियंगु के संवादों में आभिजात्य है। दूसरी ओर दन्तुल, निक्षेप, मातुल जैसे सामान्य पात्र भी शैलीय विविधता लिए हुए हैं। नाटक की भाषा में सर्वत्र अभूतपूर्व मितव्ययता दिखाई देती है।

आषाढ़ का एक दिन एक सफल कृति है। इसे 'कल्पना का यथार्थ' कहें या 'यथार्थ की कल्पना', 'भावबोध का रोमानी धरातल' कहें या 'यथार्थ का कटु अहसास', 'नियति का अंकन' कहें या 'आकांक्षा का भटकाव' सच्चाई यह है कि यह बद्ध दृष्टि में बँधनेवाला नाटक नहीं है। हर अन्वेषी दृष्टि के आगे यह अर्थ के नए स्तर खोलता है और बहुधा 'एक से अधिक संकेत' देता है। फिर भी यह नाटक मन में कई शंकाएँ

जगाता है। कालिदास की महानता का इस नाटक में कोई अहसास नहीं होता। उसका द्वन्द्व भी बहुत मुखर नहीं जिससे उसके चरित्र की ऊँचाइयों को मापा जा सकता। विलोम की तुलना में वह दुर्बल है। विलोम की आक्रामकता प्रभावित करती है जबकि कालिदास की दुर्बलता पर सिर्फ़ तरस आता है। मल्लिका को आरोपित आस्था कहा गया है पर उसे कालिदास की आस्था का विस्तारित रूप किस आधार पर माना जाए ? सम्पूर्ण नाटक की संवेदना कालिदास की न होकर मल्लिका की है। मल्लिका की मानसिकता आस्था की है या खंडित विश्वास की ? इस पर भी विचार करने की ज़रूरत है। उसकी उदारता हठ में बदलती दिखाई देती है। वह जो कुछ करती है क्षतिपूर्ति के रूप में करती है। वस्तुतः खंडित व्यक्तित्व में दिशाबोध का निरूपण यह नाटक नहीं करता क्योंकि यह भौतिक आवश्यकताओं के साथ-साथ आत्मिक आवश्यकताओं, आकुल चेतना और अभीप्सा को आन्दोलित करता नहीं लगता।

दो दीपाधारों की जलती लौ

बहुत पहले से मन में एक बिम्ब था। दो दीपाधार। एक ऊँचा, शिखर पर पुरुष-मूर्ति—बाँहें फैली हुईं तथा आँखें आकाश की ओर उठती हुईं। दूसरा छोटा, शिखर पर नारी-मूर्ति—बाँहें सिमटी हुईं तथा आँखें धरती की ओर झुकी हुईं। पहले-पहल शायद अश्वघोष का सौंदरनन्द पढ़ते हुए यह बिम्ब मन में बनने लगा था। क्यों और कैसे, यह कह सकना असम्भव है। उस काव्य का अपना बिम्ब तरंगों पर तैरते राजहंस का है या अनिश्चय में उठे-रुके एक पैर का। परन्तु मेरे लिए यह सब धुँधला दृश्य था। स्पष्ट थे दो दीपाधार जो सौंदरनन्द में नहीं थे।

—भूमिका, लहरों के राजहंस

लहरों के राजहंस (1963/63) दो बिम्बों को उभारता है। सच्चाई इतनी ही है कि दो दीपाधारों के शिखर पर स्थापित पुरुष और स्त्री-मूर्ति का बिम्ब राकेश का अपना है और लहरों पर तैरते राजहंस का बिम्ब **सौंदरनन्द** से गृहीत होने के नाते नाटक पर आरोपित है। यदि उसे हटा दें तो नाटक नहीं बनता—पूरा ढाँचा, प्रतीक तत्त्व उसी पर आधारित है। फिर भी लहरों पर तैरते राजहंस का बिम्ब नाटक की संरचना पर आरोपित है और पुरुष तथा स्त्री-मूर्ति का बिम्ब कथ्य के बीच से उभरता है। कथ्य के बीच से उभरनेवाले बिम्ब को ही राकेश ने वास्तविक बिम्ब माना है। वैसे दोनों में कोई तात्त्विक विरोध नहीं है। राजहंस पुरुष और नारी-मूर्ति के भाव को बख़ूबी व्यंजित करते हैं। पुरुष की सिमटी हुई बाँहें और नारी की झुकी हुई आँखों के बिम्ब में भी कोई ऐसा भाव नहीं जो राजहंसों के तैरने की गतिशील मुद्रा में निहित न हो ! यदि 'राजहंस नन्द और सुन्दरी के समानान्तर हैं—लहरें उनकी परिस्थितियाँ हैं' तो बाँहों के सिमटने और आँखें झुकी होने का कारण भी परिस्थितियाँ ही हो सकती हैं। पर नाटककार शायद परिस्थितियों पर उतना जोर नहीं देना चाहता जितना स्त्री-पुरुष के आपसी सम्बन्धों (की भाव-मुद्रा) पर। इसीलिए उसे पहला बिम्ब अधिक प्रिय है।

वस्तुतः नाटक दोनों बिम्बों पर निर्भर करता है। नाटक का कथ्य केवल स्त्री-पुरुष सम्बन्धों पर टिककर नहीं रह जाता और न अनिश्चय के 'ययौ न तस्थौ' वाले द्वन्द्व तक ही अपने को सीमित कर देता है। नाटक में दोनों बिम्ब परस्पर घुल-मिल जाते हैं,

एक-दूसरे के पूरक बन जाते हैं। एक व्यक्ति के आन्तरिक विभाजन का संकेत देता है तो दूसरा स्त्री-पुरुष के पारस्परिक सम्बन्धों की परणिति का। पर आन्तरिक विभाजन भी वस्तुतः पार्थिव और अपार्थिव तत्त्वों को लेकर है जिसको स्त्री-पुरुष के सम्बन्धों के सन्दर्भ में चरितार्थ किया गया है।

पुरुष और नारी-मूर्ति के जिस बिम्ब की बात राकेश ने की है, वह प्रसाद के नाटकों की याद दिलाता है। **स्कंदगुप्त** में धातुसेन की एक उक्ति आती है : 'पुरुष उछाल दिया जाता है, **उत्प्रेक्षण** होता है। स्त्री **आकर्षण** करती है।' सुन्दरी की चेतना में भी यह भाव कहीं गहराता नज़र आता है : 'नारी का **आकर्षण** पुरुष को पुरुष बनाता है तो उसका **अपकर्षण** उसे गौतम बुद्ध बना देता है।' मिलिन्द के नाटक **गौतमनन्द** में सुन्दरिका भी कुछ इसी भाव से प्रश्न करती है : 'क्या सीता के जीते-जी राम का उन्हें छोड़कर संन्यासी हो सकना सम्भव था ? क्या मेरे जीते-जी मेरे पति मुझे छोड़कर संन्यास-ग्रहण कर सकते हैं ? यह मेरी समझ में नहीं आता कि मेरी माता के रहते मेरे पिता संन्यास-ग्रहण की बात कैसे सोच रहे हैं ? सौभाग्यवती यशोधरा देवी के रहते राजकुमार सिद्धार्थ घर छोड़कर कैसे जा सके ?'[1] सम्भवतः इस संवाद ने भी राकेश को इस दिशा में सोचने का अवसर दिया हो।

जब मोहन राकेश ने कहानी के रूप में सर्वप्रथम **लहरों के राजहंस** की परिकल्पना की थी तो नारी के आकर्षण में पुरुष की पूर्णता और अपकर्षण में उसकी विसंगति का भाव उसमें भी मुख्य था। सुन्दरी की वह उक्ति भी तब उसमें विद्यमान थी। कई वर्षों बाद **रात बीतने तक** के प्रारूप में इसका जो नया संस्कार हुआ, उसमें भी यह उक्ति विद्यमान रही कि 'गौतम बुद्ध को गौतम बुद्ध बनाने का श्रेय यशोधरा को है।' दरअसल, यह विचार ही उसमें एक तुला पर रखा गया है जिसको मन और कर्म से नर्तकी चन्द्रिका विश्वासभरी चुनौती देती है। यह चुनौती बाद में सुन्दरी के अपने जीवन में खरी नहीं उतरती है। सुन्दरी के यशोधरा न होने पर भी नन्द बुद्ध का मार्ग ग्रहण कर लेता है। इस प्रकार **रात बीतने तक** में नारी के अपकर्षण से बुद्ध बनने की बात खंडित होती है और ऐसी आध्यात्मिक शक्ति का बिम्ब उभरता है जो नारी के प्रति आकर्षण पैदा करती है। फलतः जीवन को 'मादक रात' बनाकर बिताने का विश्वासी नन्द रात बीतने तक नई किरण के लिए छटपटा उठता है। यह छटपटाहट सीधी और सतही है। राकेश उसे आत्मिक गहराई नहीं दे सके। फिर भी सुन्दरी का रूप-दर्प, नन्द की आसक्ति, अलका द्वारा व्यंजित क्षणभंगुरता का अहसास दोनों नाटकों में सामान्य रूप से विद्यमान है। जो सामान्य नहीं है, वह है नर्तकी चन्द्रिका की परिणति; किन्तु सबसे बड़ी बात यह है कि **रात बीतने तक** का कथ्य वही नहीं है जो **लहरों के राजहंस** का है।

लहरों के राजहंस में दो दीपाधारोंवाली परिकल्पना किस रूप में गृहीत है, उसकी व्याख्या राकेश ने श्यामानन्द जालान को लिखे एक पत्र में की है : 'सुन्दरी, पृथ्वी के प्रतीक रूप में, पुरुष और उसकी चेतना को अपने तक बाँधे रखना चाहती है—पुरुष बँधना चाहकर भी उससे ऊपर उठना एक अपार्थिव जिज्ञासा में अपने लिए उपलब्धि

ढूँढ़ना चाहता है। बुद्ध पार्थिवकता को तिलांजलि देकर उस उपलब्धि की ओर जाते हैं—नन्द तिलांजलि नहीं दे पाता; नहीं देना चाहता।'[2] आकर्षण और अपकर्षण के इस द्वन्द्व को ही राकेश ने पार्थिव का द्वन्द्व कहा है। यह द्वन्द्व एक ओर स्त्री और पुरुष का है, दूसरी ओर दो जीवन दृष्टियों का। जहाँ तक दो जीवन-दृष्टियों का प्रश्न है, यह दो दिशाओं में अग्रसर है। एक दिशा में यह गौतम बुद्ध और सुन्दरी के बीच दिखाई देता है; किन्तु दूसरी में और भी जटिल है। यहाँ यह द्वन्द्व नन्द और सुन्दरी में ही नहीं दिखाई देता वरन् स्वयं नन्द की दो विरोधी चेतनाओं में भी मुखर हो उठता है। बुद्ध और सुन्दरी के बीच जो द्वन्द्व है वह रागमयी नारी और विरक्त होते पुरुष का द्वन्द्व है और उसका केन्द्रीय बिन्दु है कामना। सुन्दरी यही आदिम कामना लिए हुए है—उसका कामोत्सव कामना का त्यौहार है। कामना की पूर्ति में ही वह जीवन की सार्थकता पाती है। बुद्ध की अकामना को वह चुटकी में उड़ा देती है, 'उन्होंने बोध प्राप्त किया है, कामनाओं को जीता है ! परन्तु मैं जानना चाहती हूँ कि कामनाओं को जीता जाए, यह भी क्या मन की कामना नहीं ?' वह कामना को जीवन के आदिम स्तर पर स्थापित कर कहती है : 'कोई गौतम बुद्ध से कहे कि कमलताल के पास आकर कभी' राजहंसों से भी 'निर्वाण और अमरत्व की बात कहें।' बुद्ध निर्वाण और अमरत्व की बात कहने नहीं आते, पर अलका, श्यामांग, नन्द उनकी उपस्थिति का अहसास दिलाते हैं। सुन्दरी की इसी कामना का विरोध यशोधरा प्रव्रज्या ग्रहण कर करती है; अतिथि कामोत्सव में भाग न लेकर करते हैं। हमारे देश में युगों से विरागी पुरुष नारी से घृणा करते रहे हैं। भोगवादी नारी का भी विरागी पुरुष का विरोध करना स्वाभाविक है।

कुल मिलाकर सारी विसंगति के मूल में कामना है, चाहे वह बुद्ध की निष्काम होने की कामना हो, सुन्दरी की भोग की कामना हो या अन्तर्विरोध में फँसे नन्द की कामना हो। नन्द, सुन्दरी, बुद्ध इन तीन स्तरों पर कामना का टकराव दृष्टिगत होता है और बिखराव वहाँ से शुरू होता है, जहाँ चित्त-वृत्ति आस्था से विराग होने का उपक्रम करती है। किन्तु यह स्त्री-पुरुष सम्बन्ध का मामूली-सा नुक्ता है। इसमें कोई नवीनता नहीं पर यहीं एक प्रश्न यह भी उठता है—इस द्वन्द्व में सुन्दरी के रूप में पार्थिव का जैसा बिम्ब उभारा गया है, क्या बुद्ध के रूप में अपार्थिवता सही अर्थों में मुखर हो पाई है ? बुद्ध नाटक में अत्यन्त उपेक्षित पात्र है। इसलिए वह सीधे द्वन्द्व में नहीं आता। किन्तु इतना सच है कि बुद्ध की एक अदृश्य छाया सारे नाटक पर छाई है। अलका और श्यामांग प्रारम्भ से ही बुद्ध और उनकी अपार्थिवता के मुख्य प्रवक्ता हैं। उनकी मानसिकता को बुद्ध के आगमन और देवी यशोधरा के भिक्षुणी बन जाने के प्रसंग से जोड़कर देखा जा सकता है। कामोत्सव का आयोजन बुद्ध और सुन्दरी को आमने-सामने खड़ा कर देता है। श्यामांग के मन की उलझन इसी वस्तु-स्थिति की देन है : 'पिछले वसंत में आम कैसे बौराए थे !...परन्तु तब कामोत्सव का आयोजन नहीं किया गया। आयोजन किया गया है इस बार जब आम के वृक्षों ने भिक्षुओं का वेश धारण कर रखा है।...कल प्रातः देवी यशोधरा भिक्षुणी के रूप में दीक्षा ग्रहण करेंगी, और यहाँ...यहाँ

रात-भर नृत्य होगा ! आपानक चलेगा !'

एक कार्य के पीछे बुद्ध कारण हैं, दूसरे के सुन्दरी। सुन्दरी और बुद्ध जैसे चेतना के दो ध्रुवान्त हैं। नन्द और सुन्दरी के बीच एक तीसरा तत्त्व है बुद्ध। बुद्ध केवल अध्यात्म का तत्त्व मात्र नहीं है। वह स्त्री और पुरुष के कामनामय जीवन के बीच समय-असमय आ जानेवाली विरक्ति का प्रतीक भी है। सुन्दरी भावना के स्तर पर जीती है; शुद्ध इन्द्रिय स्तर पर। बुद्ध की विराग चेतना इंद्रिय स्तर की अस्वीकृति ही नहीं, चिन्तन की देन भी है। सुन्दरी चिन्तन की आवश्यकता नहीं समझती कि अग्नि काष्ठ के भीतर है या बाहर। दूसरी ओर बुद्ध की चेतना से आविष्ट श्यामांग सोचने में ही पत्तियों को उलझा देता है। फिर कमल ताल में पत्थर फेंकना, हंसों को आहत करना, छाया को देखना ऐसे प्रसंग हैं जो **बुद्ध का अहसास** और गहरा करते हैं। श्यामांग के रूप में बुद्ध का भोग-विरोधी भाव मुखर होता है। इसीलिए उस व्यक्ति को देखकर सुन्दरी को सदा उलझन होती है और वह उसके कार्य, हर विचार के सामने प्रश्न-चिह्न बनकर खड़ा हो जाता है। इसी प्रकार नन्द के माध्यम से भी बुद्ध का पक्ष ही उभरकर आता है—आखेट में अपनी ही थकान से मरते मृग के प्रति करुणा का भाव लेकर लौटा नन्द जैसा थका-थका महसूस करता है उसमें बुद्ध की करुणा के दर्शन होते हैं। नन्द को पार्थिव से अपार्थिव की ओर ले जाने में इस घटना का भी कुछ हाथ हो सकता है। यशोधरा के प्रव्रज्या ग्रहण करने के कारण अतिथि कामोत्सव में सम्मिलित नहीं होते। यह कटु अहसास भी अपार्थिव की चेतना को गाढ़ा करता है। इसलिए यह कहना ग़लत होगा कि बुद्ध या अपार्थिवता का पक्ष निरूपित ही नहीं हुआ। फिर पार्थिवता और अपार्थिवता के द्वन्द्व का प्रश्न ही नहीं उठता। पर सच्चाई यह है कि द्वन्द्व है। बात इतनी-सी है कि सब-कुछ संवेदना के स्तर पर उभारा गया है। इसलिए दोनों पक्षों को लेकर न विवाद है, न सिद्धान्त कथन, न विरोध न समर्थन। कुछ है तो बस स्थितियों और पात्रों का तनाव। किसी भी रचनात्मक कृति के लिए यही अपेक्षित भी है।

द्वन्द्व का वास्तविक केन्द्र न बुद्ध है न सुन्दरी। वास्तविक द्वन्द्व तो नन्द के मन में उभरता है। और उसी के द्वन्द्व के माध्यम से नाटक की मूल समस्या भी सामने आती है। यों भी कह सकते हैं कि नन्द में द्वन्द्व पनपानेवाला अहसास भी स्वयं बुद्ध का है—उस मन का जो मनन करता है। इसलिए नन्द का द्वन्द्व भी बुद्ध और सुन्दरी का द्वन्द्व है। किन्तु बुद्ध नन्द को आकृष्ट करता है, नन्द बुद्ध नहीं बन जाता। उसकी अपनी निजता है और निजता की अपनी तलाश है। इसलिए बुद्ध और सुन्दरी दोनों उसे अपने ढंग से प्रभावित करते हैं। यही प्रभाव उसे उलझाता है।

नन्द बुद्ध की चेतना से अभिभूत होता है; अंक के प्रारम्भ में ही नन्द श्यामांग से तादात्म्य स्थापित कर लेता है। वह बुद्ध की ओर खिंचनेवाले उसके परोक्ष व्यक्तित्व का प्रतीक बन जाता है। और यही वह स्थल है जहाँ नन्द सही शब्दों में अपना कुछ विश्लेषण कर पाता है : 'कुछ है जो चेतना पर कुंडली मारे बैठा रहता है और मुझे अपने

से मुक्त नहीं होने देता।' जैसे ही बुद्ध की छाया नन्द पर हावी होने लगती है, सुन्दरी संघर्ष में आ जाती है। और फिर एक नई मानसिकता में दोनों खंडित होने लगते हैं। सुन्दरी पहले ही अहं से ग्रस्त है। कामोत्सव की असफलता में उसे जो ठेस लगती है, उसकी क्षति-पूर्ति वह अधिकार-भावना से करती है। नन्द को दर्पण लेकर खड़े रहने की अभ्यर्थना में भी यही भावना है। किन्तु रूप में इस दासता को भिक्षुओं का 'बुद्धं शरणं गच्छामि' का स्वर बेध जाता है। दर्पण काँपता है और श्वास से धुँधला भी हो जाता है। अनन्यता का अभाव सुन्दरी को रुष्ट कर देता है। और फिर एक के स्वर के खिंचाव और दूसरे के मन के अलगाव से दर्पण गिरकर टूट जाता है। यह दर्पण बाहर का नहीं, कहीं बहुत अन्दर का है जो चटख जाता है। दर्पण का टूटना दोनों के सम्बन्धों की परिणति को मुखर करता है। सुन्दरी इस बात को अच्छी तरह जानती है : 'दर्पण का टूटना क्या अकारण था ?' उस टूटे दर्पण में वह अपना प्रतिबिम्ब देखती है—अपरूप और अनचीन्हा। पर इसी में वह अपने खंडित सीमांत, वर्जित सुहाग को चीन्ह लेती है। साथ ही नन्द के व्यक्तित्व का एक खंड टूटकर उसके सामने आ जाता है : 'आपका मन इतना कच्चा है, मैं न जानती थी।'

गौतम बुद्ध का भिक्षाटन के लिए आना और द्वार पर किसी को न पाकर लौट जाना उस मन के कच्चे आदमी को और भी विचलित कर देता है। नन्द के अपार्थिव की ओर बढ़ते चरण उसे बुद्ध की ओर ले जाते हैं। सुन्दरी का आहत अहं उसे अलगाव की स्थिति में ले जाता है। नन्द का केश कटवाकर लौटना उसे अर्थपूर्ण लगता है क्योंकि वह सोचती है कि जो नन्द लौटा है वह 'दूसरा व्यक्ति' है। वह इस दूसरे व्यक्ति को पहचान लेती है, पर नन्द इस दूसरे व्यक्ति पर आवरण डाल देता है। किन्तु तनाव की स्थिति ऐसा नहीं होने देती। फलतः पहला और दूसरा व्यक्ति स्वयं आपस में टकराने लगते हैं। नन्द का चुपचाप केश कटवाना, फिर वन में जाकर व्याघ्र से लड़ना या भिक्षु आनन्द के साथ घर लौट आना इसी स्थिति का परिचायक है। सुन्दरी को पहले व्यक्ति के आवरण में दूसरे की उपस्थिति खलती है और वह भी दुर्बल व्यक्ति के रूप में कहीं अधिक। दुर्बल और परबद्ध नन्द उसे कभी ग्राह्य नहीं रहा। यही नन्द उसका सीधी समान्तर दृष्टि से सामना नहीं कर पाता क्योंकि यहाँ सुन्दरी उसे दुर्बल नगण्य व्यक्ति के रूप में आँकती है। वस्तुतः नगण्यता का अनुभव शायद जीवन का सबसे कटु अनुभव कहा जा सकता है। उसका अहसास नन्द को बिलकुल दूसरा व्यक्ति बना डालता है। किन्तु यह दूसरा नन्द अपने तर्कों में कहीं अधिक वास्तविक लगता है जो परार्थ-जीवी नन्द से भिन्न है। वह अपनी आन्तरिक आवश्यकताओं के लिए जीता है। यह नन्द जीवन को किसी एक सिरे पर जीनेवाला नहीं; वह उसके बीचोंबीच जीता है जहाँ दोनों हैं या दोनों नहीं हैं। 'उसके पास था तो मन यहाँ के लिए व्याकुल था। अब तुम्हारे सामने हूँ तो मन कहीं और के लिए व्याकुल है।...क्योंकि **मैं यह भी हूँ और वह भी।** इनमें से कोई एक नहीं, जैसा कि तुम सब अलग-अलग से विश्वास करना चाहते हो. ..।' नन्द के दोनों व्यक्तित्व पूरक हैं और दोनों उसकी आन्तरिक आवश्यकता को प्रकट

करते हैं। पर ये विरोधी पूरक सुन्दरी को ग्राह्य नहीं। इसीलिए रंग-संकेत के अनुसार जब नन्द 'उसकी बाँह हाथ में कस लेता है तो सुन्दरी झटके से उसका हाथ अपनी बाँह से अलग कर देती है।' वह उसे 'वहीं-का-वहीं' नहीं बने रहने देना चाहती।

अश्वघोष के **सौंदरनन्द** ने मोहन राकेश को जो बिम्ब दिया था 'वह तरंगों पर तैरते राजहंस का था या अनिश्चय में उठे-रुके पैर का।' पर इस धूमिल बिम्ब के समानान्तर दो दीपाधारों का बिम्ब था। किन्तु दोनों बिम्ब धुँधले होते रहे और नए दृश्यों में परिवर्तित होते गए। ये दोनों बिम्ब कहीं आपस में जुड़ जाते हैं। राजहंस जहाँ सुन्दरी और नन्द के प्रतीक बन जाते हैं वहीं दीपाधार स्त्री-पुरुष के साथ-साथ पार्थिव-अपार्थिव के सूचक भी।

वस्तुतः दो दीपाधारोंवाले बिम्ब का उपयोग एक रंग-उपकरण के रूप में करते हुए नाटककार ने एक रूढ़ किन्तु आंशिक तथ्य को ही उजागर किया है। आधुनिक भाव-बोध से उसे पूरी तरह जोड़ना सम्भव नहीं। राकेश ने उसकी व्याख्या पार्थिव-अपार्थिव चेतना के रूप में भले ही की है, किन्तु स्त्री-पुरुष को इतने रूढ़ अर्थों में व्याख्यायित नहीं किया जा सकता।

अब दूसरा बिम्ब लीजिए। नाटक के अन्त तक पहुँचते-पहुँचते नन्द की स्थिति **सौंदरनन्द** के लहरों पर तैरते राजहंस की-सी हो जाती है—

तं गौरवं बुद्धगतं चकर्ष भार्यानुरागां पुनराचकर्ष।
सोऽनिश्चयान्नापि ययौ न तस्थौ तुरंस्तरंगेष्विव राजहंसः ॥

'ययौ न तस्थौ' की स्थिति में नाटककार दो विरोधी मनःस्थितियों की टकराहट का बिम्ब प्रस्तुत करता है। इस टकराहट से ही नाटक की सम्भावनाएँ फूट पड़ती हैं। लहरें मन और परिस्थिति की तरलता को व्यक्त करती हैं और राजहंस मन की निर्मलता, विवेकशीलता और समृद्धि सभी कुछ ध्वनित करता है। 'ययौ न तस्थौ' में मन की जड़ दशा का प्रवाह और प्रवाह की जड़ता दोनों की व्यंजना निहित है। फिर भी प्रश्न उठता है, राजहंस लहरों के अधीन है या अनिश्चय के ? श्लोक नन्द की अनिश्चय युक्त भाव-मुद्रा को व्यक्त अवश्य करता है किन्तु **लहरों के राजहंस** स्थिति की जड़ता का नाटक नहीं है। उसमें सर्वत्र—अनिश्चय में भी—स्थिति की टकराहट है और नन्द उसके बीच सक्रिय है। एक घटना-विहीन घटना निरन्तर उसके अन्दर घटती है जो मन को, मनन को, अनुभूति को एक स्तर पर तोड़कर फिर-फिर गढ़ती चलती है। फलतः नाटक घटनाओं का नाटक न रहकर प्रश्नों और प्रश्नों की तलाश का नाटक बन जाता है। परिस्थितियाँ प्रश्न उठाती हैं। प्रश्न पुराने हैं, उत्तर भी किसी-न-किसी ने दिए हैं; पर नन्द अपना उत्तर आप पाना चाहता है। प्रश्न जिसका है, उत्तर भी उसी को खोजना होगा। इसलिए समस्या पार्थिव या अपार्थिव की एकांगी स्वीकृति की नहीं—एक तलाश की है; एक चुनौती की है : 'कितने-कितने बिन्दु खोजे हैं आज तक तुमने ? जाओ एक और बिन्दु खोजो।' यह तलाश एकदम दार्शनिक नहीं है—वह उसके अन्तर की तलाश

है, एक पूरे जीवन का समर्पण है और इसमें कोई सतही निर्णय नहीं, जीवन की एक पूरी प्रक्रिया निहित है, सीमान्त की एक कँटीली बाड़ है जहाँ अकुल प्राण बार-बार आ टकराते हैं। प्रकाश के दो दावेदार उसे अंधकार की संधि-रेखाओं में भटकाते हैं। इसलिए सारी स्थिति जड़ता की स्थिति नहीं है।

यही कारण है कि सम्पूर्ण नाटक जो बिम्ब उभारता है, वह लहरों पर डोलते राजहंस से भी आगे बढ़ जाता है। दुविधा में भी एक स्थिति आती है जब परिस्थितियों से भी अधिक मन महत्त्वपूर्ण हो जाता है, जब हंस लहरों के अधीन नहीं रह जाता। इसलिए अन्ततः जो तथ्य उभरता है वह अन्वेषण की अधूरी यात्रा या रास्ते की तलाश का लगता है। आन्तरिक संवेदना के स्तर पर इसमें नन्द और सुन्दरी को स्त्री-पुरुष के पारस्परिक सम्बन्धों की राह के बीच से गुज़रना पड़ता है। ये सम्बन्ध नाटककार के लिए महत्त्वपूर्ण रहे हैं। पुरुष नारी को अनेक आन्तरिक आवश्यकताओं में एक आवश्यकता मानता है, नारी के लिए वह सम्पूर्ण उपलब्धि है। नारी की भोग्या-वृत्ति उसके जीवन की सार्थकता है, पुरुष सदा से भोक्ता रहा है। सुन्दरी ऐकांतिक भोग और समर्पण में ही अपने 'स्व' की विमुक्ति पाती है और यही 'स्व' की 'अन्य' में एकाकार हो जाने की स्थिति को जन्म देता है। यही वह रहस्यात्मक एकान्विति है जो उसे अपने अस्तित्व की पूर्णता का आभास दिलाती है। पर नन्द की स्थिति उसकी समर्पण-भावना और आस्था को खंडित करती है। इस प्रकार कल्पित प्रेम एक प्रवंचना सिद्ध होता है और यातना भोगने के लिए एक अभिशप्त नारी आश्रयहीन छूट जाती है।

वस्तुतः सुन्दरी का जीवन जिस भोग-वृत्ति पर आधारित है उसमें हर इच्छा की परिणति इसी प्रकार उसे धरती पर ला पटकने के लिए काफ़ी होती है। नन्द रसिक नायक नहीं है—वह चिन्तन-प्रिय चेतन पुरुष है। इसीलिए रूप उसे बाँध नहीं पाता और सुन्दरी के आकर्षण और अधिकार-भाव में वह अचेतन में दासता का अनुभव करता है। इसीलिए मुक्ति के लिए छटपटाता है। सुन्दरी अपने को नन्द का पहला और अन्तिम लक्ष्य मानती है; नन्द उसकी सीमाओं को समझता है। सुन्दरी सब मूल्यों से ऊपर अपने को मूल्यवान समझती है जबकि नन्द के समक्ष उससे भी भिन्न मूल्य विद्यमान हैं। सुन्दरी के लिए जगत् नन्द के माध्यम से ही उद्घाटित होता है, किन्तु नन्द की दुनिया उससे बाहर भी है।

लहरों का राजहंस मुट्ठी से रिसते क्षणों का नाटक है। इसके मुख्य पात्र नन्द और सुन्दरी क्षण की प्रतिक्रियाओं और चरम बिन्दुओं पर जीते हैं। जीवन के ताल में राजहंसों को सम जल प्रवाह की अपेक्षा होती है जिससे तैरने में अनागत की कोई आशंका न हो और प्रत्येक क्षण को समग्रता से पाया जा सके। विपरीत स्थितियों की छाया जब श्यामांग, नन्द, सिद्धार्थ और यशोधरा के माध्यम से सुन्दरी के भोगों और प्राप्य पर पड़ने लगती है तब कामोत्सव में अभ्यागतों का न आना, गौतम बुद्ध का राजभवन के द्वार पर भिक्षाटन, यशोधरा का भिक्षुणी बन जाना, नन्द का बुद्ध के पास से विशेषक न सूखने तक लौट न आना ऐसी घटनाएँ हैं जिनमें क्षण विशेष सुन्दरी की अस्मिता से

फिसल जाता है। ऐसी स्थिति में सुन्दरी के लिए मुंडित केश नन्द 'दूसरा' हो जाता है और स्वयं वह बीते क्षण को न पकड़ पाने के कारण खंडित दर्पण, सूखा विशेषक, मृत मृग का प्रतीक बन जाती है। वही उसका अभिशप्त क्षण है और वही नन्द का एक बिन्दु से दूसरे बिन्दु तक भटकाव का क्षण भी है। दोनों इस क्षण अलग-अलग व्यक्ति हो जाते हैं—एक में ठहराव आता है तो दूसरे में गति। ठहराव अहं और जड़ता लाता है तो गति संशय और प्रश्न। प्रश्न सुन्दरी में भी है—अपनी अस्मिता के, निरर्थकता के प्रश्न; किन्तु उसमें उत्तर खोजने की ललक नहीं। एक में आधुनिक मानव की जड़ता और टूटन व्यक्त हुई है तो दूसरे में आधुनिकता बोध से जुड़ा बोध।

नारी जितना ही अधिक प्रेम पाती है, उतनी ही वह अपने अस्तित्व से वंचित होती है और उतनी ही अधिक असुरक्षा और दयनीयता उसके हाथ आती है। पुरुष जितना ही लिप्त होता है उतना ही दासता को भोगता है और उतना ही तीव्र उसका विकर्षण भी होता है। स्त्री और पुरुष दोनों के लिए इसीलिए प्रेम विघटनकारी होने के लिए नियति-बद्ध है। इसमें दोनों टूटते हैं। सवाल इतना ही है कि कौन कितना टूटता है और किसकी दृष्टि कहाँ टिकती है—धरती पर या आकाश पर ? ऐसा प्रतीत होता है कि नाटक में स्त्री-पुरुष को उनके सम्बन्धों को लेकर आमने-सामने खड़ा कर देना ही मोहन राकेश का लक्ष्य रहा है। सुन्दरी की नन्द के प्रति यह उक्ति कि 'इतना ही समझ पाते हैं ये लोग' के द्वारा राकेश ने दोनों के एक-दूसरे को न समझ पाने की विडम्बना की ओर संकेत किया है। यह उक्ति उस सन्दर्भ में कही गई है जब नन्द मुंडित-केश घर लौट आता है और सुन्दरी विक्षुब्ध होती है। नन्द यह सोचता है कि उसका विक्षोभ केश कटवाने के कारण है, जबकि सत्य यह है कि सुन्दरी का क्षोभ उसके मुंडित हो जाने के कारण नहीं वरन् इस कारण से है कि नन्द ने उस मात्रा में उसके विश्वास को खंडित किया है; नन्द उसका अर्थ यही लेता है कि उसे मुंडित रूप में देखना सुन्दरी को सह्य नहीं हुआ जबकि सुन्दरी पुरुष द्वारा न समझे जाने की बात कहती है। पार्थिव के संघर्ष में नारी की पुरुष से यह शाश्वत शिकायत रही है।

यह शिकायत जैसी भी है, ठीक है, पर यह शायद **लहरों के राजहंस** में पूरी तत्परता से नहीं उभर पाती। इसका एक कारण यह भी है : नन्द का मुंडित-केश लौटना और उससे कुछ पहले घटनाक्रम का अमनोवैज्ञानिक न होकर भी उसके अपने मन के तर्क-वितर्क के बावजूद आरोपित-सा लगना। और फिर इतने बड़े द्वन्द्व के बाद यदि तथ्य का इतना-सा अणु ही यदि हाथ लगे तो सारी उपलब्धि पर शंका हो सकती है। किन्तु इस नाटक में अर्थ के और भी आयाम हैं जो नाटक की सम्पूर्ण संवेदना से जुड़कर कथ्य को गरिमा प्रदान करते हैं। वस्तुतः यह नाटक कई रूपों में कई बार लिखा गया। **रात बीतने तक** के रूप में जब इसकी पहली अवधारणा हुई थी तो **सौंदरनन्द** के प्रभावस्वरूप लहरों पर तैरते युगल हंसों का बिम्ब ही लेखक पर हावी था जिसमें विवशता का चित्रण ही मुख्य था। बाद में राकेश ने पार्थिव-अपार्थिव के द्वन्द्व का समाहार कर दीपाधार के बिम्ब को आरोपित किया और उसके माध्यम से स्त्री-पुरुष सम्बन्धों को उजागर करने

का भी प्रयास किया। इस प्रयास में कथ्य उलझ ज़रूर गया है, किन्तु नाटक फिर भी अर्थ की सम्भावनाओं से युक्त रहा।

लहरों के राजहंस भी उसी अर्थ में 'ऐतिहासिक' नाटक है जिस अर्थ में **अषाढ़ का एक दिन**। वस्तुतः उसके सन्दर्भ में ऐतिहासिकता की बात उठाना वैसे भी ठीक नहीं, क्योंकि नन्द और सुन्दरी की कथा केवल कहने भर के लिए ऐतिहासिक है, उसकी विस्तृत सूचना इतिहास में नहीं मिलती। ऐतिहासिक तथ्य सम्भवतः इतना ही है कि नवगृह-प्रवेश और राज्याभिषेक के अवसर पर नन्द भिक्षु बन गया था। बाकी कथा के सूत्र **सौंदरनन्द** में अश्वघोष की सर्जनात्मक काव्य प्रतिभा द्वारा सर्जित हुए हैं। काव्यग्रन्थ होने के नाते उसमें कवि की कल्पना का समावेश स्वाभाविक था। उस काव्य-कल्पना में नाटकीयता का समावेश राकेश के लिए अपेक्षित था। अतः 'जो स्वतन्त्रता **सौंदरनन्द** के लेखक ने उपलब्ध तथ्यों से आगे जाने में ली, वही स्वतन्त्रता यदि आज का लेखक सौंदरनन्द के तथ्यों को आगे ले जाने में लेता है तो पुराणसर्वस्व व्यक्तियों को चौंकना नहीं चाहिए।'[3]

मोहन राकेश ने **सौंदरनन्द** काव्य से कथा का ढाँचा और चारित्र्य का कुछ आधार ग्रहण किया है। नन्द की कामासक्ति (4/1), सुन्दरी का रूपगर्व (4/3), नन्द की दीक्षा (5/8), केश-कर्त्तन (7/16) आदि प्रसंग **सौंदरनन्द** के आधार पर रचे गए हैं। प्रसाधन-सम्बन्धी पूरा दृश्य तो उसी पर आधारित है। दर्पण का आविल होना (4/14), बुद्ध का भिक्षाटन के लिए उपस्थित होना और लौट जाना और स्थिति की सूचना पर नन्द का काँप उठना (4/31), इन सारे प्रसंगों का भी राकेश ने वहीं से उपयोग किया है। काम-आसक्ति और बुद्ध के अनुराग के बीच तनाव झेलती नन्द की मनःस्थिति भी **सौंदरनन्द** (4/42-44), में अंकित है। उसमें नन्द का चुपचाप बुद्ध के पीछे-पीछे चलते हुए सुन्दरी के विशेषक के सूख जाने की चिन्ता करते जाना बड़े मार्मिक ढंग से चित्रित किया गया है। **लहरों के राजहंस** में नाटकीय शिल्प और मंच की सीमा के कारण यह द्वन्द्व उस रूप में प्रत्यक्षतः चित्रित करना सम्भव न था। इसी प्रकार **सौंदरनन्द** में भी नन्द दीक्षा को पहले हृदय से स्वीकार नहीं करता, पर बाद में पूरी तरह स्वीकार कर लेता है। (इस प्रकार की निर्द्वन्द्व स्वीकृति राकेश ने **रात बीतने तक** में दिखाई है।) फिर भी **सौंदरनन्द** में द्वन्द्व है—उसमें केश-कर्त्तन कराते हुए नन्द रोता है; दो पाटों के बीच पिसता अनुभव करता है (7/16) और एक स्थल पर आकर वह घर लौटने के लिए सन्नद्ध भी दिखाई देता है (7/47)। इसमें कोई सन्देह नहीं कि **लहरों के राजहंस** का द्वन्द्व इस काव्य की ही देन है। अन्तर इतना है कि काव्य पार्थिव और अपार्थिव के द्वन्द्व के बीच अपार्थिव का समर्थन करता है। राकेश ने द्वन्द्व तो ग्रहण कर लिया, किन्तु आधुनिकता के आग्रह से उसे कोई दिशा देने से बचते रहे। आधुनिक दृष्टि से यह ठीक भी था क्योंकि आज का आदमी दुविधा की स्थिति में बँटा है और मात्र आध्यात्मिकता उसे अस्तित्व के किसी संकट का समाधान नहीं दे सकती।

आधुनिकता के आग्रह के कारण ही नन्द की ख्यात कथावस्तु नया स्वरूप ग्रहण कर लेती है। स्वयं मोहन राकेश के शब्दों में, 'नाटक का नन्द इतिहास के नन्द की भाँति आचरण नहीं करता।...मैंने इस इतिहास-कथा का उपयोग इसलिए किया, क्योंकि इस कथा के माध्यम से विशेष प्रकार की व्याख्या प्रस्तुत की जा सकती थी। बुद्ध और नन्द की पत्नी सुन्दरी के बीच होनेवाले संघर्ष का भी मैं उपयोग करना चाहता था। और यही दो बातें उस स्थिति का, जिसमें मैं अपने आपको आज पाता हूँ अर्थात् दो शक्तियों के बीच विभाजित होने की नियति का—प्रतीक बन गईं। वस्तुतः **अपने भीतर के संघर्ष को मैं चित्रित करना चाहता था। आप सार्त्र के 'लुसिफर एंड द लॉर्ड' में भी इसी बात को पाएँगे।** यदि आप यह कहें कि यह केवल ऐतिहासिक नाटक है तो मैं समझता हूँ आप मनुष्य के प्रति न्याय नहीं करेंगे।'[4] स्पष्ट है कि इतिहास के आधार को एक ओर लेखक ने अपनी आत्मपरकता (सब्जेक्टिविटी) और दूसरी ओर जीवन-दृष्टि को व्यक्त करने के लिए स्वीकार किया है।

यही कारण है कि **लहरों के राजहंस** में घटनाएँ नहीं, घटनाओं का समष्टिगत यौगिक प्रभाव मुख्य हो जाता है। उसमें घटना-क्रम अधिक नहीं है; क्रिया-व्यापार है जो पात्रों की निजता और आत्मिक स्वरूप को उद्घाटित करता है। सारे नाटक में घटना-क्रम इतना ही है—कामोत्सव की योजना, बुद्ध का भिक्षाटन के लिए उपस्थित होना और लौट जाना, नन्द का उनके पीछे-पीछे जाना, बुद्ध का उसे दीक्षित करना, नन्द का वन में जाकर व्याघ्र से जूझना और आनन्द के साथ घर लौटने पर सुन्दरी की उपेक्षा पर फिर घर से निकल जाना। घटना-क्रम पर आधारित किसी परम्परागत नाटक के लिए कथावस्तु का यह ढाँचा शायद स्वीकार्य नहीं होता; किन्तु मोहन राकेश ने साधारण घटनाओं से असाधारण अर्थवत्ता उभारी है। और सारा नाटक आत्म की खोज का एक सफल माध्यम बन गया है। राकेश के लिए यह सारी यात्रा अपनी इस खोज में अँधेरी गुफ़ा में प्रवेश करने के समान रही है।

और यह सारी यात्रा विडम्बना पर आधारित है। घटना के स्तर पर एक ओर कामोत्सव का आयोजन सामने आता है, दूसरी ओर देवी यशोधरा का प्रव्रज्या ग्रहण; एक ओर नन्द और सुन्दरी की काम-प्रवृत्ति और दूसरी ओर बुद्ध का भिक्षाटन; एक ओर सुन्दरी का केश-प्रसाधन और दूसरी ओर 'बुद्धं शरणं गच्छामि' का समवेत स्वर और नन्द के मुंडित केश। ये विरोधी स्थितियाँ नाटक की कथावस्तु में द्वन्द्व को पुष्ट करती हैं और कहीं-कहीं तो पात्रों के बीच से ही नाट्य स्थितियाँ और क्रिया-व्यापार उभरता है। कथानक में जो द्वन्द्व है, उससे कहीं अधिक आमने-सामनेवाली स्थिति पात्रों में दिखाई देती है। श्यामांग-श्वेतांग, नन्द-सुन्दरी, श्यामांग-सुन्दरी, अलका-सुन्दरी, बुद्ध-सुन्दरी विरोधी बिम्बों को उभारते हैं। नाटक का वास्तविक द्वन्द्व नन्द और सुन्दरी में नहीं, सुन्दरी और बुद्ध में है।[5] इस प्रकार **लहरों के राजहंस** क्रिया-व्यापार और चेतना के द्वन्द्व को लेकर चलता है। इसका आधार व्यक्ति का द्विधाग्रस्त होना है—दो तरह से सोचना और दोनों को एक साथ ठीक या ग़लत मानना। यह द्विधा आधुनिक साहित्य

में मानवीय नियति के रूप में स्वीकार की गई है और अन्वेषण की दिशा में इसका महत्त्वपूर्ण योगदान है। राकेश की सफलता इस बात में है कि मानसिक द्वन्द्व केवल स्वीकार या अस्वीकार का संवाद-भर बनकर नहीं, रह जाता। भावात्मक द्वन्द्व मानसिक ऊहापोह में नहीं, क्रिया में परिणत हो जाता है—मानसिक द्वन्द्व स्वयं क्रिया-व्यापार बन जाता है और क्रिया-व्यापार स्वयं बाहर की अपेक्षा अन्तःक्षेत्र की ओर मुड़ जाता है। इस प्रकार कथानक चरित्रांकन में परिणत हो जाता है। और पात्र कथानक के वाहक बन जाते हैं।

नाटक का क्रिया-व्यापार तीन अंगों में विभक्त है : इसका पहला चरण कामोत्सव के लिए की जानेवाली साज-सज्जा, श्यामांग की उलझन और किसी भी अतिथि के कामोत्सव में उपस्थित न होने से सम्बन्धित है। पहले अंक की इस कथावस्तु में क्रिया और भावना का, कथावस्तु और चरित्र में निहित द्वन्द्व का सही आभास मिलने लगता है। यहीं से परस्पर विरोधी स्थितियाँ और चरित्र लहरों के राजहंसों को तरंगायित करते दीखने लगते हैं। समरस स्थिति पहले चरण के अन्त में टूटने लगती है, किन्तु वास्तविक त्रासद स्थिति का बीज दूसरे चरण में पनपता है। श्यामांग जिस द्वन्द्व को पहले अंक में मुखर करता है दूसरे अंक में वह चरम पर ले जाता है। इसी अंक में बुद्ध का भिक्षाटन के लिए आना और अनदेखा चला जाना—एक ऐसी घटना के रूप में घटित होता है जो नन्द के लिए द्वन्द्व का मार्ग खोलती है। पहला अंक सुन्दरी और बुद्ध के द्वन्द्व पर आधारित है। दूसरे अंक में भी सुन्दरी मुख्य है; किन्तु तीसरे अंक में नन्द की अस्तित्ववादी चेतना मुख्य हो जाती है। इस अंक का सबसे बड़ा दोष यह है कि वह द्वन्द्व की अभिव्यक्ति परोक्ष रूप में करता है (जिसकी अलका की भाँति कल्पना-मात्र की जा सकती है : 'ओह कैसे अन्तर्द्वन्द्व के क्षण रहे होंगे कुमार के लिए !'), उससे द्वन्द्व एक लम्बे संवाद के रूप में विवरण बनकर रह जाता है। मंच के पीछे के इस द्वन्द्व के बावजूद आन्तरिक झंझावात की स्थिति के बीच नाटक चरम पर पहुँचता है।

सारा नाटक द्वन्द्व के दो स्तरों पर आगे बढ़ता है—द्वन्द्व का एक स्तर वह है जहाँ मुख्य पात्र स्वयं अपनी विरोधी मनःस्थितियों को झेलता है और दूसरा वह है जिसमें वह औरों के साथ विरोध में आता है। पर दोनों में इतना अन्तःसूत्र अवश्य विद्यमान है कि दूसरों के साथ का विरोध भी उसके अपने अन्तर्विरोध की ही बाह्य परिणति है। नाटककार इसकी सूचना नाटक के प्रारम्भ में ही श्यामांग-श्वेतांग तथा अलका जैसे गौण पात्रों के संवादों से देता है। वे आनेवाले संकट की अनायास ही सूचना देते हैं और बाद में जब सुन्दरी और फिर नन्द का प्रवेश होता है तो वे उन्हीं के छाया रूप बन जाते हैं। दूसरे अंक में आगे जो घटित होता है वह नन्द और सुन्दरी के द्वन्द्व की उलझन और खुले में लाकर रख देता है।

इसमें कोई सन्देह नहीं कि एक अंक की दूसरे में सही परिणति हुई है। पहले अंक के बाद दूसरे में द्वन्द्व की उतनी तीव्रता नहीं और तीसरे में वह बिखर ही जाता है; फिर भी प्रत्येक अंक एक सुविचारित उत्कर्ष पर पहुँचकर समाप्त होता है। यह ठीक है कि

तीसरे अंक पर पहुँचते-पहुँचते लेखक की पकड़ ढीली हो जाती है और ऐसा लगने लगता है कि लहरों पर तैरनेवाले हंस लहरों पर तैर नहीं, डोल मात्र रहे हैं। नाटक का अन्त करते हुए लेखक स्वयं इस स्थिति का शिकार रहा है। एक स्थान पर पहुँचकर वह नाटक को दो बार छोड़ चुका था। वह क्या करता ! सुन्दरी की हत्या कर सकता था—उसकी मृत्यु से समस्या का हल हो सकता था; पर बार-बार मन में प्रश्न उठता : 'सुन्दरी मर कैसे सकती है ?' तो पराजय ?—नहीं 'वह मर सकती है, पराजय स्वीकार नहीं कर सकती।'[6] एक अन्त यह भी हो सकता था जो **रात बीतने तक** में राकेश कर चुके थे अर्थात् नन्द का भिक्षु बनना ! एक और संभावित अन्त यह भी हो सकता था कि नन्द बुद्ध के मार्ग का पुरजोर विरोध करता और सुन्दरी के साथ रहकर सुखोपभोग करता ! पर नन्द दोनों नें से कोई भी मार्ग नहीं चुनता, त्रिशंकु बनकर रह जाता है। अस्तित्ववादी वैचारिक आग्रह और कुछ निजी दुविधाग्रस्त मानसिकता के कारण[7] मोहन राकेश नाटक को द्वन्द्व में ही छोड़ देते हैं जिसमें नन्द और सुन्दरी का 'आमने-सामने होना और एक-दूसरे तक अपनी बात न पहुँचा पाना, यही उसकी वास्तविकता' बन जाती है। इसीलिए नाटक का अन्त निगति से नहीं होता।

दुविधा एक स्तर पर बुरी बात नहीं है, वरन् विश्वसनीयता और सत्य का स्रोत है जो कभी-कभी चुनाव की ओर ले जा सकती है। आख़िर नाटक के बाहर भी तो एक ज़िन्दगी है। इससे भी बड़ी बात यह है कि व्यक्ति से उसके चिन्तन की प्रक्रिया को अलग नहीं किया जा सकता। नन्द नाटक में जिस प्रकार का व्यक्ति है वह तलाश में है, सत्य की ओर उन्मुख नहीं। उन्मुख शायद इसलिए भी नहीं क्योंकि सत्य उसके भौतिक जीवन के लिए शायद उपादेय तत्त्व भी नहीं। अनन्त जिजीविषा से युक्त नन्द बुद्ध के त्याग के लिए तैयार नहीं। उसने ज्ञान और विराग दोनों नहीं हैं, तृप्ति का सन्तोष भी नहीं। वह चेतना के उस निचले छोर पर स्थित है जहाँ से द्वन्द्व का दूसरा छोर पकड़ना उसके लिए सम्भव नहीं। इसीलिए भोक्ता से द्रष्टा नहीं बन पाता। फलतः वह अपने अस्तित्व का मूल्य राग और विराग की दोनों स्थितियों के अनुभव से चुकाता है। अस्तित्व का यह द्वन्द्व कौन नकार सकता है, पर द्वन्द्व की प्रसव-पीड़ा में ही जीना काफ़ी नहीं और उसमें जीना निर्णय की ओर नहीं ले जाता।

वस्तुतः विवशता का अनुभव मूल्यवान ही हो, ऐसा भी नहीं है। मूल्यवान तो वह अर्थ है जो उस अनुभव से व्यक्ति निकालता है। जहाँ तक नन्द का प्रश्न है, वह यंत्रणा के गर्भ से जीवन का कोई अर्थ नहीं निकाल पाता। इसका कारण यह है कि उसके जीवन में ऐसी उभयमुखता की जो स्थिति आती है, वह अर्थों की प्रक्रिया के बीच से पैदा नहीं हुई है। उसका गौतम बुद्ध की ओर जाना इस मामूली-सी बात पर आधारित है : 'उन्हें कैसा लगा होगा; उन्होंने मन में क्या सोचा होगा ?' यही नहीं, कामासक्ति के आरोप में वह चुपचाप सिर मुड़वा लेता है और सुन्दरी के द्वारा 'बाहर का व्यक्ति' कहलाए जाने पर घर से बाहर निकल आता है। नन्द घर से बाहर इसलिए नहीं निकलता कि उसे किसी सत्य की तलाश है, वरन् इसलिए कि सुन्दरी उसे कटे केशों में स्वीकार

नहीं करती, उसे ग़लत समझती है। वह जो कुछ करता है गहरे अन्तर्द्वन्द्व से उद्भूत नहीं है। लगता है जैसे एक शाश्वत समस्या उभरते-उभरते रह गई है। यहाँ लेखक अपने बचाव के लिए पार्थिव और अपार्थिव के द्वन्द्व की बात करने लगता है, पर वे जीवन्त क्षण नाटक में कहाँ हैं जहाँ दार्शनिक अर्थ उजागर होते हैं ?

पार्थिव और अपार्थिव एक-दूसरे से जुड़े तत्त्व हैं। उनकी पारस्परिक क्रिया को अस्वीकार कर उन्हें द्वन्द्व के स्तर पर एक-दूसरे के विरुद्ध खड़ा करना कोई दार्शनिक अर्थ नहीं रखता। वास्तव में यह दुहरी दृष्टि मनुष्य की आत्मिक और भौतिक ज़रूरतों के बीच खाई खड़ी करती है, उसे विभाजित करती है। एक ज़रूरत को दूसरी ज़रूरत के—एक आन्तरिक आवश्यकता को दूसरी आन्तरिक आवश्यकता के विरुद्ध खड़ा करना दोनों के पारस्परिक आश्रयत्व को झुठलाना है। राकेश स्त्री-पुरुष की सक्षमता को ही इस सन्दर्भ में काफ़ी समझते हैं; किन्तु सक्षमता दो व्यक्तियों की महत्त्वपूर्ण नहीं; महत्त्वपूर्ण केवल सत्य की सक्षमता ही है। और नन्द सत्य की सक्षमता अस्तित्ववादी ढंग से ही सिद्ध कर पाता है। इस रूप में द्वन्द्व सही और ग़लत के बीच नहीं, सही और सही के बीच है। किन्तु विचित्र बात यह है कि यह द्वन्द्व प्रबलता से नाटक में उभरता नहीं। इसका मुख्य कारण यह है कि द्वन्द्व को जन्म देनेवाली घटनाएँ नेपथ्य में घटती हैं और साथ ही दोनों का वैचारिक पक्ष उभरकर सामने नहीं आता।

वैसे भी नाटक की कथावस्तु इतनी संक्षिप्त है कि नाटककार को उसे पल्लवित करने के लिए अतिरिक्त श्रम करना पड़ा है। प्रायः पात्रों के एक युग्म के बाद दूसरा युग्म कथानक को आगे बढ़ाता है। उन्हीं से स्थितियाँ बनती और आगे बढ़ती हैं और इस प्रकार एक के बाद दूसरी स्थिति आ जुड़ती है और कथानक उनका योग बन जाता है। इसी प्रकार कुछ पाश्चात्य परम्परा के मोह में आकर भी नाटककार को प्रतीकों, अतिप्राकृत तत्त्वों, विद्रूप तथा खंडित चरित्रों की योजना से कथानक को संकुलता प्रदान करनी पड़ी। **लहरों के राजहंस** में राकेश ने प्रतीकों की व्याख्या की है जैसे लहरें जीवन के आवेग और विक्षोभ की द्योतक हैं। वे नन्द और सुन्दरी की परिस्थितियाँ हैं; कमल ताल संसार का, सुन्दरी भोग का, बुद्ध वैराग्य का, नन्द अन्तर्मन या अचेतन का, श्यामांग नन्द के मन की आकुलता का और **दर्पण सुन्दरी के रूप दर्प का** प्रतीक है। यह प्रतीक तत्त्व नाटक का प्रस्तुतीकरण करते हुए मोहन राकेश को अपने निर्देशक श्यामानन्द जालान को समझाना पड़ा—यह प्रतीकों के आरोपित होने का संकेत देता है। विलियम नाइट (प्रिंसिपल्स ऑफ शेक्सपीयरियन प्रोडक्शन) में लिखते हैं कि कविता में प्रतीक प्रभावशाली हो जाते हैं पर नाटक में अनर्थ पैदा करते हैं। फिर भी आलोच्य नाटक में कुछ प्रतीक ऐसे अवश्य हैं जिनके प्रतीकार्थ की खोज नहीं करनी पड़ती। प्रायः आलोचकों ने नाटक में प्रतीकों की भरमार को अच्छी दृष्टि से नहीं देखा है। राजहंस, छाया, सूखा सरोवर, मृग, दर्पण, श्यामांग आदि कई प्रतीकों के साथ सबसे बड़ी कठिनाई यह होती है कि उन पर अर्थ का आरोप कभी-कभी लेखक अपनी इच्छानुसार करने

लगता है। तब वे बोधगम्य नहीं रह जाते हैं। उदाहरण के लिए हमारे यहाँ हंस नीर-क्षीर विवेक के लिए ही ख्यात है। राकेश ने उन्हें आदिम मानव के उन्मुक्त इन्द्रिय-संवेद्य जीवन का प्रतीक माना है। राकेश के यह प्रतीक चेख़व के **सी गल** से मिला हो तो आश्चर्य नहीं। इसमें एक स्थान पर (अंक 2) त्रिगोरिन कहता है : (झील के पास रहनेवाली हंसिनी के समान युवती की ओर संकेत करता हुआ) 'लेकिन अचानक वहाँ एक आदमी आ जाता है और उसकी हत्या कर डालता है।' **लहरों के राजहंस** में भी हंसों पर आघात होता है। उसका दोष छाया को दिया जाता है। श्यामांग कमल ताल में डरावनी लम्बी छाया देखता है जिसका अति प्राकृत रूप में विस्तार से वर्णन किया गया है। उसको लगता है जैसे 'वह छाया सबको लील जाएगी, राजहंस के जोड़ों को भी।' वैसे यह छाया श्यामांग के अपने मन की हो सकती है। डॉ. सुरेश अवस्थी ने इसे बुद्ध की कल्पित छाया माना है।[8] राकेश ने स्वयं इसे परोक्ष की छाया कहा है जिसके चेतन रूप से श्यामांग बचना चाहता है।[9]

एक और प्रतीक के रूप में मृत्यु का प्रसंग नन्द के मुँह से उभरता है। मृत्यु का अहसास अपने में विलक्षण है। मृत्यु समय पर ही आती है, पर उसका अहसास जीवन में हर समय आदमी में बना रहता है। मृत्यु एक प्रकार का अनस्तित्व है जिसका बोध जीवन की निस्सारता की ओर ले जाता है। मृग की मृत्यु का दृश्य नन्द को अपने मृत्यु-भाव की याद दिलाता है। और यह उसके विराग के लिए पर्याप्त भूमि तैयार करता है। एक शव-यात्रा को देखकर बुद्ध भी तो विरक्त हुए थे। पराई मौत का प्रत्यक्षदर्शी होना एक मनःस्थिति का स्वीकार है। नन्द की वह मनःस्थिति नाटक पर बहुत दूर तक छाई रहती है। इसी मनःस्थिति में वह जंगल में जाकर व्याघ्र से युद्ध करता है। यह भी एक विचित्र अनुभव है कि मृग नन्द के बाण से नहीं, अपनी की क्लान्ति से मरता है। असल में वह मृग की क्लान्ति में अपने ही मन की क्लान्ति देखता है। नन्द की क्लान्ति वैचारिक उलझन की क्लान्ति है; खंडित व्यक्ति के अन्दर परस्पर जूझनेवाली प्रवृत्तियों से उत्पन्न क्लान्ति है। क्लान्ति केवल कार्य करने से ही नहीं होती—स्वयं अस्तित्व भी क्लान्तिजन्य हो सकता है। और अस्तित्व की पीड़ा से कतराने की स्थिति भी क्लान्ति लाती है। क्लान्ति रिक्तता का बोध कराती है और आशा का हनन कराती है। मृग की क्लान्तिजन्य मृत्यु के माध्यम से राकेश ने ऊब और थकान से घिरे आधुनिक मानव की नियति को व्यक्त किया है। वस्तुतः एक स्थूल प्रतीक के रूप में आहत मृग-शावक के प्रतीक का प्रयोग राकेश पहले ही **आषाढ़ का एक दिन** में कर चुके थे। वहाँ वह राजसत्ता का शिकार था; यहाँ वह नन्द की प्रवृत्ति और बुद्ध की निवृत्ति का शिकार बनता है। किन्तु क्लान्त मृग नन्द ही नहीं, सुन्दरी भी है। नन्द सुन्दरी की ओर संकेत करते कहता है : 'उस समय तुम कितनी सुन्दर लग रही थीं उस मृग की तरह...।' इसमें कोई सन्देह नहीं कि मृग का प्रतीक सुन्दरी और नन्द दोनों के लिए प्रयुक्त है और दुहरा अर्थ देता है।

नन्द का जंगल में जाकर व्याघ्र से युद्ध करना एक गढ़ी हुई-सी घटना लगती है।

पर उसकी सार्थकता उसमें निहित प्रतीकार्थ पर निर्भर करती है। वस्तुतः उसका व्याघ्र से लड़ना—उसकी आक्रामक प्रवृत्ति—उसकी आत्मविध्वंसक प्रवृत्ति की ही बाह्य परिणति है। दूसरों के प्रति आक्रामकता अपने प्रति आक्रामकता का ही प्रतिरूप है। प्रत्येक व्यक्ति के भीतर ऐसा बहुत कुछ (विसंगत) होता है जिससे वह जूझना चाहता है। नन्द भी अपने अन्दर की विसंगति से जूझता है। इस व्याघ्र-युद्ध से उसका दुविधाग्रस्त स्थिति में वरण न कर सकने का विवशताजन्य रोष व्यक्त हुआ है। साथ ही इसके द्वारा उसका आहत अहं अपनी निरंकुश, शूर-प्रतिमा को प्रश्रय देता है; एक ऐसी प्रतिमा जो वास्तविक न होकर मिथ्या मात्र है। एक ऐसा मुखौटा है जो उसके चेहरे को पूरी तरह ढक नहीं पाता। नन्द जंगल में जाता है—यह उसके अपने त्रास की प्रतिक्रिया है। मनुष्य का जगत् में होना त्रास की स्थिति है। यही त्रास बचाव के लिए आदमी को एकांत—जंगल—की ओर ले जाता है। इसमें आत्मरक्षा का भाव निहित है। किन्तु नन्द वन में जाकर व्याघ्र से लड़ने में आत्म-रक्षा और आत्म-विनाश इन दोनों अन्तर्विरोधों के बीच एक साथ जीता है। यही उसकी विसंगति है जो उसे अलगाव देती है। उसी अलगाव की विडम्बना है कि जो नन्द मुंडित-केश होते वक़्त निरीह जन्तु की तरह कुछ प्रतिकार नहीं करता, वही वन में जाकर व्याघ्र से लड़ता है। फिर भी व्याघ्र से लड़ना और उसकी मृत्यु-कामना में एक प्रकार का पारस्परिक सम्बन्ध भी होता है। अपना पौरुष दिखाने के लिए नन्द आक्रामक बनता है। सुन्दरी की दृष्टि में वह नीचे गिर गया है, इससे वह अपनी पौरुषहीनता का संकेत ग्रहण करता है और उसे स्वीकार करना एक प्रकार की मृत्यु मान लेता है। नन्द ऐसा स्वीकार करने के लिए उद्यत नहीं है। इसीलिए वह मृत्यु से लड़ता है—मृत्यु के मुँह में जीता है। मृत्यु की कामना और मृत्यु का आक्रामक मुक़ाबला पश्चिम के आधुनिक नाटकों का प्रिय विषय है। उनमें पात्र जीवन की त्रासदी का सामना शहीदों की तरह करते हैं; साथ ही जिजीविषा के क्षीण सूत्रों को भी टूटने नहीं देते। मोहन राकेश ने पाश्चात्य नाटकों की देखा-देखी इस प्रकार के प्रकरणों की अवतारणा की है !

दर्पण का प्रकरण भी अपने सम्पूर्ण प्रतीक तत्त्व के साथ नाटक की अर्थवत्ता का प्रसार करता है। नन्द एक स्थान पर अपने को सुन्दरी का दर्पण कहता है। सुन्दरी की आत्मसज्जा और नन्द का सामने दर्पण लेकर खड़ा रहना भी दोनों के सम्बन्धों को व्याख्यायित करता है। प्रसाधन की क्रिया के बीच यशोधरा के भिक्षुणी बन जाने के प्रसंग के साथ जब 'बुद्धं शरणं गच्छामि' का स्वर उभरता है तो दर्पण हिल जाता है और नन्द के अन्तर्जगत की हलचल के कारण उसकी साँसों से धुँधला हो जाता है। दर्पण का हिलना, आविल होना और फिर टूट जाना नन्द की आन्तरिक स्थिति का सही प्रतिनिधित्व करता है। किन्तु इससे भी अधिक, दर्पण सम्बन्धी यह प्रसंग एक दार्शनिक तथ्य की अभिव्यक्ति के लिए प्रयुक्त किया गया है। नन्द का—सही अर्थों में सुन्दरी का नन्द को अपना दर्पण कहना एक मनोवैज्ञानिक और दार्शनिक तथ्य है। विचार या परावर्तन (रिफलेक्शन) चेतना का महत्त्वपूर्ण कार्य है। जिस प्रकार दर्पण में उस वस्तु

के अतिरिक्त कुछ नहीं होता जो उसमें परावर्तित होती है, उसी प्रकार चेतना में परावर्तित वस्तु के अतिरिक्त उसकी कोई अन्तर्वस्तु नहीं होती। नन्द की चेतना भी कुछ इसी प्रकार की है—उसमें सुन्दरी अपने को देखती है और अपने को प्रतिम्बिबित पाती है। जब दर्पण टूटता है तो सुन्दरी को अपने ही रूप के आकर्षण की टूटन दिखाई देती है। वह नन्द को अपने ही आकर्षण से मापती रही—वही उसकी पहचान बना हुआ था।

नन्द सुन्दरी के लिए एक दर्पण है—वह अपने रूप, अहं, आकांक्षा को उसमें ही साकार पाती है। वह उसके अस्तित्व के लिए नितान्त आवश्यक है, क्योंकि वह उसके अहं को आश्रय देता है, तुष्ट करता है। अपनी सत्ता और भविता को वह उसी के माध्यम से व्यक्त करती है। दूसरी ओर दर्पण का जिस तरह अपना अलग से महत्त्व नहीं—वह माध्यम होने के नाते अपनी अपेक्षा देखनेवाले के लिए महत्त्वपूर्ण है—कुछ ऐसी ही स्थिति नन्द की भी है। उसका अपना अस्तित्व प्रामाणिक नहीं, परावर्तित मात्र है। किन्तु एक समय आता है जब नन्द की स्थिति बदल जाती है। उसके आन्तरिक उलझाव से गीली साँसों से दर्पण धुँधला हो जाता है। अब चेतना में भाव या विचार जुड़ जाता है तो वह अशुद्ध हो जाती है—उस पर भव की छाप पड़ जाती है और विषय और विषयी में द्वैत उत्पन्न हो जाता है। अस्मित का अहसास नन्द को सुन्दरी से अलग कर देता है जो एक त्रासद अनुभव है। इसीलिए नन्द काँप उठता है और दर्पण का टूटना उस सम्बन्ध का टूटना बन जाता है जो दोनों के बीच कभी विद्यमान था।

दर्पण के टूटते ही वह प्रवंचना भी समाप्त हो जाती है जिसका वह दर्पण अब तक प्रतीक था। मोहभंग की इस स्थिति में दर्पण के टुकड़े-टुकड़े नन्द के मन में विभाजन को प्रकट करते हैं। पर यह विभाजन केवल नन्द का ही नहीं, सुन्दरी के अहं का भी है। इसीलिए वह उससे कहती है : 'इसमें मुझे देखिए, कैसा लगता है मेरा टूटा प्रतिबिम्ब ?...यह दो भागों में बँटा चेहरा—खंडित मस्तक, खंडित सीमान्त !' अपरूपता का यह अहसास उसके अन्तर्मन की देन है। सार्त्र के **नौसिया** में भी रॉक्वेंटिन अपने को दर्पण के जितने नज़दीक लाता है उतना ही अधिक विरूप (सुर-रियलिस्टिक) स्थिति में पाता है। अपनी ही छवि अपरूप तब दीखने लगती है जब व्यक्ति अपने को सन्दर्भ से कटा पाता है। सुन्दरी सन्दर्भों से ही नहीं कटती, यथार्थ के प्रति भी जागरूक हो उठती है। यह स्थिति उसे प्रामाणिक जीवन की ओर ले जाती है क्योंकि वह स्वयं और जगत् को एक निर्धारित दृष्टि से देखने के अभ्यास से अपने को मुक्त करती है। इसीलिए नाटक में अलका तब टूटे दर्पण के बिखरे टुकड़ों को हटाते हुए दिखाई गई है। दरअसल, नन्द द्वारा सुन्दरी के सामने रखा दर्पण नियति की विडम्बना से वैयक्तिक अन्तर्बोध का कारण बनता है और विष्यगत तथा विषयीगत परावर्तन का सही अन्तर प्रस्तुत करता है। सुन्दरी इसी से विषयीगत सत्य से विषयगत सत्य की ओर उन्मुख होती है। इसके अतिरिक्त राकेश ने कई रंगमंचीय प्रतीक भी उभारे हैं। उदाहरण के लिए श्यामांग का क्रियाकलाप इसका सुन्दर उदाहरण है। रंगमंचीय उपकरणों में दीपाधार, मत्स्याकार आसन, झूला, दर्पण, चील, कठफोड़वा, सुरसुराती हवा आदि के प्रतीक

किसी-न-किसी रूप में नाटक को रंग-तत्त्व और अर्थ-गौरव प्रदान करते हैं।

प्रायः यह कहा जाता है कि **लहरों के राजहंस** में निहित यह प्रतीक-योजना कथा-वस्तु पर टिप्पणियाँ लगाने या उसकी रिक्तता की पूर्ति करने-जैसा प्रयास है। किन्तु सच्चाई यह है कि ये प्रतीकात्मक प्रसंग नाटक को वैचारिक पृष्ठभूमि प्रदान करते हैं। कुछ लोग श्यामांग सम्बन्धी प्रसंग को नाटक के 'सहज प्रवाह में बाधक' मानते हैं; किन्तु इस बात को अस्वीकार नहीं किया जा सकता कि नाटक की वस्तुयोजना की महत्त्वपूर्ण अर्थवत्ता इसी प्रकरण पर निर्भर करती है। नाटक के कृशकाय ढाँचे को पूर्णता देने में इसका योग अपरिहार्य है। डॉ. जगदीश शर्मा ने ठीक ही कहा है कि 'प्रथम अंक के उपरान्त श्यामांग की विक्षिप्तता कथागति की शिथिलता की क्षतिपूर्ति में महत्त्वपूर्ण योग देती है। प्रथम अंक में कथा-गति की क्षिप्रता पाठक की चेतना को बाँधे रखती है।...प्रथम अंक के उपरान्त एकाएक नाटकीय प्रभाव की क्षति घातक सिद्ध हो सकती थी, किन्तु राकेश ने श्यामांग की विक्षिप्तता के माध्यम से भावोद्वेग की सशक्त अभिव्यक्ति की है जो तनाव को बनाए रखती है।[10] इसमें कोई सन्देह नहीं कि श्यामांग का प्रसंग मूल कथावस्तु की ख़ाली दरारों को भरता है। पर वह सहसा मूल कथावस्तु से विलग भी नहीं, उसी का अंग है। यह कहना ठीक नहीं कि वह मूल कथावस्तु को खंडित करता है। नन्द की चेतन और अचेतन, आक्रामक और वशवर्ती प्रवृत्तियों में वह उसकी अचेतन और आक्रामक प्रवृत्ति का प्रतिनिधित्व करता है। नन्द श्यामांग को भी अपने में जीता है। वह नन्द के भीतर का ही 'दूसरा व्यक्ति' है। इसलिए उससे सम्बद्ध कथावस्तु नन्द की कथावस्तु का ही पूरक अंग है।

चरित्र-सृष्टि की दृष्टि से विचार करें तो ऐसा लगता है जैसे 'अस्तित्व और अनस्तित्व के बीच चेतना को प्रश्न-चिह्न' बनाकर पात्रों की प्रतिमाएँ गढ़ी गई हैं। नन्द की प्रतिमा इसका एक उदाहरण है; किन्तु सुन्दरी भी उससे मुक्त नहीं। फिर भी वह नन्द की तरह द्वन्द्व में नहीं जीती। उसकी दृष्टि एकपक्षीय है और वह केवल देह के स्तर पर जीती भोगमयी नारी का बोध कराती है। उसका भावात्मक विपक्षी स्वरूप किंचित् अलका में मुखर हुआ है जो बुद्ध की परोक्ष पक्षधरता निभाती है। सुन्दरी की वैचारिक चेतना पुष्ट नहीं है। इसीलिए उसकी चेतना में विकृत मानसिक प्रतिक्रियाएँ और पूर्वग्रह जुड़े हुए हैं। ये उसे आत्म-प्रवंचना के भाव से ग्रस्त करते हैं जो जीवन का एक निषेधात्मक मूल्य है। इन्द्रिय स्तर पर जीनेवाली यह नारी स्वकेन्द्रित होने के नाते सचेतन चुनाव नहीं कर पाती। वह अपने अस्तित्व में भौतिक जगत् से बद्ध है, पर यह बद्धता ज्ञान के स्तर पर नहीं, द्वेष के स्तर पर है—यह वह आदिम स्तर है जहाँ चेतना मनुष्य का निर्माण नहीं करती, केवल भोग ही उसके द्वारा होता है। इसके साथ ही उसमें अहं की प्रबलता है जिसके कारण वह सह-भू (बींग विद अदर्स) बनकर रहने की अपेक्षा व्यक्तिहीन इकाई बनकर रह जाती है। वह न बुद्ध और यशोधरा से जुड़ती है, न प्रजाजनों से; जुड़ती

है तो अहं और सुखवादी दृष्टि से। इसी से उसके व्यक्तित्व की मूलभूत सम्भावनाएँ समाप्त हो जाती हैं। अप्रामाणिक तथा असत् आस्था से प्रेरित होने के कारण नन्द को व्यक्ति नहीं वस्तु बना डालती है। वह सबको अपनी कामना का अस्त्र समझती है और लक्ष्य के संधान में असफल होने पर स्वयं अपनी प्रतिमा पर खीझती है; वस्तुओं को तोड़-फोड़ डालती है और टूटकर भी बाहर से ठोस बनी रहती है और कहती है : 'दुर्बलता कहीं थी तो मुझमें नहीं।'

सुन्दरी उस प्रकार के व्यक्तित्व से सम्पन्न है जिसे कीर्केगार्द ने 'ऐस्थेटिक' के अन्तर्गत रखा है। यह बच्चे के स्तर पर जीना है जिसमें वह अयथार्थ उड़ानें भरता है। सुन्दरी जीवन के प्रति अपूर्ण दृष्टि रखती है जिसे वह पूर्ण समझकर जीती है। इसीलिए जितना तीव्र उसका मोह है, उतनी ही बड़ी चोट भी खाती है। कीर्केगार्द के शब्दों में, 'जब सुख की पुकार बहुत आग्रहपूर्ण होती है तो हृदय में दुख भी बड़ी तीव्रता से जागता है।'[11] प्रायः इन्द्रियपरक भोगवादी जीवन की प्रक्रिया व्यक्ति को अध्यात्म की ऊँचाइयों तक ले जाती है; सुन्दरी वहाँ तक पहुँच नहीं पाती है, वैसा उपक्रम भी नहीं करती। केवल उसकी तरलता ठोस रूप धारण कर लेती है। वह स्थिति का अतिक्रमण नहीं कर पाती। सत्य उसे जितना पीड़ित करता है, उतना ही झूठे अर्थों में उसे ठोस और आत्मनिष्ठ भी बनाता है। इसीलिए नन्द के चले जाने पर या सामने होने पर वह न रोती है, न झगड़ती है, न उसे वापस चाहती है, और न सहानुभूति की माँग करती है। हताशा का शिकार होकर सिर्फ़ बर्फ की तरह जम-सी-जाती है। वह तीव्र वेदना भोगती है। वेदना भोगना चेतना का मार्ग है—इसी के द्वारा व्यक्ति आत्मचेता बनता है और आत्मचेता बनना अपनी सीमाओं को समझना है। सुन्दरी इसी क्रम से अपने अस्तित्व की मूक परिभाषा प्रस्तुत करती है।

नन्द की स्थिति कहीं सुन्दरी के समानान्तर और कहीं उससे भिन्न है। नन्द का द्वन्द्व सुन्दरी जैसा नहीं। सुन्दरी द्वन्द्व को पी जाती है। नन्द दो यथार्थों से जूझता है जबकि सुन्दरी का मौन यथार्थ और अयथार्थ के बीच है। दोनों में समानता यह है कि दोनों भावना में एक समय भोग्य और भोगी के स्तर पर सुखपूर्ण क्षणों के लिए जीते हैं। इस समानान्तर स्थिति के बावजूद भोग्या और भोगकर्त्ता के बीच द्वैत उभरकर आता है। नन्द के सामने एक ओर निजता का प्रश्न है, दूसरी ओर अपनी निजता को सुन्दरी की निजता में विलीन कर देने का। सुन्दरी उस पर हावी है, किन्तु साथ ही उसके प्रति समर्पित भी। वह ग्रहणशील है और साथ ही आक्रामक भी। उसकी दृष्टि में प्रेम एक आदान-प्रदान नहीं, बल्कि एक आपूर्ति मात्र है। वस्तुतः नन्द और सुन्दरी के बीच जो प्रेम-सम्बन्ध है उसमें सहजता नहीं है—एक खिंचाव मात्र है। अतः प्रेम के नाम पर ऐसी स्थिति आती है जो अस्तित्व का एक नया संकट पैदा करती है। नन्द आसक्ति और अनासक्ति के द्वन्द्व में पड़ जाता है। विडम्बना यह है कि 'आन्तरिक जीवन का भयंकर और त्रासद सूत्र यह है कि या तो न्यूनतम प्रेम के साथ परम सुख की स्थिति ही प्राप्त की जा सकती है और या फिर अधिकतम प्रेम के साथ न्यूनतम सुख। एक या दूसरे

के बीच कोई चुनाव करना नहीं होता है।'[12] सुप्रसिद्ध लेखक यूनामुनो ने यह भी कहा है कि प्रेम में दो व्यक्ति एक-दूसरे की तृप्ति के लिए अस्त्र होते हैं। फलतः दोनों प्रतिस्पर्धा में नृशंस और दासता से ग्रस्त होते हैं।'[13] प्रेम की सार-भावना में वह विध्वंसक तत्त्व विद्यमान रहता है जो शरीरों को मिलाकर भी आत्माओं को विच्छिन्न कर देता है। नन्द और सुन्दरी का प्रेम-सम्बन्ध भी ऐसी ही विषमता पैदा करता है : 'तुम समझती हो तुम्हीं वह केन्द्र हो जिसके वृत्त में मैं एक नक्षत्र की तरह घूमता हूँ, जिसका कहीं वृत्त नहीं, जिसकी कोई धुरी नहीं।' स्वतन्त्रता की चाह, यथास्थिति की ऊब, और लभ्य से अलभ्य की ओर संक्रमण नन्द के व्यक्तित्व के एक अंश को उससे अलग कर देता है। यहीं वह निस्सार अस्तित्व की विडम्बना भोगता हुआ पार्थिव और अपार्थिव के द्वन्द्व में फँस जाता है।

इस द्वन्द्व के कारण प्रायः यह कहा जाता रहा है कि नन्द इतना दुर्बल है कि निर्णय से पलायन करता है। ऐसा कहते हुए यह न भूलना चाहिए कि यह कहना मुश्किल है कि मनुष्य शरीर-मात्र है या यह कि वह केवल आत्मा है। न दोनों की एकता सिद्ध की जा सकती है और न दोनों का द्वैत। जैसा कि टेनिसन ने अपनी प्रसिद्ध कविता **द एशियंट सेज** में कहा है, 'कुछ भी सिद्ध या असिद्ध नहीं किया जा सकता है।'[14] आस्था के अभाव में मानव के पास द्वन्द्व के अतिरिक्त रह क्या गया है ? जिस संशय से नन्द ग्रस्त है वह हमारे वैज्ञानिक और तकनीकी युग की देन है। आज मानवता ऐसे कगार पर खड़ी है जिसके आगे राह नहीं है। एक राह अध्यात्म अवश्य है; पर उस पर बरसों से रूढ़ियों का मलवा पड़ा है। दो ध्रुवों के बीच खड़ा मानव आज पूछ रहा है : ततः किम् ? और उत्तर मिलता है : नान्य पंथा।

वस्तुतः 'नन्द एक प्रामाणिक जीवन जीना चाहता है, अपनी पूरी अस्मिता के साथ वह पूर्व-धारणाओं को स्वीकार नहीं करता—न बुद्ध को, न सुन्दरी को। वह प्रयोग करना चाहता है, निर्णय नहीं। वह अनुभव से निर्णय पर पहुँचना चाहता है, निर्णय से अनुभव की राह पर नहीं चलना चाहता है। जब कोई पराई धारणा स्वीकार बन जाती है तो सत्य की तलाश बन्द हो जाती है। उसके मन में प्रश्न-ही-प्रश्न है; उत्तर भी वह अपने ही चाहता है। बुद्ध के पास अब प्रश्न नहीं, उत्तर-ही-उत्तर हैं। नन्द बुद्ध के उत्तरों को अपना उत्तर नहीं मानना चाहता। दूसरे की प्रेरणा से कुछ बन जाना महत्त्वपूर्ण नहीं है, महत्त्वपूर्ण है होना। जीवन का परम सत्य सहज होता है, चेष्टा से प्रेरित नहीं। इसीलिए एक वास्तविक अहसास के लिए नन्द बेचैन है, उभयमुखता से ग्रस्त है। उभयमुखता की यह विसंगति घृणा और प्रेम की समकालिक भावना की देन है एक को स्वीकार और दूसरे को नकारने में भय, संत्रास, अधूरेपन और विवशता का अहसास आड़े आता है। फलतः नन्द चुनाव नहीं करता। चुनाव कर भी लेता तो भी किसी एक पक्ष की स्वीकृति उसके लिए संघातक सिद्ध हुए बिना न रहती ! मनुष्य हर हालत में विसंगति से ग्रस्त होने के लिए बाध्य है। यह एक ऐसा यथार्थ है जिसको झुठलाना सरल नहीं। आधुनिक जीवन में मानव का इतना अवमूल्यन हो चुका है कि नन्द जैसा है वह जीवन

का विकृत प्रतीक बन गया है। इस रूप में नन्द कितना त्रासद और साथ ही कितना हास्यास्पद लगता है ! जब वह मुंडित-केश होकर भिक्षु आनन्द के साथ घर आकर अपनी कैफियत देता है तो उसके शब्द उसका मजाक उड़ाने लगते हैं। चौराहे पर खड़े नंगे व्यक्ति के रूप में वह करुणा नहीं जगा पाता—करुणा जगाती है तो सुन्दरी। या फिर करुणा जगाता है श्यामांग। सुन्दरी की कारुणिक स्थिति उसके अहं की परतों के नीचे दबी चेतना है। वस्तुतः राकेश के इस नाटक में इस परत में छिपे द्वन्द्व ही पात्र को प्रतीक बना देते हैं चाहे वह सुन्दरी हो, श्यामांग या स्वयं नन्द।

श्यामांग जो व्यक्ति नहीं, अहसास बनकर नाटक में आया है। उसके चरित्र के बारे में कई प्रश्न उठाए गए हैं, जैसे—श्यामांग पात्र नहीं, प्रतीक बनकर रह गया है; वह ख़ाली जगह को भरता है; नाटक की सीम्ति सामग्री को विस्तार देने के लिए उसका उपयोग हुआ है, या उसका प्रलाप नाटकीय स्थापत्य को शक्ति नहीं देता। इसमें कोई सन्देह नहीं श्यामांग अपने ढंग का पात्र है और उसकी सृष्टि में राकेश को अपेक्षित कौशल दिखाना पड़ा है। पहले अंक में उसे हटाना चाहा तो हट न सका;[15] दूसरे अंक में आकर नेपथ्य के स्वरों में वह नाटक पर छा गया और अपनी चेतना में नन्द से जैसे एकाकार हो गया। पुराने संस्करण के तीसरे अंक में वह 'मुझे एक किरण ला दो, बस एक किरण, केवल एक किरण' कहता हुआ उपस्थित दिखाई देता है। पर संशोधित संस्करण में उसे नाटककार ने तीसरे अंक से निकाल बाहर ही कर दिया, क्योंकि तब उसकी आवश्यकता ही नहीं रह जाती। फिर भी नाटक के कथ्य को मुखर करने के लिए उसका प्रयोग वहाँ भी संभव था।

श्यामांग के चरित्र की सार्थकता इस बात में है कि नन्द को उसकी 'बातों में अपने अन्तर्मन की छाया झलकती दिखाई देती है।' श्यामांग नन्द के अन्दर का वह दूसरा व्यक्ति है जो सहज नहीं है। व्यक्ति के अन्दर का वह व्यक्ति द्वन्द्व की एक स्वाभाविक स्थिति को प्रकट करता है जिससे वह अपने सहवर्ती को व्याख्यायित करता है। इसमें कोई सन्देह नहीं कि श्यामांग प्रतीक पात्र है। वह नन्द के खंड व्यक्तित्व का प्रतिनिधि मात्र है। पाश्चात्य नाटकों में इस प्रकार के प्रयोग होते रहे हैं और ऐसे पात्रों की विश्वसनीयता पर कभी इतना बल नहीं दिया जाता जितना प्रतीकात्मकता पर। वह जिस अतिरंजित रूप में चित्रित है, वह विसंगति को उभारने के लिए अनिवार्य है। इसीलिए उसका उन्मादी, अति प्राकृत, अयथार्थ होना सार्थक है। वह नन्द की काल्पनिक चेतना को जीता है जिसमें वह अयथार्थ को ही यथार्थ मान लेता है। फलतः राजहंसों की छाया द्वारा आहत होना, चील के मँडराने आदि को वह यथार्थ मानता है और उन्हें अपनी मनःस्थिति के अनुरूप अर्थ देता है। एकाकीपन, भय और त्रास से आवृत्त उसकी ज़िन्दगी विसंगति से परिपूर्ण है जिसे ज्वर के माध्यम से व्यक्त किया गया है। उसके प्रलाप में प्रतिभा और चिन्तन में स्फुलिंग हैं; पर वह बौद्धिक चेतना का कोई बड़ा संकेत नहीं देता। उसकी बौद्धिकता नाटक की अर्थवत्ता के लिए एक बहुत बड़ी शक्ति सिद्ध हो सकती थी; पर जैसे राकेश के और पात्र बौने हैं, श्यामांग भी बौना है। सनकीपन

में वह जो बौद्धिक उत्कर्ष नाटक को दे सकता था, वह रह गया। फिर भी वह नाटक का प्रमुख स्वर बनता है और यथेष्ट वातावरण की सृष्टि करता है। उसका चरित्र नाटकीय है। यह कहना ग़लत होगा कि वह नाटकीय प्रभाव की क्षति करता है। सच्चाई यह है कि उसके संवाद नाटक को अर्थ और वातावरण प्रदान करते हैं। यह कहना भी ग़लत होगा कि अलका का 'अत्यन्त प्रयोजनशील और स्पष्ट चरित्रांकन' उसी के कारण 'विकसित नहीं हो पाता।' वस्तुतः अलका सूत्रवाही पात्र है। वह प्रारम्भ में श्यामांग का ही पूरक चरित्र है जो बुद्धि की चेतना को प्रतिबिम्बित करता है। राकेश के शब्दों में, 'आरम्भ में अलका और श्यामांग का प्रेम स्त्री और पुरुष के बीच की भावना के दूसरे स्तर को प्रस्तुत करता है—नन्द और सुन्दरी के आवेशपूर्ण और वासनात्मक प्रेम से अलग।' बाद में श्यामांग और अलका का सम्बन्ध पूर्णता को नहीं पहुँचता। श्यामांग से अलग नाटक के अन्त में अलका नन्द और सुन्दरी दोनों के त्रास को झेलती दिखाई देती है।

इसी प्रकार एक शंका बुद्ध के सम्बन्ध में उठाई गई है। नाटक का नायक नन्द है, पर जो अहसास नाटक पर छाया हुआ है, वह स्पष्टतः गौतम बुद्ध का है। इसका मुख्य कारण यह है कि वास्तविक संघर्ष बुद्ध और सुन्दरी के बीच हैं, नन्द तो उसका शिकार मात्र है। नाटककार नन्द की अपेक्षा बुद्ध का प्रयोग करना चाहता था : 'इस नाटक में मैं नन्द के नाम का उपयोग नहीं करना चाहता था। मैं बुद्ध के नाम का उपयोग करना चाहता था, हालाँकि नाटक में बुद्ध नाम का कोई पात्र नहीं। अतः मैंने इस नाटक के लिए बुद्ध का नाम और मनुष्य की उस विशिष्ट तलाश का उपयोग किया।'[16] इसीलिए उसे पात्र न सही एक वैचारिक छाया के रूप में बुद्ध को महत्त्व देना पड़ा। पर नन्द के स्थान पर यदि बुद्ध को अपेक्षाकृत अधिक महत्त्व मिलता तो नाटक कुछ और ही होता ! इसीलिए, उन्हें मंच पर नन्द जैसा महत्त्व मिलना चाहिए था—यह तर्क कोई अर्थ नहीं रखता नाटक बुद्ध का नहीं, नन्द का है। एक भावात्मक उपलब्धि के लिए बुद्ध का मूक आभास ही पर्याप्त है। सुन्दरी बुद्ध पर जो प्रहार करती है, उससे उनकी परम्परागत प्रतिमा खंडित होती है, यह सोचना भी व्यर्थ है। 'नारी का अपकर्षण पुरुष को गौतम बुद्ध बना देता है'—यह कटाक्ष मात्र है, नाटकीय अर्थ नहीं। यह कथन बुद्ध पर चरितार्थ होता ही नहीं, क्योंकि नाटक का घटनाक्रम स्वयं इसका खंडन करता है और इस धारणा को लेकर सुन्दरी स्वयं खंडित हो जाती है। इसी प्रकार बुद्ध नन्द पर प्रव्रज्या जिस रूप में थोपते हैं, उसे भी अमानवीय कहना ठीक नहीं। जैसा कि **सौंदरनन्द** में कहा गया है, इस कृत्य में उस भिषग की सदाशयता निहित है जो बलपूर्वक रोगी को कड़वी दवा पिला देता है।

कुल मिलाकर रूपबन्ध की आंतरिक बुनावट तनावों और द्वन्द्वों पर आधारित है। कथानक और पात्रों का निर्धारण इसी को लेकर हुआ। सभी पात्र कथानक के निर्माण में एक निश्चित प्रयोजन निभाते हैं। वे घटना, प्रतीक, द्वन्द्व, मानसिक वृत्ति और रंगतत्त्व के साथ जुड़े उपादान सभी कुछ हैं। इसमें नन्द और सुन्दरी के तनावों के बीच

नाटककार की एक दृष्टि का भी पता चलता है। यह भी एक उपलब्धि है कि 'दो तरह से सोचने के बावजूद' राकेश के राजहंस 'हमारे मन में कई प्रश्न उठाते हैं', भले ही अंधी गली में पहुँचाते हैं पर जीवन जीना चाहते हैं—छटपटाते हुए सही ! एक बात और, नन्द आसक्ति और विरक्ति, अस्तित्व और अनस्तित्व के बीच जिस तरह भी जीता है, वह अधूरेपन का अहसास लिए हुए है

यह भी कहा जाता रहा है कि नन्द अस्पष्ट पात्र है और उसका द्वन्द्व अनाकर्षक है—उसका बौद्धिक अन्तर्द्वंद्व उभरता नहीं और भावात्मक अन्तर्द्वंद्व हृदय को छूता नहीं। इस बात में भी कुछ सत्य दिखाई देता है कि मुंडितकेश नन्द को देखकर सुन्दरी के मन में जो प्रतिक्रिया होती है, वह गर्विता स्त्री का मामूली घरेलू झगड़ा बन गया है—वह किसी गहन आध्यात्मिक संकट की द्योतक बन सकता था। इस प्रकार की स्थिति के गहन दार्शनिक पैठ न होने के कारण नाटक अन्तिम अंक में नन्द और सुन्दरी के बीच चिन्तन की सम्भावनाएँ समाप्त होने से वह अपना प्रभाव नहीं छोड़ता। उसका सारा क्रिया-कलाप और स्थापत्य नाटक की भाव-सघनता समाप्त कर देता है।

लहरों के राजहंस की कुछ सीमाएँ हैं। ये सीमाएँ कथानक की हैं या लेखक की दृष्टि की, या मंचीय विधान की। जहाँ तक लेखक का प्रश्न है, उसने पहले ही कह दिया है : 'मैं केवल इतना जानता हूँ कि कुछ ऐसा है जो इस नाटक में होना चाहिए था और नहीं है। यह क्या है, यह मेरे मन में स्पष्ट नहीं।'[17] फिर भी कथ्य के स्तर पर यह राकेश का सबसे सफल नाटक है। वैचारिक संकेतों की ऐसी परिपूर्णता उनके किसी नाटक में उपलब्ध नहीं। इसके अतिरिक्त कवित्व का हलका-सा स्पर्श इस नाटक को संवेदना के स्तर तक उठाता है। काव्यात्मक दार्शनिकता से परिपूर्ण यह नाटक आधुनिक भाव-बोध को भी पूरी सामर्थ्य के साथ अभिव्यक्त करता है। हो सकता है भविष्य में आधुनिक मानव को उसकी नियति से मुक्ति दिलाने के लिए कोई और समर्थ चिन्तन उभरे। किन्तु अभी जब तक नन्द के द्वन्द्व का निदान नहीं होता, तब तक **लहरों के राजहंस** श्रेष्ठ नाटक माना जाएगा।

घर की तलाश

उस सारी रात राकेशजी एक ही बात बोलते रहे—मुझे घर चाहिए अन्ना, घर ! मुझे ज़िन्दगी में और सब-कुछ मिला है, सिर्फ़ एक घर ही नहीं मिला। मैं कहाँ-कहाँ इसके लिए नहीं भटका ? क्या-क्या इसके लिए नहीं किया ? लेकिन पता नहीं क्यों 'घर' नाम की चीज़ मुझसे हमेशा रुसवा रही। दो बार मैंने इसे पाने का अपने में विश्वास भरा और दोनों ही बार मुझे ख़ुद ही उससे भाग जाना पड़ा।...क्या तुम मेरे लिए एक ऐसा घर बना सकोगी जो मेरे, सिर्फ़ मेरे अनुकूल ही हो—मैं बहुत ही दुखी आदमी हूँ अन्ना—एक बहुत ही थका हुआ आदमी हूँ। मैं चाहता हूँ कि मुझे अब तुम सँभाल लो, मुझे और मेरे घर को...।

—अनीता राकेश : चन्द सतरें और, पृ. 75

घर की यह तलाश मोहन राकेश के सभी नाटकों में है, पर **आधे अधूरे** (1969) पूरी तरह एक घरेलू नाटक है। वैसे आनुषंगिक रूप से यह नाटक अर्थ की कई छायाएँ उजागर करता है—नारी की मुक्ति-भावना, विघटनशील जीवन-मूल्य, वैवाहिक सम्बन्धों की विडम्बना और पुरुष का अधूरापन, किन्तु नाटक के केन्द्र में दो ही बातें मुख्य हैं—घर और घर में रहनेवाले लोगों का अधूरापन। 'घर' है तो ऐसा जिसमें बिखराव और गर्द है। उसमें किसी चीज़ का कोई ठौर-ठिकाना नहीं और उनमें कोई संगति नहीं। संगति इसलिए नहीं कि संगति बिठानेवाला कोई नहीं है। घर है पर 'घर में कोई नहीं।' यदि कोई एक घर में होता भी है तो दूसरा नहीं होता और जब सब होते भी हैं तो उनमें संवाद नहीं होता : 'स्कूल से आई तो घर पर कोई नहीं था। और अब आई हूँ तो तुम भी हो, डैडी भी हैं, बिन्नी भी है—पर सब लोग ऐसे चुप हैं जैसे...।' सबको इस घर से शिकायत है। चुप्पी इस घर में रहने की मजबूरी है और लड़ना उससे बाहर निकलने का बहाना। इसे कोई अपनाने को तैयार नहीं। लड़का सवाल करता है : 'इसे घर कहती हो तुम ?' बड़ी लड़की को जब सावित्री यह कहती है कि 'तेरा अपना घर है' तो वह चौंक उठती है : 'मेरा अपना घर ?' गृहस्वामी के सामने एक चुनौती है कि घर उसका है। इसीलिए वह कहता है : 'तो मेरा घर नहीं है। यह कह दो नहीं है ?'

उसको बार-बार यह अहसास दिलाया गया है कि वह एक कीड़ा है जो उस घर को खा गया है। अकेली सावित्री उसे किसी तरह थामे हुए है। वह भी जानती है कि 'मेरे करने से इस घर का जो कुछ हो सकता था, इस घर का हो चुका आज तक। मेरी तरफ़ से अब यह अन्त है उसका, निश्चित अन्त।' इस अन्त को लानेवाली एक 'हवा' है जो इस घर पर हावी है। यही हवा बाहर की गर्द इस घर में लेकर आती है। 'अधिकार, रुतबा, इज़्ज़त—यही सब बाहर के लोगों से मिल सकता है इस घर को !' आपसी प्रेम, सद्भाव और आदर यह घर स्वयं अपने को नहीं दे पाता। ऊब, घुटन और तनाव के अभिशप्त जीवन का प्रतीक यह घर नरक-जैसा अनुभव देता है।

नाटक के प्रारम्भ में ही घर के घर न होने का आभास होने लगता है। इसमें गृहस्वामिनी का जो रूप सामने आता है, वह हर वक़्त आमादा रहनेवाली नारी का है। दूसरी ओर गृहस्वामी एक मूक निरपेक्ष पशु-तुल्य-अस्तित्व झेलता दिखाई देता है जिस पर एक चुप्पी थोप दी गई है। अपने ही अकर्मण्य जीवन की परिणति को वह निरीह होकर पत्नी की लताड़ों के बीच भोगना दिखाई देता है। वह नकारा, दब्बू और निठल्ला है और भावना के स्तर पर निर्जीव और निर्वीर्य, किन्तु अन्दर से बहुत गहराई में कहीं तल्ख़ी और तुर्शी को छिपाए हुए है। और ऐसे घर में सन्तान भी माता-पिता की कार्बन-कॉपी है। बड़ी लड़की बिन्नी माँ के पूर्व-प्रेमी मनोज के साथ भागकर इस अहसास को सार्थक करती है कि वह अपने मायके से एक 'हवा लेकर गई है'। दूसरी लड़की किन्नी छोटी उम्र में काम-सम्बन्धी चर्चा में जो रस लेती है, वह भी उसकी माँ के व्यक्तित्व का हो एक अंग लगता है। इसी प्रकार लड़का अपने बाप का ही प्रतिरूप जीवन बिताता है। सब निष्क्रिय, परजोवी, कामुक अस्तित्व को दुहराने लगते हैं। घर की यह हालत बनाने में किसका हाथ है, यह स्वीकार करने के लिए कोई तैयार नहीं। सब एक-दूसरे पर दोषारोपण करके रह जाते हैं। सावित्री सारा दोष महेन्द्रनाथ पर थोप देती है। महेन्द्रनाथ अपने ऊपर लगाए लांछनों से बिद्ध हुआ एक आहत संवेदना में घुलता दीखता है : '(लड़के की ओर इशारा करके) यह आज तक बेकार क्यों घूम रहा है ? मेरी वजह से। (बड़ी लड़की की तरफ़ इशारा करके) यह बिना बताए एक रात घर से क्यों चली गई थी ? मेरी वजह से। (स्त्री के बिल्कुल सामने आकर) और तुम भी...।' वह अपने को रबर-स्टैम्प, घरघुसना, आरामतलब, कीड़ा न जाने क्या-क्या महसूस करता है। फिर भी वह और उसी की तरह और लोग उस घर से चिपके हैं क्योंकि उनकी 'हड्डियों में जंग लगा है' और वे अन्दर से आरामतलब हैं। घर किसी-न-किसी रूप में उन्हें सुविधा और सुरक्षा देता है। इसलिए इस अहसास के बावजूद कि 'कोई मुझे वजह बता सकता है, एक भी वजह, कि क्यों मुझे रहना चाहिए इसमें' वे वजह जानते हुए भी उस घर में रहते हैं। सुविधा के मूल्य पर उनके लिए विद्रोह अकल्पनीय बन जाता है। फलतः यथास्थिति कहीं नहीं टूटती।

जैसी मनःस्थिति में राकेश ने नाटक लिखा है उसमें यथास्थिति को तोड़ना सम्भव भी नहीं था। जैसा कि उनकी पत्नी के संस्मरणों से स्पष्ट है कि घर को बनाने और

सँभालने की ज़िम्मेदारी वे दूसरे की समझते थे—'मैं चाहता हूँ, मुझे अब तुम सँभाल लो—मुझे और मेरे घर को।' इसमें समर्पण का भाव है; किन्तु अपने को दूसरे की कृपा पर छोड़ देने और दूसरे के मूल्य पर सुविधा-भोग का दर्शन भी छिपा है। इस दर्शन को लेकर किसी को 'घर' मिल जाए यह दूर की सम्भावना है। वस्तुतः राकेश की घर की मायावी तलाश पुरुष की आकांक्षा को व्यक्त करती है जिसमें वह स्त्री को अपने लिए उपयोगी होते देखना चाहता है। ऐसे में पुरुष की स्थिति प्राथमिक हो जाती है और स्त्री की गौण। राकेश अपने जीवन में भी स्त्री को दूसरा भी नहीं, तीसरा स्थान देने की बात करते हैं। तीसरे स्थानवाली नारी पुरुष को 'घर' कहाँ दे सकती है ? पुरुष उसके साथ घर बसा भी ले तो वह नारी का घर नहीं होगा। घर एक भावात्मक रचना है—एक शरण स्थल। पर स्त्री-पुरुष घर की आकांक्षा लेकर भी किस प्रकार घर नहीं बसा पाते—इस तथ्य को राकेश ने अपने नाटक और जीवन में सही रूप में समझा है। राकेश की तरह महेन्द्रनाथ भी इस सत्य को भूल गया कि घर की ज़रूरत उसी को नहीं, सावित्री को भी है। केवल अपने अनुकूल घर चाहने का अर्थ है कि वह दूसरे के अनुकूल न हो। यही कारण है कि सावित्री और महेन्द्रनाथ के हाथ वह घर लगा जो उन्हें उनका नहीं लगता। मनहूसियत के इसी तत्त्व को वे खोज नहीं पाते। 'तुम बता सकती हो ममा, क्या चीज़ है वह ? और कहाँ है वह ? उस घर के खिड़कियों-दरवाज़ों में ? छत में ? दीवारों में ? तुममें ? डैडी में ? किन्नी में, अशोक में ? कहाँ है वह मनहूस चीज़ ?' नाटक के मध्य भाग में वह मनहूसियत अपने को व्याख्यायित करती है और अन्त में दायित्व-निर्धारण के प्रयास में परस्पर दोषारोपण ही हाथ लगता है—घर ज्यों का त्यों रह जाता है। महेन्द्र और सावित्री दो समानान्तर रेखाओं की तरह कभी नहीं मिल पाते। इसका कारण है—दोनों की टूटन। सावित्री के टूटने का कारण है महेन्द्रनाथ और महेन्द्रनाथ के टूटने का कारण सावित्री। दोनों के जीवन दो अँधेरे बन्द कमरे हैं, जिनसे वे बाहर निकलना चाहकर भी निकल नहीं पाते। घर तब केवल लड़ने, कुढ़ने और झींकने के लिए रह जाता है !

यह नाटक आज के शहरी क्षेत्र के मध्यमवर्गीय पारिवारिक ज़िन्दगी की विसंगति को उभारता है। शहरों में भीड़ के फैलाव के साथ आदमी सिकुड़ता जा रहा है और जितना वह सिकुड़ता जाता है, उतनी ही उसकी यह ज़रूरत बढ़ती जाती है कि कोई उसे अपना समझे, कोई तो उसकी अलग से पहचान करे, कहीं तो वह इंसानियत का रिश्ता क़ायम करे; किन्तु रिश्ता जुड़ने के बजाय आदमी द्वीप की तरह अलगाव से घिर जाता है—फिर कहाँ रिश्ते टूट जाते हैं और कहाँ से अन्तरंगता शुरू होती है—इसका अहसास ही नहीं होता। ऐसा लगता है कि जैसे हर चीज़ दुहराई जा रही है; जो भोगा जा चुका है, उसी को भोग रहे हैं। और फिर इसी ऊब, घुटन और तनाव में छोटी-छोटी ख़ुशियाँ महत्त्वपूर्ण हो जाती हैं। महेन्द्रनाथ जुनेजा के पास, सिंघानिया सावित्री के पास, सावित्री

जगमोहन के पास इन्हीं की खोज करती है। लड़का लड़कियों की तस्वीरें बटोरने और छोटी लड़की कच्ची उम्र में ही सेक्स की बातें करने में ऐसी ही ख़ुशी पाती है। साधनों और सीमाओं से परे सावित्री और उसका परिवार एक सुविधापरक रँगीली ज़िन्दगी जीना चाहता है। किन्तु असफलता और निराशा के बीच सारी ख्वाहिशें जमकर ठंडी हो जाती हैं।

सारी कथावस्तु महेन्द्रनाथ और सावित्री के टूटते वैवाहिक सम्बन्धों पर आधारित है। साथ रहने के बावजूद उनके बीच उपेक्षा, घृणा, खीझ और उपालम्भों की ऐसी दीवारें हैं जो उन्हें परस्पर जुड़ने नहीं देतीं। सम्भवतः जुड़ानेवाले तन्तु एकदम टूटे नहीं हैं; किन्तु दोनों के बीच आपसी समझदारी और सम्प्रेषण का अभाव आड़े आता है। यदि उनका कोई मानवीय रिश्ता उभरता है तो रात-दिन एक-दूसरे की जान नोचने का। और इसमें भी एक कोण से महेन्द्रनाथ ज़िम्मेदार लगता है और दूसरे कोण से सावित्री। सावित्री महेन्द्रनाथ को अधूरा आदमी मानकर उनसे खिन्न रहती है और महेन्द्रनाथ सावित्री के व्यवहार को देखकर मन मसोसकर रह जाता है। स्त्री-पुरुष को अर्थपूर्ण पारस्परिक सम्बन्ध में बाँधनेवाला एक तत्त्व प्रेम या सेक्स माना जा सकता है। प्रेम या सेक्स की एक शर्त पूर्ण आत्मसमर्पण या आत्मविस्मरण भी है। सावित्री में उसका अभाव है। वह ठीक अपने नाम के विरुद्ध पति से विमुख है और अनेक पुरुषों से सम्पर्क बनाती है। वह सोचती है कि महेन्द्रनाथ के स्थान पर यदि दूसरा पुरुष होता तो शायद जीवन में सब-कुछ ठीक चलता। इसीलिए वह एक के बाद दूसरे पुरुष की खोज करती है। यह खोज पहले मन के स्तर पर शुरू होती है, फिर शरीर के स्तर पर उतर आती है। हर तलाश पर लगता है जैसे आकाश बाँहों में भर लिया है; पर हर बार हाथों में आकर जैसे वह टूटकर खंड-खंड हो जाता है। यह अनुभव उसे दुनियादार बना देता है। जिन-जिन पुरुषों की ओर वह मुखातिब होती है उनसे वह एक ओर सेक्स की पूर्ति करती है, दूसरी ओर अतिरिक्त अनुग्रह जुटाने की व्यावहारिक बुद्धि भी दिखाती है। यद्यपि चारों पुरुषों से उसकी अपेक्षाएँ अलग-अलग हैं, फिर भी सेक्स और स्वार्थ-साधन की प्रवृत्ति उसमें मुख्य दिखाई देती है। इस रूप में वह सह-अपराधी और शिकार दोनों है। परिस्थितियाँ उत्तरदायी हो सकती हैं; पर उसका दोष इतना तो है ही कि वह एक 'रोल' स्वीकार कर लेती है और उस पर चलकर ज़िन्दगी गुज़ार देती है। जब वह जागरूक निर्णय लेना भी चाहती है तो बहुत विलम्ब से। इसीलिए वह उसे कहीं नहीं ले जाता।

सावित्री उस मध्यवित्तीय मानसिकता की शिकार है जो धन को जीवन में अतिरिक्त महत्त्व देती है। यही वृत्ति उसे विभिन्न पुरुषों से जोड़ती है और इसी के वशीभूत होकर वह अपनी देह का अस्त्र की तरह प्रयोग करती है। कुछ वैयक्तिक लक्ष्यों के लिए वह अपने को व्यक्ति से वस्तु बना डालती है। इसी से वह अपने को छलती है—जिस पूरेपन की तलाश उसे अपने अन्दर करनी चाहिए थी, उसकी तलाश वह अपने से बाहर औरों में करती है। इस तलाश में उसे वे सब पुरुष अधूरे दिखाई देते हैं, जिनके सम्पर्क में वह आती है। उनके माध्यम से वह उन रोमानी क्षणों से बँध जाती है जो उसके यथार्थ जीवन के सत्यों को झुठलाते हैं और उन्हें आकर्षक और स्वप्निल बनाते हैं। वह दूसरों

के इरादों से शोषित होना इस मोहान्धता में स्वीकार कर लेती है कि वह स्वयं उनका शोषण करने का इरादा रखती है। उस उपलब्धि के लिए वह क्षण को स्वीकार कर लेती है और क्षण के बीतते ही नए क्षण की तलाश शुरू करती है। यह भावना से विहीन सम्बन्ध-सूत्र शर्तों के साथ जुड़ा होने के कारण आन्तरिकता का विकास नहीं होने देता। इसके अभाव में पुरुषों के साथ उसका सम्बन्ध वस्तुगत या समूहगत-सा बनकर रह जाता है। यह एक विलक्षण बात है कि पुरुषों के साथ अपने सम्बन्धों को वह 'ऐडवेंचर' के रूप में लेती है और विरोधों के बीच जीती है—क्षणिक वासना और स्थायी प्रेम की युगपत् आकांक्षा उसे घर या घाट कहीं का नहीं रहने देती। यह स्थिति प्यार की अवधारणा के विरुद्ध जाती है।

मोहन राकेश ने इस तथ्य को जीवन में पाया था। एक पत्र में एक बार उन्होंने लिखा : 'एक बात कहूँगा, अप्राप्त की प्राप्ति कल्पना में नहीं हो सकती—भावना में होती है, क्रिया में होती है।...**तुम जो हो वह भी बनी रहना चाहो और अतिरिक्त भी लेना चाहो, यह क्यों कर होगा** ? यह नहीं कि उजली धूप में भीगी घटियों तक जाने के रास्ते नहीं हैं। लेकिन रास्ते के अतिरिक्त चरण भी तो चाहिए। यों **अपनी समर्थता को दूसरों की देन** मान लेने से असन्तोष कुछ हलका भले ही हो जाए, परन्तु सिद्धि क्या होगी ?.. .**दूसरे को उपादान के रूप में कभी मत ग्रहण करो—पुरुष हो, भावना हो या पत्थर।** अपने से बाहर उसके स्वतन्त्र अस्तित्व को स्वीकारो।...**प्यार भी लो, लेकिन किसी रेखा के बीच, भीतर या बाहर नहीं क्योंकि भावना में कोई रेखा नहीं होती...।**'[1]

'किसी एक के नाम' लिखे राकेश के पत्र का यह अंश किसी भी नारी के नाम हो सकता है। मल्लिका के नाम न सही तो सुन्दरी के नाम हो सकता है और कम-से-कम सावित्री के नाम तो निश्चयतः हो सकता है। सावित्री में वे सब बातें हैं जिनको ऊपर राकेश ने प्रेम की अप्राप्ति के सन्दर्भ में लिखा है। अमूर्त चाह, अप्राप्त की प्राप्ति के लिए ललक, भावना और क्रिया का अभाव, दूसरों को उपादान समझने की भावना और अपनी असमर्थता को दूसरों को उपादान समझने की भावना और अपनी असमर्थता को दूसरों की देन समझने की प्रवृत्ति सब उसमें हैं। उसमें उजली भीगी घाटियों तक जाने के लिए कामना तो है पर चरण नहीं हैं। वह स्वयं आधी-अधूरी है। इस कोटि की आधी-अधूरी नारियों का संकेत स्ट्रिंडबर्ग ने **मिस जूली** की भूमिका में देने का प्रयत्न किया है। उनके अनुसार ऐसी आधी औरतें अपने को अधिकार, यश, धन आदि के लिए पराए हाथों बेच देती हैं और कई पीढ़ियों तक एक त्रास और ह्रास का कारण बनती हैं। भ्रष्ट पुरुष ही ऐसी स्त्रियों को चुनते हैं और इस प्रकार ये ऐसी सन्तति को जन्म देती हैं जो यन्त्रणा की भागी बनती हैं।[2]

एक सुखी वैवाहिक सम्बन्ध को धराशायी करने में सबसे बड़ा हाथ उन अपेक्षाओं का होता है जो स्त्री-पुरुष को एक-दूसरे से होती है। **अन्तराल** उपन्यास में राकेश ने इस तथ्य को बड़ी बारीकी से उभारा है : 'कोई भी सम्बन्ध एक-साथ दो-दो अपेक्षाओं की पूर्ति से ही निभ सकता है और ये अपेक्षाएँ भी एक सीमा तक ही पूरी हो सकती

हैं क्योंकि हर व्यक्ति एक भरे-पूरे बाज़ार की तरह है जिसके सब-कुछ की तुम प्रशंसा कर सकते हो, पर वह सब-कुछ तुम अपने लिए नहीं ले सकते। तुम उसमें से वही लो जिसे लेने की सामर्थ्य तुममें है। इसकी चिन्ता मत करो कि शेष कहाँ जाता है, कौन लेता है ?'[3] वस्तुतः सब-कुछ पा जाना और एक-साथ पा जाना, अपनी अपेक्षाओं के आधार पर दूसरे के जीवन को निबन्धित करना अच्छे पारिवारिक सम्बन्धों में विसंगति पैदा करता है। इस तथ्य को राकेश ने बख़ूबी समझा है।

सावित्री का अधूरापन कुछ उसके व्यक्तित्व और कुछ परिस्थितियों का परिणाम है। उसका व्यक्तित्व आक्रामक नारी का व्यक्तित्व है। दूसरी ओर महेन्द्रनाथ दब्बू (दब्बू शायद वह पहले नहीं था) है। आक्रामक स्त्री जब प्रतिस्पर्द्धा करती है तो पुरुष की श्रेष्ठता स्वीकार नहीं कर पाती और जब वह उसे प्रतियोगिता के अयोग्य पाती है तो भावात्मक सुरक्षा के लिए पूरे आदमी की तलाश उसके लिए आवश्यक हो जाती है।[4] वैवाहिक असन्तोष का एक मूल कारण यह भी है। जब एक प्रकार की सम्बन्धहीनता पति-पत्नी में पनपने लगती है, तब स्त्री-पुरुष विरोधी गुणों को स्वीकार करने के लिए बाध्य होते हैं। युंग का विचार है कि स्त्री और पुरुष दोनों में ही दोनों के सजातीय गुण होते हैं। जिस पुरुष का नारीत्व पक्ष पत्नी के रूप में परावर्तित (प्रोजेक्ट) नहीं हो पाता, वह थोथी भावनाओं और अज्ञात मनःस्थितियों का दास हो जाता है। इसी तरह वे स्त्रियाँ जिनका पुरुष-भाव पति में सन्निहित नहीं हो पाता, वे आक्रामक और सिद्धान्तवादी बन जाती हैं और पुरुष के लिए वे बड़ी कष्टकारक होती हैं। आधा-अधूरा ऐसे ही लोगों का व्यक्तित्व रह जाता है।[5]

इस कोटि की स्त्रियाँ यद्यपि पुरुष से पौरुष की माँग करती हैं, किन्तु मनोवैज्ञानिक कारणों से आज्ञाकारी और दब्बू पुरुष से ही ब्याह करती हैं।[6] मूलतः उन्हें अधूरे पुरुष की ही तलाश होती है और वे ही उन्हें मिलते भी हैं। एक मनोवैज्ञानिक तथ्य यह भी है कि पुरुष के प्रति उनके आकर्षण का मूलाधार अप्राप्त की कल्पना भी होती है। सावित्री अधूरे पुरुष से बँधी ही नहीं रहती, बल्कि जिन-जिन पुरुषों से बँधती है इसी अप्राप्त की प्राप्ति के कारण। जुनेजा और सावित्री के संवादों से इस बात का कुछ अनुमान लगाया जा सकता है। जुनेजा को वह कभी इसलिए प्यार करती थी—'सिर्फ़ इसलिए कि मैं जैसा भी था, जो भी था, महेन्द्र नहीं था।' सावित्री की सारी प्रणय-यात्रा इस 'नहीं' से शुरू होती है—प्राप्त से अप्राप्त की ओर : 'पहले कुछ दिन जुनेजा एक आदमी था तुम्हारे सामने। तुमने कहा तब तुम उसकी इज़्ज़त करती थीं ...जुनेजा के बाद जिससे कुछ दिन चकाचौंध रहीं तुम, वह था शिवजीत। एक बड़ी डिग्री, बड़े-बड़े शब्द और पूरी दुनिया घूमने का अनुभव। पर असल चीज़ वही कि वह जो भी था, और ही कुछ था—महेन्द्र नहीं था।. ..उसके बाद सामने आया जगमोहन। ऊँचे सम्बन्ध, ज़बान की मिठास, टिपटाप रहने की आदत और ख़र्च की दरियादिली। पर तीर की असली नोक फिर उसी जगह पर कि उसमें जो कुछ भी था जगमोहन का-सा था, महेन्द्र का-सा नहीं था... ।'

वस्तुतः सावित्री के ये सम्बन्ध 'कितना कुछ एक-साथ होकर, कितना कुछ

एक-साथ पाकर और कितना कुछ एक-साथ ओढ़कर जीने' को चरितार्थ करते हैं। उसकी यह लिप्सा उस बच्चे की मनोवृत्ति को व्यक्त करती है 'जो मुट्ठी में एक-साथ कितना-कुछ भर लेना चाहता है, पर उसमें जो था, वह भी बाहर फिसल जाता है।' प्राप्ति के नाम पर यह क्षति स्वाभाविक है। पति को समर्पित जिस्म जब औरों की वासना का माध्यम बनता है तो सुख के बदले वह एक दूषण का आधार बनता है। पर-पुरुष-सम्बन्ध सावित्री के लिए एक ऐसा प्रलोभन सिद्ध होता है जिसमें आन्तरिक पूर्ति और प्राप्ति का कामना ही अधिक कारगर होती है। सावित्री को अपनी मनःस्थिति के कारण इन सम्बन्धों में जो अधूरा सुख मिलता है उसके कारण ही उसे लोग अधूरे लगते हैं। वस्तुतः अभाव और अधूरापन मानव चेतना की देन है। चेतना ही वस्तुतः अभावग्रस्त होती है। पूर्णता की सम्भावना/कल्पना अभाव को जन्म देती है और पूरे के सन्दर्भ में ही अधूरेपन की प्रतीति होती है। अभाव और अधूरेपन की प्रतीति में व्यक्ति की अपनी एषणाएँ का बड़ा हाथ होता है। यही एषणा सावित्री को विभिन्न पुरुषों के पास ले जाती है और सदा उससे प्राप्त का अतिक्रमण करवाती है। अभाव और एषणा मन के मूल्यों की वासना जगाते हैं क्योंकि मूल्य उसकी पूर्ति में सहायक होते हैं। सावित्री अपने लिए जो मूल्य गढ़ती है वे उसके लक्ष्य के अनुकूल हैं। ये लक्ष्य ऐसे हैं, इतनी असत् आस्था से परिपूर्ण हैं कि उनसे सुखद वर्तमान क्या, भविष्य का निर्माण भी नहीं हो सकता।

सावित्री की तलाश इसीलिए मानव के सम्पूर्ण अस्तित्व की तलाश नहीं कही जा सकती। तलाश के लिए न उसमें जिज्ञासा है, न बौद्धिक और आत्मिक बल। केवल एक असन्तुष्ट दृष्टि और अतृप्त संवेग को लेकर वह महेन्द्रनाथ तथा अपने सम्पर्क में आए अन्य पुरुषों पर अधूरेपन का आरोप लगाती है। वह पुरुष को जैसा है उस रूप में स्वीकार नहीं करती। मसलन एक स्थिति में जुनेजा, सिंघानिया या जगमोहन को इसलिए स्वीकार करती है कि वे महेन्द्रनाथ नहीं हैं। शायद वह महेन्द्रनाथ से इसलिए घृणा करती है कि वह सिंघानिया या जगमोहन या जुनेजा नहीं है। यह नकारात्मक दृष्टि व्यक्ति के व्यक्तित्व को नकारती प्रतीत होती है। कोई भी व्यक्ति अपनी निजता को त्यागकर दूसरा व्यक्ति नहीं बन सकता। महेन्द्रनाथ, जैसा भी है, अपने को महेन्द्रनाथ बनाए रखता है। सावित्री का उसे अपनी निजता में न जीने देना उसे पंगु बना देता है और स्वयं उसके जीवन में बाहरी लोग इतने प्रमुख हो जाते हैं कि उनके बिना वह अपने अस्तित्व को सिद्ध ही नहीं पर पाती। फिर भी वह घर का फालतू आदमी है जिसमें उसको फालतू बनाने में सबका हाथ है। स्वयं सावित्री जो उस पर इतने दोष मढ़ती है, एक कारण है। सच्चाई यह है कि एक-दूसरे को आगे-पीछे वैसा बनाने में दोनों का हाथ है। इस प्रकार नाटक की मूल समस्या—स्त्री-पुरुष सम्बन्ध—में दोनों के अधूरेपन का दोष साथ जुड़ जाता है और नाटक का आयाम कुछ व्यापक और दार्शनिक-सा हो जाता है। पर अधूरापन अपने में कोई दर्शन नहीं है। सच्चाई यह है कि वह उसके अपने भटकाव का ही एक प्रतिरूप है। इसीलिए उसके जीवन

में द्वन्द्व नहीं, 'क्राइसिस' है। और **आधे अधूरे** सही अर्थों में क्राइसिस का जीवन-नाटक है जो एकांगी दृष्टि से पैदा होता है।

सावित्री को घर का पुरुष जब सुरक्षा नहीं दे पाता तो वह एक के बाद एक बाहर के पुरुषों को आजमाती है पर अन्ततः सबको आधा-अधूरा करार देती है। सच्चाई यह है कि कभी कोई आदमी पूरा नहीं होता—आपसी सम्बन्धों की पूर्णता ही उसे पूरा बनाती है। यौन सम्बन्धों की दासता और आत्मिक इच्छा-शक्ति के अभाव में उसके आगे पूर्णता का अहसास कभी उभरता ही नहीं। किन्तु अधूरी होकर भी दूसरों का अधूरापन देखना उसका दोष नहीं—किसी भी स्त्री के मन में यह आकांक्षा होती है कि पुरुष में उसे वह पूर्णता मिले जो स्वयं उसमें नहीं। चूँकि महेन्द्रनाथ में (और उसके साथ ही जुनेजा, जगमोहन, सिंघानिया आदि में) वह अपने भावों की पूर्ति नहीं पाती, अतः वह उन्हें अधूरा कहकर लताड़ती जाती है। नायिका के अवचेतन में पुरुष का जो बिम्ब निहित है वह उसके अपने संप्रत्यय का ही परिवर्तन मात्र है। वह जिन पुरुषों के सम्पर्क में आती है, सोचती है कि वे उस बिम्ब के अनुरूप होंगे। किन्तु जब वे अनुरूप सिद्ध नहीं होते तो वह प्रवंचित अनुभव करती है और उन्हें छोड़ती जाती है। उसके लिए ऐसी स्थिति में पुरुष 1, 2, 3, 4 को एक ही नज़र से देखना आसान हो जाता है : 'सबके-सब एक-से। अलग-अलग मुखौटे, पर चेहरा सबका एक ही।' सभी चेहरे उसे एक-से दिखाई देते हैं क्योंकि सबको वह एक-सी दृष्टि और एक-से पहलू से देखती है। वस्तुतः वह सबसे एक ही पहलू से जुड़ी है। इस भटकाव में क्या उसका दोष नहीं ? केवल सुरक्षा चाहने के नाम पर उसे दोषमुक्त नहीं किया जा सकता क्योंकि वह स्वयं अर्थ उपार्जित करती है। एक महत्त्वपूर्ण बुनियादी सुरक्षा उसके पास है।

इससे भी भयंकर भ्रम वह अपने जीवन में यह पाले हुए है कि वह परिवार के लिए हलाल हो रही है; घर को किसी तरह सँभाले हुए है या जो कुछ घर के बाहर कर रही है, घर के लोगों के लिए कर रही है। लड़का इस धारणा पर तीव्र प्रहार करता है : 'मैं पूछता हूँ, क्यों करती हो ? किसके लिए करती हो ?' जिनके लिए भी वह करती है वे इस लायक़ नहीं—यह अहसास भी वही दिलाता है। फलतः सावित्री तिलमिला उठती है : 'आज से मैं सिर्फ़ अपनी ज़िन्दगी को देखूँगी—तुम लोग अपनी-अपनी ज़िन्दगी ख़ुद देख लेना।' और फिर 'सिर्फ़ अपनी ज़िन्दगी' जीने के लिए वह एक बार फिर जगमोहन के पास पहुँचती है। जाते हुए वह एक पूरे जीवन की कल्पना लेकर जाती है, किन्तु लौटती है तो उसके हाथों का लिजलिजा पसीना लेकर। असल में जो सम्बन्ध कभी शरीर का था, वह आत्मा का कैसे हो सकता था ? राकेश ने सावित्री के इस ओढ़े हुए जीवन का सुन्दर विश्लेषण किया है। जिसे वह घर के लिए खपना कहती है, वह एक प्रकार से दूसरों की नज़र बचाकर अन्दर-ही-अन्दर सड़ना है, स्वाभाविकता के प्रति विश्वासघात है। इसीलिए दुहरी ज़िन्दगी को तोड़ने का उपक्रम अन्त में जब वह करती है तो विवशता ही हाथ लगती है। इस सब में उसकी अपनी भूमिका के प्रति वह जागरूक नहीं प्रतीत होती। वह वंचिता तो महसूस करती है—किन्तु सबसे अधिक वंचित व्यक्ति महेन्द्रनाथ

है—यह उसे कहीं महसूस नहीं होता। राकेश नारी की इस एकांगी दृष्टि की ओर संकेत करना नहीं भूलते हैं। सावित्री सारी शिकायतें औरों पर लादकर जीना चाहती है।

वस्तुतः महेन्द्रनाथ, सिंघानिया, जगमोहन, जुनेजा चारों पुरुष उसके लिए किसी-न-किसी रूप में फूहड़ तो हैं ही। सिंघानिया का फूहड़पन उसके संवादों में पूरी तरह मुखर है। जगमोहन में स्वार्थ का एक विलक्षण ठंडापन है और जुनेजा में आक्रामक बेरुखी। इनमें वास्तविक विसंगत व्यक्ति महेन्द्रनाथ है। वह निकम्मा, अकेला और अस्तित्वहीन जीवन जीता दिखाया गया है। असफलताओं के कारण वह दुश्चिन्ता से ग्रस्त है। वास्तव में वह निजात्म स्वरूप से निर्वासित लगता है। 'दूसरों' की दुनिया उसे घेरे हुए है जिसमें वह अपमान महसूस नहीं करता। इसीलिए वह परजीवी है—एक ही जिए जीवन को फिर-फिर जीता है, उसी तरह उसी क्रम में। घर में जो कुछ घटता है, उसमें उसका कोई दख़ल नहीं और स्वयं जो उसके घटना-विहीन जीवन में घटता है, उसे वह रोक नहीं पाता जिससे वह निरन्तर टूटता जाता है। वह अपने और अपने परिवार के अस्तित्व का साक्षी मात्र बनकर एक अजनबी की तरह उस घर में रहता है जो उसका होकर भी उसका नहीं। सारी दुनिया में जुनेजा ही एक ऐसा व्यक्ति है जो उसका पक्षधर है, किन्तु सावित्री के अनुसार वही उसे पराश्रित बनाने का ज़िम्मेदार भी है। विचित्र बात यह है कि परिवार में जो उसके प्रति उपेक्षा का भाव है उसने उसे निर्द्वन्द्व होकर स्वीकार कर लिया है। वह अपने इस अप्रामाणिक जीवन में दुःख को इस तरह पचा लेता है कि उसमें न आन्तरिक अस्वीकार है और न मोहभंग की स्थिति। इसीलिए उसके लिए वरण की गुंजाइश नहीं । वह जीवन की उस सीढ़ी पर है जहाँ पैरों तले ज़मीन नहीं, केवल मृत्यु का अहसास है। फलतः आत्मग्लानि और अपराध-बोध से घिरा हुआ कभी वह असह्य स्थिति में घर से बाहर निकल भी पड़ता है तो 'हर मंगल-शनीचर को' वापस लौट आता है क्योंकि अन्तिम निर्णय से बचना उसके लिए ज़रूरी हो गया है। इसीलिए वह घर में बन्दी रहकर जीना बाहरी जगत् को तुलना में अधिक सुरक्षित मानने लगता है। महेन्द्रनाथ आधुनिक संस्कृति की देन है। वह आज का आदमी है जिसने अपने स्वत्व को खो दिया है और जिसकी इच्छाशक्ति उसकी कुंठित मानसिक स्थिति की शिकार बन गई है। आधुनिक साहित्य में ऐसे आदमी का हवाला अक्सर दिया जाता रहा है जिसकी आस्थाएँ मिट गई हैं। महेन्द्रनाथ ऐसा ही आधुनिक मानव है।

यह विसंगति मेहन्द्रनाथ या तीन अन्य पुरुषों में ही नहीं, परिवार की लड़कियों और लड़कों में भी है। अशोक, बिन्नी, किन्नी आदि महेन्द्रनाथ की विसंगति के ही कुछ और अधूरे रूप हैं। वस्तुतः राकेश ने एक ही मूल समस्या को लेकर दो पीढ़ियों के नग्न सत्य को उघाड़कर रखा है। कार्ल यास्पर्स ने ठीक ही कहा है कि 'जब जीवन एक कार्य-व्यापार मात्र रह जाता है तो उसकी ऐतिहासिक विशेषता ख़त्म हो जाती है जिससे विभिन्न वयों के लोगों में एक-स्तरीयता आ जाती है। यौन आकांक्षा जो युवा-वर्ग की अपेक्षित भूख है वह सबकी भूख बन जाती है और एक ख़ास उम्र के बीत जाने पर भी बड़ी उम्र के लोग उस ओर अपना रुझान दिखाने लगते हैं।...लोग

ऐसा व्यवहार करने लगते हैं जैसे वे सब एक ही उम्र के हों। बच्चे बड़े लोगों की तरह हो जाते हैं। जब बूढ़े जवान होने का बहाना करें तो जवान लोगों के मन में उनके लिए कोई आदर नहीं रह जाता।'[7] **आधे अधूरे** में दो पीढ़ियों के बीच अधूरेपन की यही विसंगति है। बड़ी उम्र के लोग सावित्री, सिंघानिया, जगमोहन सब युवाओं जैसा विकृत व्यवहार करते हैं। ऐसी हालत में तेरह वर्ष की उम्र में किन्नी वयस्कों की तरह सेक्स की चर्चा में आनन्द लेने लगती है; बड़ी लड़की माँ के प्रेमी के साथ भाग जाती है। और लड़का नंगी तस्वीरें काटने में दिन गुज़ारता है। 'ज़िन्दगी की उस बेबसी का वह क्या करे ! किसी रचनात्मक आयोजन में विसर्जित न कर पाने की स्थिति में उसकी सृजन शक्ति अपने ही मूल उत्स की ओर लौट-लौट आती है और उसके लिए ज़िन्दगी में सेक्स ही सबसे बड़ी समस्या बन जाती है—एक स्वस्थ आदमी की यौन मुक्ति के रूप में नहीं—एक घुटे आदमी की यौन विकृति के रूप में।'[8] अशोक एक रुकी हुई ज़िन्दगी का शिकार है तो बिन्नी और किन्नी एक अवरोध में क़ैद आक्रोश के। किन्नी में तीव्र आक्रोश है और सेक्स की जागरूकता। बिन्नी मनोज के साथ अपने दाम्पत्य सम्बन्ध को अपनी माँ की तरह ही दुहरा रही है। दोनों के बीच ऐसी 'हवा' है जिसका अहसास उन्हें अजनबी महसूस कराता है। यह स्थिति उसे सावित्री की तरह आक्रामक बना डालती है : 'मन करता है आसपास की हर चीज़ को तोड़-फोड़ डालूँ। कुछ ऐसा कर डालूँ जिससे...' किन्तु 'धीरे-धीरे हर चीज़ उसी ढर्रे पर लौट आती है।' लौटकर फिर उसी स्थिति में आ जाना—आगे की ओर न बढ़ना ही वह अधूरापन है जो सारे परिवार की तरह उसे भी घेर लेता है।

अशोक के अधूरेपन के और भी आयाम हैं। वह किसी ज़िम्मेदारी को स्वीकार करने के लिए तैयार नहीं। एक 'आउट साइडर' की तरह वह उस घर में रहता है और बाप की ज़िन्दगी को दुहराता है। फिर भी नई पीढ़ी के प्रतिनिधि के रूप में उसमें व्यंग्य और विरोध का माद्दा है। यह यथास्थिति को ख़त्म करने में 'तेज़ औजार' की तरह काम देता है।

इस प्रकार नाटक के सारे पात्र या तो आधे हैं या अधूरे। नाटक इस अधूरेपन के कारणों के कुछ संकेत देता है। उसमें आर्थिक, मनोवैज्ञानिक कारण मुख्य हैं। रागात्मक सम्बन्ध का अभाव भी एक बड़ा कारण है पर सबसे बड़ा कारण है सिर्फ़ अपने लिए जीना। पर-निर्भरता, यौन विकृति भी उनमें समाहित हैं। कुछ बातें सब में सामान्य हैं पर विडम्बना यह है कि उनमें दूसरे के लिए न मानवीय भावनाएँ हैं न समझ। टकराव उनके जीवन का सत्य है। उसके विषय अलग-अलग भले हों पर सबके केन्द्र में सावित्री है। इसलिए सारी कथावस्तु की इकाइयों में एक सूत्रता भी है। पर सबका अधूरापन एक-सा है—यह नहीं कहा जा सकता। हर पात्र का अपना अधूरापन दूसरे से मिलता-जुलता अवश्य है। सारे पात्र जहाँ अपने को दुहराते हैं, वहाँ अधूरेपन की विभिन्न स्थितियों का संकेत भी देते हैं। अशोक कल का महेन्द्रनाथ है और किन्नी-बिन्नी स्वयं सावित्री के पूर्ववर्ती विकास-क्रम को द्योतित करती हैं। महेन्द्रनाथ भी पाँच व्यक्तियों में बँटा एक व्यक्ति है या यों कहिए कि उन सबका जोड़ है। अधूरापन अंग-अंगी रूप में

कहाँ नहीं है ? फिर भी वह जिस रूप में नाटक में चित्रित है, वह व्यक्ति या व्यक्तियों का होकर भी पूरे समाज की सामान्य समस्या नहीं है, आधुनिकता के बोध से जुड़े हुए नए चिन्तन से सम्बद्ध भले ही मान लिया जाए।

अधूरापन एक आधुनिक धारणा है। एक युग था जब धर्म और विज्ञान पर अलग-अलग आस्था रखनेवाले लोग मुनष्य की पूर्णता की सम्भावनाओं के प्रति आश्वस्त थे। तब महामानव की धारणा थी, वैज्ञानिक प्रगति और देवत्व के विश्वास के आधार पर मानव के सम्बन्ध में मसीहाई घोषणाएँ हुई थीं। किन्तु दो महायुद्धों के बाद आशावादी स्वर इतने मन्द पड़ गए कि मानव को महामानव की प्रतिष्ठा दिलाना तो दूर रहा, उसकी लघुता और अपूर्णता को ही उसकी पूँजी मान बैठना युगचिन्तन का महत्त्वपूर्ण अंग बन गया। दरअसल पूँजीवादी व्यवस्था, विज्ञान और टेक्नोलॉजी के परिणाम मानव के लिए इतने भयंकर सिद्ध हुए कि आज वह साध्य न बनकर साधन मात्र रह गया है। वैज्ञानिक यन्त्रवाद की यन्त्रणा ने इस विश्वास को समूल उखाड़कर फेंक दिया कि मनुष्य पूर्ण है या हो सकता है।[9] इसके साथ ही इतिहास और परम्परा अथवा आदर्शवाद के विरुद्ध जो प्रवृत्तियाँ आज सक्रिय हैं वे मानव का अवमूल्यन करने के लिए कटिबद्ध हैं। अतः युग के इन सत्यों से आँखें मूँदकर कौन आशावादी बना रह सकता है ? जो कुछ भी आज संहारोन्मुख वर्तमान और प्रतीक्षारत भविष्य के बारे में कहा जा रहा है, वह एक भयंकर सच्चाई है।[10] समाज जिस प्रकार की अव्यवस्था का शिकार होता जा रहा है उसमें मनुष्य की इयत्ता नहीं रह गई है—सब लोग जैसे एक-से असहाय, बेनाम, बेचेहरा हो गए हैं। ऐसी स्थिति में अधूरापन, नंगापन उसके जीवन के आधारभूत सत्य हो गए हैं जिनके बीच ही अब उसे जीना होगा। इसीलिए पूर्णता की खोज अब उसे अधूरेपन के बीच ही करनी होगी।

किन्तु इसके साथ ही मानवीय अस्तित्व की सार्थकता की खोज भी अनिवार्य हो जाती है। राकेश ने अपने नाटकों में आधा-अधूरापन तो खोज निकाला, पर पूर्णता की तलाश के लिए आकुलता उस अनुपात में नहीं दिखाई। वस्तुतः आदमी न जानवर है न मशीन का पुर्ज़ा और न ही अपने अस्तित्व का निश्चेष्ट साक्षी; वह नियामक भी है और सम्भावनाओं का स्रष्टा भी। **आधे अधूरे** में राकेश मानव को एक ऐसे बिन्दु से जोड़ना भूल गए हैं जहाँ से जीवन की सार्थकता नज़र आने लगती है। वस्तुतः अधूरापन एक साक्षेप स्थिति है जिसका अहसास केवल पूरेपन के सन्दर्भ में ही सम्भव है। आश्चर्य की बात यह है कि एक पूरी ज़िन्दगी जीने का हिमायती नाटककार[11] आधे-अधूरों की दुनिया का पहरेदार बनकर रह जाता है ! स्वयं राकेश ने बन्धन तोड़े और बनाए। फिर **आधे अधूरे** सिर्फ़ भँवर में घूमते परिवार का नाटक बनकर क्यों रह गया ?

राकेश आधे पुरुष की बात करते हैं। किन्तु प्रश्न यह उठता है कि पूरा पुरुष क्या होता है ? यह सब राकेश ने नहीं बताया है। एक उत्तर यह हो सकता है जो बातें अधूरे पुरुष में नहीं होतीं, वे ही पूरे पुरुष की पहचान बन सकती हैं। साधारणतः जिस व्यक्ति में ह्रासोन्मुख प्रवृत्तियाँ या कुंठाएँ न हों और जिसमें भावात्मक, मानसिक और शारीरिक

शक्तियों का सकारात्मक समन्वय हो जो उसे व्यक्तित्व की ऊँचाइयों तक ले जा सके उसे पूर्ण पुरुष कह सकते हैं। जीवन की गुणता, मानवीय व्यवहार, सर्जनात्मकता, ऊर्जा, कार्यक्षमता भी उसके गुण माने जा सकते हैं। प्रगाढ़ता में जीना, नई राहों का अन्वेषण, संवेदनशीलता आदि भी कई विशेषताओं को गिनाया जा सकता है। पर पूर्ण पुरुष की पहचान कराना मुश्किल है, केवल कुछ अनुमान लगाए जा सकते हैं। नाटक में ऐसे पुरुष का संकेत नहीं। ऐसा पुरुष संसार में बना-बनाया नहीं आता, उसे बनाना पड़ता है और इसके लिए रूपान्तरण की अपेक्षा होती है। यह ठीक है कि आज का आदमी विवश है और सब पूर्ण हो भी नहीं सकते। फिर भी उसके लिए राकेश नाटक में ललक या आकुलता नहीं पैदा कर सकते थे ?

इस प्रकार **आधे अधूरे** अर्थ के दो आयाम प्रस्तुत करता है। घर-परिवार के विघटन की समस्या इसके यथार्थवादी स्वर को उद्घाटित करती है और पात्रों का अधूरापन आधुनिक भाव-बोध को। कुछ आलोचकों ने इसे 'समकालीन यथार्थ जीवन का महत्त्वपूर्ण दस्तावेज़' कहा है। किन्तु कोई भी नाटक यथार्थ या समकालीन जीवन का दस्तावेज़ इसलिए नहीं बन जाता क्योंकि वह समकालीन है या उसमें किसी युग-पुरुष का समाजिक यथार्थ चिचित्र है। **आधे अधूरे** में जैसा भी यथार्थ है, वह पारिवारिक स्थूल यथार्थ है जिसके ऊपर आभासी दर्शन का आरोपण किया गया है। इस यथार्थ से जीवन की एक नंगी तस्वीर अवश्य उभरती है किन्तु चिन्तन की मुद्रा के बावजूद कोई सूक्ष्म तथ्य हाथ नहीं आता। सूत्रधार के द्वारा व्यक्ति और व्यक्ति के बीच के समीकरण की बात सुनकर भी जो कथानक ज़ेहन में आता है वह महेन्द्रनाथ, सावित्री और उसके परिवार का भले ही हो सकता है, सबका नहीं हो सकता। न सब पुरुष महेन्द्रनाथ हैं और न सब स्त्रियाँ सावित्री हो सकती हैं। महेन्द्रनाथ और सावित्री दोनों 'एब्नॉर्मल' हैं, किसी रुग्ण मानसिकता के शिकार हैं। फिर भी यदि स्त्री-पुरुष के बिगड़ते सम्बन्धों और उनके बीच सन्तान की बिगड़ती हालत को लें, तो यह बहुत बड़े वर्ग की कहानी कही जा सकती है। यह बात दूसरी है कि यदि राकेश की कथन-शैली को ध्यान में लाएँ तो यह बहुत सामान्य और स्थूल कहानी है जो न जाने कब से घरों में घटती रही है। राकेश ने उसे शैली, शिल्प, भाषा और संवाद की गरिमा अवश्य दी, पर कथ्य को गहराई नहीं दे पाए। केवल उसे महसूस-भर किया है जैसा कि पहले किसी ने महसूस नहीं किया था। फिर भी पति-पत्नी के बीच टकराहट के जो मुद्दे हैं वे आयामहीन हैं। उनके बारे में ठीक ही कहा गया है कि मोहन राकेश उसे किसी गहरे संकट, संवादहीनता, सम्बन्धहीनता या किसी का अहसास नहीं कराते न किन्हीं प्रश्नों की तह में ले जाते हैं। वे मानव मात्र की समस्या को नहीं उभारते।

जो नाटक में उभरकर नहीं आता उसकी पूर्ति राकेश ने विचार के माध्यम से करने का प्रयत्न किया है। इस नाटक को विशिष्ट सर्जनात्मक क्षमता प्रदान करनेवाला तत्त्व

इसकी ओढ़ी हुई वैचारिक पृष्ठभूमि में है। किन्तु इसके बावजूद इसकी कथावस्तु में इतनी सामान्यता है कि उसके माध्यम से गहरी संवेदनाओं और प्रश्नों तक पहुँचना सम्भव भी न था। इसीलिए नाटक जहाँ से शुरू हुआ है, वही पहुँचकर रह जाता है। बड़ी लड़की, लड़के, सावित्री, जुनेजा के बावजूद एक असहाय स्थिति अपना वृत्त पूरा करती है : 'अंकल/हाँ बेटे/सचमुच कुछ नहीं हो सकता क्या ?/एक दिन के लिए हो सकता है शायद, दो दिन के लिए हो सकता है, पर हमेशा के लिए नही।'/'तो इस हालत में यही बेहतर नहीं कि...।' लेकिन बेहतर क्या है—इसका नाटककार के पास कोई जवाब नहीं। शायद बेहतर उसके पात्रों की नियति में ही नहीं क्योंकि वे 'मिट्टी के लौंदे' हैं। उनके लिए अब कोई चुनाव रह ही नहीं गया है : 'दाएँ से हटकर बाएँ, सामने से हटकर पीछे, इस कोने से हटकर उस कोने में। क्या सचमुच कोई चुनाव नज़र आया है तुम्हें ?' और इस अहसास के साथ ज्यों ही नाटक का अन्त होता है 'अँधेरा अधिक गहरा होता जाता है।'

इसमें कोई सन्देह नहीं कि महेन्द्रनाथ, सावित्री, सिंघानिया, बिन्नी, किन्नी, अशोक—सब अँधेरे के किसी यथार्थ को प्रकट करते हैं; किन्तु इस यथार्थ की विडम्बना यह है कि इसे सामान्यीकृत नहीं किया जा सकता—यह व्यक्ति का यथार्थ है, व्यक्ति-व्यक्ति का यथार्थ नहीं; विभाजित यथार्थ है, समाज के सम्पूर्ण ढाँचे का यथार्थ नहीं। इसीलिए का.सू.वा. सूत्रधार के रूप में इस बात को साफ़ कर देता है : 'जो मैं इस मंच पर हूँ वह यहाँ से बाहर नहीं हूँ।' इस विभाजित यथार्थ में भी वह एक सर्वसामान्य यथार्थ की घोषणा करता है : 'जहाँ इस समय मैं खड़ा हूँ वहाँ मेरी जगह आप भी हो सकते थे—दो टकरानेवाले व्यक्ति होने के नाते आपमें और मुझमें बड़ी **समानता** है। यह समानता आपमें और उसमें, उसमें और उस दूसरे में, उस दूसरे में और मुझमें...।' पर यह ज्यामितीय सम्बन्ध वैयक्तिक यथार्थ को झुठलाता लगता है।

ओम शिवपुरी ने एक स्तर पर इस नाटक को 'अनुभव की समानता का दिग्दर्शन' कहा है। समानता की जो बात का.सू.वा. कहता है, वह अगर अभिधा में ली जाए तो बड़ी अटपटी लगती है क्योंकि पुरुष 1, 2, 3, 4 और सावित्री या उसकी सन्तान के अपने-अपने व्यक्तित्व हैं जिनका अपना-अपना वैशिष्ट्य है। उनमें एक के स्थान पर दूसरे को नहीं रखा जा सकता, न वे ऐसे व्यक्ति हैं जो फुटपाथ पर टकरा जाते हैं। इसी आधार पर कुछ लोगों ने यह कहने का प्रयत्न किया कि का.सू.वा. के कथन और वास्तविक कथ्य में कोई संगति नहीं है। यह कैसे कहा जा सकता है कि महेन्द्रनाथ, जुनेजा, सिंघानिया, सावित्री, बिन्नी, किन्नी, अशोक सब एक-से हैं और सावित्री यदि उनमें से किसी भी पुरुष के साथ रहती तो वह वैसी ही होती जैसी अब है। सामान्यतः यह माना जाता है कि परिस्थितियों के बदल जाने के साथ आदमी भी बदल जाता है; पर का.सू.वा. कहता है—'परिवार दूसरा होने पर भी परिस्थितियाँ बदल जातीं, **मैं वही रहता।**' इसी विचारधारा के अनुरूप नाटक में अन्ततः सब पात्र वही और वहीं रहते हैं। यह सब चुनाव का निषेध करनेवाली धारणा है। किन्तु इसका एक दूसरा

चिन्तन-पक्ष भी है जिसको का.सू.वा. के द्वारा व्यक्त किया गया।

वस्तुतः आज का व्यक्ति भीड़ में खो गया है और उसके व्यक्तित्व की निजता नई परिस्थितियों और व्यवस्था में निचुड़ गई है। जैसा कि यास्पर्स ने कहा है, मानव का अस्तित्व आज समूहगत अस्तित्व हो गया है और वह एक अलग इयत्ता को समाप्त कर सामूहिक चेतना का अंग हो गया है।[12] इसी को आधुनिक चिन्तकों ने 'संगठन मनुष्य' (ऑरगनाइजेशनल मैन) की संज्ञा दी है। एक स्थिति यह भी है कि एक का सत्य दूसरे का सत्य हो सकता है। महेन्द्र, जुनेजा, सिंघानिया या जगमोहन कहीं पर 'एक' हो सकते हैं—कहीं का तात्पर्य है किसी एक सन्दर्भ में। ये सब सावित्री के सन्दर्भ में समान हैं, अन्यथा भिन्न-भिन्न व्यक्ति हैं। जब एक व्यक्ति दूसरे जैसा लगता है तो इसका अर्थ यह है कि दोनों में शाश्वत या सार्वजनीन तत्त्व है। व्यक्ति की निजता सदा शाश्वतता की ओर प्रवाहित होती है क्योंकि व्यक्ति शाश्वत की आधारशिला है; अन्यथा शाश्वतता का कोई अर्थ नहीं, यदि वह व्यक्ति के प्रयोजन के लिए नहीं।[13] ऐसी स्थिति में का.सू.वा. जिस समानता की बात करता है वह वैचारिक यथार्थ है और एक बार व्यक्ति-व्यक्ति की समानता को स्वीकार कर लेना परिस्थिति के यथार्थ को स्वीकार करना है, व्यक्ति को नहीं। व्यक्ति महेन्द्रनाथ की तरह कभी परिस्थितियों के दबाव से यों जम जाता है कि अतीत उसकी चेतना पर हमेशा-हमेशा के लिए अंकित हो जाता है—फिर वही वर्तमान बन जाता है, वही भविष्य। फलतः परिवर्तन की सारी सम्भावनाएँ नष्ट हो जाती हैं। स्वयं राकेश के शब्दों में, 'मैं जो बात बताना चाहता हूँ वह यह है कि अपनी परिस्थिति के लिए व्यक्ति अकेला ज़िम्मेदार नहीं होता क्योंकि स्थितियाँ कुछ भी होतीं, उसे बार-बार उसी का चुनाव करना पड़ता। ज़िन्दगी में व्यक्ति कुछ भी चुने उसमें एक विशेष 'आइरनी' होती है, क्योंकि परिस्थितियाँ फिर वही बन जाती हैं।[14] पर नाटककार सत्य के किसी एक पहलू से आश्वस्त नहीं लगता है : इसलिए वह निर्णय नहीं करता कि मुख्य भूमिका पात्र की है या परिस्थिति की, या दोनों की ? ऐसी स्थिति में वह समानता की बात को मानता भी है और झुठलाता भी है।

आधे अधूरे एक सुरचित नाटक है जिसमें कथावस्तु की रेखाएँ सुनिश्चित हैं और नाटक का अनिश्चित अन्त और पात्रों की अनिश्चित भूमिका की सफ़ाई देने के बावजूद भी सब-कुछ सुनिश्चित है। कहा जाता है कि **आधे अधूरे** राकेश की पत्नी अनीता राकेश की एक कहानी पर आधारित है। **चंद सतरें और** पढ़ने पर यह सन्देह और भी पुष्ट होता है। **आधे अधूरे** की सृजन-प्रक्रिया में राकेश का अपना व्यक्तित्व और उनकी भोगी हुई 'एक और ज़िन्दगी' बहुत सहायक रही है, किन्तु इसके साथ अनीता राकेश का भोगा हुआ यथार्थ भी उक्त कहानी या कहानी के आधारभूत जीवन से जुड़ा प्रतीत होता है। **चंद सतरें और** के प्रारम्भिक पृष्ठ **आधे अधूरे** के कथानक की मूल संवेदना तक ले जाते हैं। इस अर्थ में बीज नाटकों को उनका बीज मान लेना कोई अर्थ नहीं रखता क्योंकि बीज नाटक **आधे अधूरे** से भी आगे की चीज़ है।

आधे अधूरे की कथावस्तु एक परिवार तक सीमित है। इस सीमा के कारण नाटक की संवेदना का क्षेत्र भी छोटा हो जाता है। सारी कथावस्तु 'अनेक चेहरों के पीछे जिस एक चेहरे' को उभारती है वह जीवन के किसी सार्थक प्रश्न का उत्तर नहीं प्रस्तुत करता। सारा नाटक एक दूसरे स्तर पर अभावजनित असन्तोष की अभिव्यक्ति का माध्यम अवश्य बन जाता है, पर इस तथ्य में कोई नयापन नहीं। फिर भी कथानक का संयोजन राकेश ने बड़े कौशल के साथ किया। नाटक में घटनाएँ कम हैं, मनःस्थितियाँ ही अधिक हैं जो एक समान चेतना के आधार पर परस्पर जुड़ जाती हैं। सबका मूल सूत्र संघर्ष है। सभी स्थितियों में संघर्ष के भी कई आयाम हैं। मुख्य संघर्ष सावित्री और महेन्द्रनाथ का है, किन्तु अन्य पुरुषों के साथ भी सम्बन्ध-भावना के अन्तःसूत्र में संघर्ष विद्यमान है, यद्यपि सारा संघर्ष पुंसत्वहीन है। नाटक का प्रारम्भ मन्थर गति के साथ होता है और अन्त तक उसी गति का निर्वाह है। गति का मूलाधार संवाद हैं जो नाट्य स्थितियों को विकसित करने में सहायक होते हैं। चरम सीमा के बिना ही नाटक समाप्त होता है जैसे सब-कुछ पहले जैसा दुहराने के लिए। कहा जाता है कि नाटक जहाँ से शुरू होता है, आकर वहीं ख़त्म भी होता है। सारा नाटक एक भावात्मक वातावरण पर निर्भर है। बीच-बीच में छोटी-मोटी सूच्य घटनाएँ नाटक के स्थापत्य को निर्धारित करती हैं। पर जहाँ तक कलात्मक उपलब्धि का प्रश्न है, नाटककार प्रस्तावना, संवाद और रंग उपकरणों से सँभालता है और अन्त में एक सार्थक बिम्ब से रचना को भावपूर्ण बनाने का यत्न करता है—

पुरुष चार के चेहरे पर व्यथा की रेखाएँ उभर आती हैं और उसकी आँखें स्त्री से मिलकर झुक जाती हैं। स्त्री एक कुरसी की पीठ थामे चुप खड़ी रहती है।

बड़ा लड़का : (**बड़ी लड़की से**) जल्दी से निकाल दे छड़ी क्योंकि...।

बड़ी लड़की : (**अन्दर के दरवाज़े की तरफ़ बढ़ती**) अभी दे रही हूँ। (**जाकर दरवाज़ा खटखटाती है**) किन्नी, दरवाज़ा खोल दे जल्दी से।

छोटी लड़की : (**अन्दर से**) नहीं खुलेगा दरवाज़ा।

बड़ी लड़की : (**ज़ोर से खटखटाती**) किन्नी !

सहसा हाथ रुक जाता है। बाहर से ऐसा शब्द सुनाई देता है जैसे पाँव फिसल जाने से दरवाज़े का सहारा लेकर किसी ने अपने को बचाया हो।

पुरुष चार : यह कौन फिसला है ड्योढ़ी में ?

लड़का : डैडी ही होंगे। (**दरवाज़े से निकलता**) आराम से डैडी, आराम से...!

पुरुष चार : (**एक नज़र स्त्री पर डालकर दरवाज़े से निकलता**) सँभलकर महेन्द्रनाथ, सँभलकर...।

प्रकाश खंडित होकर स्त्री और बड़ी लड़की तक सीमित रह जाता है। स्त्री स्थिर आँखों से बाहर की तरफ़ देखती आहिस्ता से कुरसी पर बैठ जाती है।...लड़के की बाँह थामे पुरुष एक की धुँधली आकृति अन्दर

आती दिखाई देती है।

लड़का : देखकर डैडी, देखकर...।

कथावस्तु को एक नई क़िस्म की सार्थकता देनेवाला प्रयोग इस नाटक में प्रस्तावना के रूप में किया गया है। कुछ आलोचकों ने इसे व्यर्थ का प्रयोग माना है जिसकी मूल नाटक के साथ संगति नहीं बैठती। अगर सतही दृष्टि से देखें तो सूत्रधार के रूप में का.सू.वा. जो कुछ कहता है वह नाटक पर पूरा नहीं उतरता। किन्तु एक का.सू.वा. इसे अनिश्चित नाटक कहता है। वह स्वयं को भी अनिश्चित मानता है। अनिश्चित होने का कारण बताते-बताते वह रह जाता है। नाटक के अनिश्चित होने का भी कोई अर्थ नहीं जबकि वह यह घोषित करता है कि परिस्थितियों के बदलने के बावजूद वह वही रहता और 'इसी तरह सब-कुछ निर्धारित करता है'। यही नहीं, प्रारम्भ में ही 'मैं कौन हूँ' की जिज्ञासा जगाकर वह 'अनिश्चित आदमी' एकदम वह आदमी बन जाता है 'सड़क की फुटपाथ पर चलते अचानक आप जिस से टकरा जाते हैं।' वह मानता है 'जहाँ इस समय मैं खड़ा हूँ, आप भी हो सकते थे। बात इतनी-सी है...कि मैं किसी-न-किसी अंश में आप में से हर व्यक्ति में हूँ।' यह प्रेक्षक और नाटक के पात्रों के बीच समानता घोषित करने का एक तरीक़ा है जिससे पात्रों के बीच में समानता भी व्यंजित है। एक व्यक्ति से चार व्यक्तियों की भूमिका करवाने के पीछे यह तर्क हो सकता है किन्तु वे अपने वैशिष्ट्य में कहीं एक नहीं लगते। यदि वे कहीं एक हैं तो अपनी वासना में। व्यक्ति के किसी अंश के कारण उसे दूसरे व्यक्ति के समान मात्र कैसे कहा जा सकता है। सवाल समानता का नहीं। स्त्री-पुरुष के स्वभाव और व्यवस्था का है जिसमें कलह होती है। चीज़ों को उनके सहज रूप में अंगीकार न करने से भी दुख होता है और संघर्ष भी। जो घर अनाग्रह से परिपूर्ण हो या आशा से अधिक माँग करता हो या जो उस स्थिति में सम्भव न हो उसको चाहता हो, वह किसी का भी नहीं हो सकता। फिर भी यदि हम प्रस्तावना की सारी उक्तियों को परस्पर जोड़कर देखें तो उनके अर्थपूर्ण होने के बावजूद उनमें सम्पूर्ण सत्य की कमी दीखती है। नाटक ज्यों-ज्यों आगे बढ़ता है, उसके अपने सत्य उभरते हैं और कुछ अंश नाटक की दृष्टि से बेकार लगते हैं। वस्तुतः यह नाट्य युक्ति पश्चिम से ग्रहण की गई है और उसी की विचारधारा की देन कही जा सकती है।

एक भिन्न दृष्टि से सोचने पर सारी प्रस्तावना अवश्य सार्थक लगती है। अभिव्यक्तिवादी नाटककारों की भाँति इस नाटक ने अनाम पात्रों का प्रयोग किया है, जैसे पुरुष 1,2,3,4 स्त्री, बड़ी लड़की, छोटी लड़की, लड़का आदि। राकेश अपने पात्रों को प्रतीकात्मक, अन्योक्तिपरक या सामान्य रूप में प्रस्तुत करते हैं जिसके लिए प्रस्तावना यथेष्ट आधारभूमि बनती है। इसी प्रकार मंच पर एक ही व्यक्ति चार भूमिकाएँ निभाता है। इस पर भी कुछ लोगों ने आपत्ति की है। उनका कहना है कि

चारों पुरुष आपस में बहुत भिन्न हैं और एक व्यक्ति दूसरा तभी हो सकता है जब वह उसका प्रतिनिधित्व या अभिनय कर रहा हो। किन्तु राकेश के मन में शायद दूसरी ही बात थी। उन्होंने जो प्रयोग किया है उसमें व्यक्ति की निश्चित भूमिका और व्यक्तित्व को अस्वीकार किया गया है। यह दृष्टि अस्तित्ववादी विचारधारा पर आधारित है जिसके अनुसार मनुष्य की कोई निर्धारित प्रकृति नहीं है, जो कुछ है वह परिस्थिति है। आदमी-आदमी में अन्तर इसलिए होता है क्योंकि उनकी परिस्थितियों में अन्तर होता है। नाटक में चूँकि परिस्थिति एक है इसलिए पात्र भी एक-से होने को बाध्य हैं।[15] यह पाश्चात्य धारणा है कि मनुष्य अपनी निजता से विलग होकर अनाम और अरूप हो गया है। इसलिए सबकी अलग-अलग भूमिका नहीं है—सब एकरूप हो गए हैं। मनुष्य आज अपना नहीं; 'संगठन' का जीवन बिता रहा है। हमारी सभ्यता आज व्यक्ति को नहीं, 'संगठनमूलक मनुष्य' को पैदा कर रही है। यही उसकी अनिश्चित भूमिका का कारण भी है। सार्त्र, स्टीफेन स्पेंडर आदि चिन्तकों ने इस बात को कई बार दुहराया है। उनके अनुसार मनुष्य का अस्तित्व न अतीत पर निर्भर है और न भविष्य पर—निरन्तर गतिशील स्थिति उसे परवर्ती अस्तित्व की ओर ले जाती है जिसमें कहीं अन्तिम मंजिल नहीं आती। मनुष्य अपने को स्वयं परिभाषित करता है और अनेकानेक सम्भावनाओं के बीच चुनाव करता हुआ आगे बढ़ता है; पर कोई चुनाव अन्तिम नहीं होता। का.सू.वा. की उक्ति को इसी सन्दर्भ में समझा जा सकता है। भूमिका का अनिश्चित होना इस बात पर निर्भर है कि मनुष्य का परिस्थितियों पर अधिकार नहीं है; इसीलिए केवल दर्शक मात्र रह गया है। यही आज औसतन देशवासी मनःस्थिति है।

प्रस्तावना के माध्यम से राकेश ने एक बड़े सत्य को छोटे दायरे में पकड़ने का प्रयास किया है। वे यह भी मानते हैं कि नाटक में पात्र की और जीवन में व्यक्ति की भूमिका एक-सी होती है। पर वस्तुतः व्यक्ति की भूमिका नाटक में सुनिर्धारित होती है; जीवन में निर्धारित नहीं होती। इसलिए नाटक के पात्र और वास्तविक जगत् के व्यक्ति की स्थिति में अन्तर होता है। राकेश नाटक और जीवन को समकक्ष मान लेते हैं; नाटक और जीवन जैसे एक हो जाता है। इसीलिए का.सू.वा. कहता है : 'यह नाटक भी अपने में मेरी तरह ही अनिश्चित है।' अनिश्चितता जीवन की हो सकती है, नाटक की नहीं। किन्तु जब नाटककार जीवन और नाटक को एक मान लेता है तो दोनों अनिश्चित हो सकते हैं।

अतः इस सन्दर्भ में प्रस्तावना सर्वथा निरर्थक ही लगती हो, ऐसा नहीं कहा जा सकता। सच्चाई यह है कि जैसी सामान्य स्थितियों का यह नाटक है, उसमें अर्थ और शिल्प का सौष्ठव यही प्रस्तावना प्रस्तुत करती है। प्रस्तावना और एक व्यक्ति द्वारा चार पुरुषों की भूमिका निभाने की रंगमंचीय युक्ति के बिना यह नाटक बहुत सामान्य नाटक होता। इससे यह नाटक कुछ-कुछ रूपवादी ज़रूर बन गया है; किन्तु यह उसे कलात्मकता प्रदान करता है। इस युक्ति में चाहे और कुछ न हो, एक प्रकार की काव्यात्मक कल्पना का समाहार अवश्य है। यह एक नए यथार्थ—वैचारिक यथार्थ का

सृजन करती है। सबसे बड़ी बात यह है कि यह रंगमंच और प्रेक्षक के बीच सीधा सम्बन्ध क़ायम करती है।

इसके अतिरिक्त कथावस्तु का ताना-बाना बुनने में लेखक ने कई और नाटकीय युक्तियों का भी प्रयोग किया है। घर की चीज़ों का बिखराव, पुरुष का फाइलें झाड़ना, अख़बार पढ़ना, लड़के का तस्वीरें काटना, पैड पर कार्टून बनाना, मोटर की बैटरी डाउन होना, डिब्बे का खोलना, पैंट में कीड़ा घुसने का नाटक करना, छोटी लड़की का हकलाना जैसी अनेक बातें नाटक में समाहित हुई हैं। ये युक्तियाँ नाटक की कृश काया में भराव का काम करती हैं। इस भराव का सबसे हीन उदाहरण नाटक के अन्त में सावित्री और जुनेजा का विवाद है। आलोचकों की इस बात से हम पूरी तरह सहमत हैं कि ये सब घिसी-पिटी रंगचर्याएँ हैं। शोभना भूटानी ने ठीक ही कहा है कि 'मंच निर्देशों में प्रतीकों का सहारा लेकर इस दोहरी रिक्तता को भरने की जो कोशिश की गई है, वह भी अतिरिक्तता मात्र लगती है। स्थूल चरित्रों की भावनात्मक स्थितियों का मुहावरा प्रतीक हो ही नहीं सकते।...दूसरे, नाटक में संकेतों को इतना खींचा गया है कि वे सपाट बन जाते हैं।...नाटककार प्रतीकों को इतना खोंचता है कि वे प्रतीक रह ही नहीं जाते।' फिर भी इन युक्तियों का सौन्दर्य और सदुपयोग निर्देशक पर निर्भर करेगा। एक अच्छा निर्देशक उन्हें रूढ़-रूप में प्रस्तुत करने के बजाय कलात्मक रूप देने का प्रयत्न करे, यह सम्भव है जिससे दोष का परिहार हो सकता है। निश्चयतः यह दोष नाटक के रूपबन्ध के कारण आया है। अन्विति के लिए महेन्द्रनाथ और सावित्री के पिछले भोगे हुए जीवन को सूच्य ही बनाया जा सकता था, उसे मंच पर ला पाना सम्भव न था। इससे प्रभाव में कमी आ जाना स्वाभाविक है। वस्तुतः सावित्री और महेन्द्रनाथ की पूर्वस्थितियों की सपाटबयानी खलती है। नाटक का तीसरा अंक भी बोझिल हो गया है।

आधे अधूरे के पात्र जीवन्त हैं। फिर भी उनमें एक विचित्र एकरसता है। उन सबमें यथास्थिति का मोह है। इसीलिए एक प्रकार की स्थिरता है—कहीं गति, स्वतन्त्रता और वरण की तीव्र आकांक्षा नहीं। वे एक अप्रामाणिक जीवन जीते हैं और व्यवस्था तक सुविधा से समझौता किए रहते हैं। सब एक प्रकार का नकारात्मक जीवन जीते हैं। इस दृष्टि से सारा परिवार एक ही मनःस्थिति का शिकार है। वे अपने व्यवहार में एक साँचे में ढले हैं। स्थिति का ठहराव, सम्बन्धहीनता, स्वतन्त्र इच्छाशक्ति का अभाव इसके कारण बनते हैं किन्तु सबसे बड़ी बाध्यता अन्तर की है। सावित्री, जगमोहन और सिंघानिया ही नहीं अशोक, बिन्नी, किन्नी भी अपने छलावे में जी रहे हैं। यह स्थिति सब पात्रों को एक जगह पर खड़ा कर देती है। इस प्रकार सारे पात्रों का सामान्यीकरण हुआ है। यहाँ तक कि पात्र परिचय में नाटककार उन्हें पुरुष 1, 2, 3, 4, लड़का, लड़की, स्त्री के रूप में अंकित करता है। अपनी व्यक्तित्वहीनता और दुर्बलताओं के बावजूद सभी पात्र सहानुभूति जगाते हैं। नाटक में सावित्री से सहानुभूति होती है क्योंकि उसे पूरा आदमी नहीं मिला; और महेन्द्रनाथ से भी सहानुभूति होती है क्योंकि वह अपने ही अधूरेपन से पीड़ित है।

आधे अधूरे के पात्र रेखाओं में उभरते हैं, किन्तु सीधी रेखा में नहीं। उनके व्यक्तित्व में स्वभाव और व्यवहार की कई विचित्रताएँ भरी पड़ी हैं। इसलिए रूढ़ होने के नाते वे अपनी समग्रता का आभास देते हैं। नाटक में स्थितियाँ कम हैं, इसलिए पात्र स्थितियों में कम और प्रतिक्रियाओं में अधिक उभरते हैं। हर पात्र क्राइसिस में है और झल्लाहट के बीच उभरता है। कथावस्तु की स्थूलता के कारण भी चरित्र केवल स्थितियों और मनःस्थितियों में जीते हैं। सबका नपा-तुला अस्तित्व नज़र आता है। पर सभी मुखर हैं। एक-दूसरे में समाहित होते हुए भी वे परस्पर विभाजित हैं और कहीं इतने आत्मबद्ध लगते हैं कि एक-दूसरे की ओर उन्मुख ही नहीं लगते—विशाल समुद्र में तैरते हिमखंडों की तरह जो कभी एक-दूसरे की तरफ़ मुखातिब ही नहीं होते, केवल टकरा-भर जाते हैं। उनके अवचेतन में बहुत कुछ भरा है। इसीलिए वे चाहे रेखाओं में अंकित हों, या 'कैरिकेचर' हों, जाने-पहचाने लगते हैं और अपना प्रभाव बख़ूबी डालते हैं। किन्तु इतना निश्चित है कि यह नाटक अलग-अलग स्थितियों और मनःस्थितियों का नाटक बनकर रह गया है। उसमें प्रश्न उठाए गए हैं, पर नाटक प्रश्नों से सीधे जूझता नहीं है और न किसी गहरे अनुभव की ओर ले जाता है। इसीलिए जहाँ से नाटक शुरू होता है, वहीं पहुँचकर अन्त भी हो जाता है—आरम्भ और अन्त की स्थिति में कोई फ़र्क़ नहीं पड़ता। जो पहले था, सो अब भी है—एक बिना पड़ाव का सफ़र है जो अपने को दुहराता जाता है। वैचारिक स्तर पर भी **आधे अधूरे** इकहरा नाटक है। सब चेहरों का एक होना, परिस्थितियाँ बदल जाने पर भी आदमी वही रहना—ये सब अधूरे और इकहरे तथ्य हैं।

आधे अधूरे की सिद्धि समसामयिक कथ्य के साथ-साथ उसकी रंगानुभूति और भाषिक संरचना में निहित है। शब्द और दृश्य का इस नाटक में बहुत ही सफल प्रयोग हुआ है। राकेश एक ओर मंच की दृश्यात्मक क्षमता का उपयोग करते हैं, दूसरी ओर शब्द की सर्जनात्मक शक्ति का। कार्य और भाव के सहयोग से वे एक विशिष्ट वातावरण का निर्माण करते हैं। जहाँ तक भाषा का प्रश्न है, राकेश इस नाटक में बोलचाल की भाषा का ऐसा प्रयोग करते हैं कि शब्द अपनी सामर्थ्य से भी अधिक अभिव्यक्त करते हैं। यही बात संवादों के विषय में भी कही जा सकती है। कुल मिलाकर अपने कथ्य, वातावरण, चरित्र सृष्टि और भाषिक क्षमता में यह नाटक अपनी एक छाप छोड़ जाता है। फिर भी यह नाटक कई प्रश्न खड़े करता है—क्या यह नाटक समग्र मध्यवर्ग की त्रासदी है ? इसकी प्रस्तावना में कितना सत्य है ? क्या दूसरी स्त्री या दूसरा पुरुष आधे-अधूरे घर को इसी प्रकार झेलता ? क्या यह मानव जीवन के घरेलू सम्बन्धों का शाश्वत नाटक है ? क्या सामान्यीकरण—सबको एक लाठी से हाँकना ठीक है ? एक बड़ा प्रश्न यह उठता है कि क्या **आधे अधूरे** की नारी अधुनातन नारी है ? इस नाटक के लिखे जाने के इतने वर्षों बाद सावित्री पुरानी पड़ गई है। नई नारी झगड़ों, घुटन, व्यभिचार, व्यतिक्रम, विरोध जैसे व्यवहारों को किसी हथियार के रूप में नहीं मानती। उसके सामने क़ानूनी अधिकार, तलाक, नैतिक मूल्य, विवेक, स्वतन्त्रता जैसे कई मूल्य हैं। विरोध का निबटारा बाहरी सम्बन्धों से भी सम्भव नहीं। आज की नारी

के सामने अपनी स्त्री सुलभ लालसाओं, दुर्व्यवहारों का सामना करने के लिए नैतिक मार्ग भी हैं। किन्तु राकेश के इस नाटक के सम्बन्ध में प्रायः मध्यमवर्ग की दुहाई दी जाती है। कहा जाता है कि यह परिवार विशेष की कहानी नहीं, मध्यमवर्ग के ह्रासोन्मुख जीवन मूल्यों की घुटन और निष्क्रिय अनैतिक सड़न और उसे सहन करने की कहानी है, पर केवल पश्चिम के अनुकरण पर मध्यमवर्गीय परिवारों और उनके जीवन मूल्यों के विघटन की बात करना काफ़ी नहीं। नए युग की सावित्री आज परम्परा के विरुद्ध जी रही है; पर इसमें सावित्री का ही दोष नहीं, राकेश की नाट्य-दृष्टि का भी दोष है जिसने अपने नाटकों में चुनौती के बजाय यत्र-तत्र लघुता को प्रश्रय दिया और वह भी एक विचारधारा का खोल ओढ़कर। कुछ समय बाद यही उनके नाटकों की कमज़ोरी बनकर सामने आ सकती है।

रचना दृष्टि

रचना दृष्टि : बिन्दु से बिन्दु तक

कितने-कितने बिन्दु खोजे हैं आज तक तुमने ?...जाओ एक और बिन्दु खोजो। कितने-कितने शब्दों से ढाँपा है उन बिन्दुओं को ?...जाओ कुछ और शब्द ढूँढ़ो ! परन्तु अन्त में कहाँ रह जाते हैं तुम्हारे ये शब्द ? फिर भी वही के वही बने हो तुम। वही...वही...वही।

—लहरों के राजहंस : नन्द के प्रति सुन्दरी की उक्ति

आज युग के सामने अनेक प्रश्न उभर आए हैं। आदमी सवालों की नोक पर टँगा है और सारी तलाश के बावजूद उसे यदि कुछ हासिल हुआ है तो नंगे ज़ख़्म, जो हमारी पीढ़ी को विरासत में मिले हैं। इन सवालों की नोक पैनी है—उनकी धार हमारे अस्तित्व के 'क्या' और 'क्यों' से बनी है। और जिस तरह सवाल शाश्वत हैं, उसी तरह उनकी नोक पर उत्तरों के बिन्दु टिकाने का प्रयत्न भी शाश्वत है। वह भी कोई एक बिन्दु नहीं, न जाने कितने बिन्दु हैं। यही चिन्तन की नियति है। ऊपर नन्द के प्रति सुन्दरी की जो उक्ति उद्धृत की गई है, वह सही अर्थों में मोहन राकेश की रचना-दृष्टि को व्यक्त करती है।

राकेश उन साहित्यकारों में से हैं, जिन्होंने ज़िन्दगी को बहुत गहराई से महसूस किया है। उन्होंने एक प्रामाणिक ज़िन्दगी भोगी थी; साथ ही उन्हें एक संवेदनशील और भाव-प्रवण हृदय भी मिला था। लेखक लेखक बाद में होता है, आदमी वह पहले होता है। राकेश के जीवन में अस्तित्व का संकट कई बार आया था। शारीरिक भूख और आन्तरिक प्यास की जैसी आकुलता उन्होंने जीवन में झेली थी, वह उनके व्यक्तित्व की सहज शक्ति की परिचायक थी। अनेक विरोधी गुण जिस व्यक्ति के स्वभाव में हों और जिसने जीवन और जगत् से समझौता न किया हो, वह एक विचित्र नियति भोगने के लिए सदा से बाध्य रहा है। इसलिए राकेश अन्दर से बहुत चिन्तनशील व्यक्ति रहे। उस चिन्तन में अनुभूति की प्रामाणिकता भी समाहित है। उनका साहित्य बहुत कुछ अस्वीकार का साहित्य है, किन्तु उसमें एक ज़बर्दस्त स्वीकार भी है और यह स्वीकार है आन्तरिकता का, अनुभवों का और अन्तरात्मा के संकट का। वे एक तरह से हृदय के अंधलोकों के राही रहे हैं और आलोक की खोज में बहुत गहरे उतरे हैं। एक-एक

बिन्दु की खोज उनके लिए एक उपलब्धि रही है।

राकेश की रचना-दृष्टि की जड़ें अस्तित्ववादी चिन्तन में खोजी जा सकती हैं। हिन्दी में इस चिन्तन के कुछ ऊपरी तत्त्वों को आज़माया गया है और कुछ अचेतन को लेकर मनोविज्ञान का सहारा लिया गया है। मनोविज्ञान मानवीय दुर्बलता का सकारण समर्थन करता है और अस्तित्ववाद जीवन को अवश, निरुपाय, रिक्त और निरर्थक मानता आया है। इसके अतिरिक्त वह मानव के स्वयं-निर्माण पर तो विश्वास करता है पर उसे क्षुद्र भी मानता है। मूलतः उसकी दृष्टि में मानव की लघुता ही रही है। अस्तित्ववाद जहाँ अस्तित्व की सारहीनता की बात करता है, वहीं उसे सम्पूर्ण वस्तु के रूप में नहीं देखता है। उसके लिए शुभ और अशुभ का अन्तर कृत्रिम है। जीवन में चुनाव है पर उसे अर्थ देने की असमर्थता भी साथ लगी है। यदि कुछ प्रामाणिक है तो वह आभ्यन्तर अनुभव है जिसमें वह जीता है। राकेश किसी न किसी रूप में इस तरह के विचारों से प्रभावित थे।

वैसे किसी भी रचनाकार के लिए रचना ही मूल लक्ष्य होती है। राकेश की दृष्टि भी रचनाकार की है; भले ही रचना के अन्दर वे अपने को दुहराते मिलते हैं। कथावस्तु, चरित्र-सृष्टि, नाट्य स्थितियों और नाटकीय शैली—सबकी उनके नाटकों में उनकी आन्तरिक वैचारिक परिणति दिखाई देती है। वे अपने विचार, भाव-बोध और दृष्टि को जीवन के एक सम्पूर्ण चित्र के साथ रला-मिला देते हैं। इसीलिए कहीं-कहीं राकेश अपने नाटकों में दार्शनिक मुद्रा धारण कर लेते हैं। पर यह दार्शनिक मुद्रा उनकी रचना-दृष्टि की देन है। उनके नाटकों में विचार भी हैं और अनुभूति भी; पर अलग-अलग तत्त्वों के रूप में नहीं। दोनों का समन्वय उनकी रचना-प्रक्रिया का महत्त्वपूर्ण अंग है। यही कारण है कि उनकी नाट्य कृतियाँ हृदय और बुद्धि दोनों की अपेक्षाओं की पूर्ति करती हैं। अनुभूति उनको अद्भुत भाव-क्षमता प्रदान करती है और बौद्धिकता उस गरिमा से युक्त करती है जिसका प्रायः हिन्दी नाटकों में अभाव रहा है। फिर भी उनके नाटकों को विचार-नाटक नहीं कहा जा सकता। विचार अलग से नहीं, उनके नाटकों की संवेदना में से उपजते हैं। वस्तुतः राकेश विचार का उपयोग मात्र करते हैं जिसमें विचार भाव का और भाव विचार का अनुसरण करता है। वस्तुतः वे अपने नाटकों में कई प्रश्न उठाते हैं, मनुष्य के अस्तित्व को परिभाषित करते हैं और उसकी नियति की तलाश करते हैं। और यह कार्य न अकेला विचार कर सकता है और न भावना। दोनों के योग से राकेश अपने नाटकों को सारतत्त्व प्रदान करते हैं और उससे पाठक/प्रेक्षक की रसग्राही मानसिकता का निर्माण करते हैं।

राकेश में एक बद्ध रचना-दृष्टि का आभास मिलता है। उनके प्रायः सभी नाटकों में एक निश्चित भाव-बोध और विचार-तत्त्व की पुनरावृत्ति मिलती है। कामू का यह कथन कि मनुष्य कुछ ही परिचित अनुभवों के बीच जीता है और सारे जीवन-भर उन्हीं का रूपान्तर या परिष्कार करता है,[1] राकेश पर पूरी तरह चरितार्थ होता है। राकेश के सभी नाटकों में एक-सा अनुभव निहित है, केवल बल और मात्रा का अन्तर है। यहाँ तक कि कहानियों और उपन्यासों में भी उन्होंने वे ही समस्याएँ उठाई हैं जो उनके

नाटकों में विद्यमान हैं।

तब एक समस्या स्वाभाविक रूप से उठती है : लेखक का अनुभव क्या है ? अस्तित्व की किसी स्थिति में वह अन्दर-ही-अन्दर क्या महसूस करता है ? इन प्रश्नों का उत्तर आधुनिक जीवन-सन्दर्भों में ही खोजा जा सकता है। हमारा युग मानव-संकट का युग है, जिसमें मशीन बनाकर आदमी स्वयं मशीन की चपेट में आ गया है। विघटन की राह से वह ऐसी पंगु स्थिति से गुज़र रहा है जहाँ सन्त्रास, पीड़ा, सम्बन्धहीनता और आत्मनिर्वासन उसकी नियति बन गए हैं। कभी ख़यालों या आदर्शों की दुनिया आदमी को बहलाने के लिए काफ़ी थी, किन्तु आज व्यावहारिक जीवन का यथार्थ अस्तित्व की दारुण पीड़ा के बीच उसे सवालों के जंगल में भटका देता है। इसीलिए लेखक के सामने आज कुछ बुनियादी सवाल महत्त्वपूर्ण हो गए हैं। आज नई परिस्थितियों ने ऐसे कुछ नए अनुभव दे डाले हैं कि वे हमारे अस्तित्व के अंग बन गए हैं। ये हैं—अस्तित्व की पीड़ा, सन्त्रास, अलगाव, निजता की खोज, संघर्ष, स्वाधीनता, वरण, स्त्री-पुरुष सम्बन्ध और त्रासद अनुभव। और इन अनुभवों के बीच मनुष्य अपने स्वत्व की खोज में अन्ततः स्वयं अपने लिए एक बहुत बड़ा सवाल बन गया है।

राकेश स्वयं यह मानते थे कि 'मेरे लिए अनुभूति का सीधा सम्बन्ध परिवेश से है और यथार्थ है मेरा समय और परिवेश।' अनुभूति और परिवेश तथा उनका यथार्थ मिलकर उनके नाटकों का संरचनात्मक स्वरूप निर्धारित करता रहता है। आदमी के अस्तित्व के संकट में आज परिवेश का बहुत बड़ा हाथ हो गया है। जीवन सत्व है और जीवन का सत्व इस बात में है कि व्यक्ति सदा एक ऐसे परिवेश में रहता है जो उसे अनचाहे-अनजाने ही मिला है। इस आत्मपरक स्थिति और परिवेश की स्वीकृति—इन दो आयामों पर ही मनुष्य का जीवन निर्भर करता है। मनुष्य जीवन नहीं है, मनुष्य और उसके परिवेश के बीच जो कुछ घटता है, वही जीवन है। मनुष्य और उसके परिवेश के बीच का यह सम्बन्ध ही भावना, विचार, स्थिति, समस्या और संकट को जन्म देता है, जिनसे वह चारों ओर अपने को घिरा पाता है। ऐसी स्थिति में संघर्ष एक अनिवार्यता बन जाता है, जिसमें मनुष्य की अपनी निजता की तलाश महत्त्वपूर्ण हो जाती है। राकेश के नाटक व्यक्ति और परिवेश के इसी द्वन्द्व में जन्म लेते और विकसित होते हैं और उन्हीं के बीच से जीवन का यथार्थ मुखर होता है।

राकेश के नाटकों में परिवेश इस दृष्टि से सक्रिय है कि आज के युग में व्यक्ति के ऊपर सामूहिक परिवेश का बहुत बड़ा दबाव है—समूह, संगठन, व्यवस्था सब व्यक्ति पर हावी हैं, जिससे उसकी वैयक्तिकता के लिए बड़ा ख़तरा उठ खड़ा हुआ है। राकेश के नाटक समूह की इसी भूमिका के विरुद्ध व्यक्ति को स्थापित करते हैं। पर व्यक्ति परिवेश के नीचे दबा लगता है। **लहरों के राजहंस** में नन्द और सुन्दरी समूह की शक्ति को महसूस करते हैं, जब कोई भी अतिथि कामोत्सव में उपस्थित नहीं होता। **आषाढ़ का एक दिन** में ग्रामीण समाज का सारा सत्व कालिदास के विरुद्ध है और उज्जयिनी में जाकर तो परिवेश ही उसकी विनाशकारी स्थिति का कारण बनता है। व्यक्ति और

समाज का ही नहीं, व्यक्ति का निजी सम्बन्ध भी एक परिवेश का निर्माण करता है। **आधे अधूरे** में यही परिवेश अधिक मुखर है, जो व्यक्ति के कार्य-कलाप को नियन्त्रित करता है। राकेश अपने नाटकों की संरचना व्यक्ति और उसके परिवेश के सन्दर्भ में करने के कारण कथावस्तु और चरित्र के तन्तु अन्तर्द्वन्द्व से रचते हैं। उनके अपने जीवन की आकुलता, विकल्पहीनता और भटकाव की उसमें महत्त्वपूर्ण भूमिका रही है। हर बात को दो तरह से सोचना, दोनों तरह से ठीक पाना, स्वीकार करना, पर स्वीकार न कर पाना उनका स्वभाव रहा है जो उनकी नाट्य संरचना का आधार बना।

यही आन्तरिक और बाह्य जीवन-स्थितियों में परस्पर विरोध पैदा करता है। एक ओर स्थापित मूल्यों और जागतिक वस्तुओं का निर्मित परिवेश है; दूसरी ओर व्यक्ति की अपनी स्वतन्त्रता, वरण और मानवीय चेतना। मनुष्य को एक ओर बाह्य परिवेश का मुक़ाबला करना पड़ता है, दूसरी ओर उसे अपना भी सामना करना पड़ता है। दोनों चीज़ें मिलकर परिस्थितियाँ बन जाती हैं। राकेश ने अपने नाटकीय पात्रों को उन्हीं दुहरी परिस्थितियों में खड़ा किया है। उनके नाटकों का जीवित संसार उन व्यक्तियों का है, जो परिस्थितियों से घिरे हैं; उनकी टकराहट में आते हैं और टूट जाते हैं। **आषाढ़ का एक दिन** में व्यक्ति और राज्य, **आधे अधूरे** में यथार्थ और असत् आस्था टकराहट में आती है। **लहरों के राजहंस** में यह टकराहट व्यक्ति के एक खंड और दूसरे में दिखाई देती है। किन्तु परिवेश में रहना, टकराना या टूटना ही जीवन के लिए काफ़ी नहीं होता। एक स्थिति मोहभंग की भी आती है। अपने ही अन्दर भावनाओं का एक अहसास इस आन्तरिक अपेक्षा को भी जन्म देता है कि वह अपने अन्दर जो हो रहा है उसको पहचाने। यह पहचान जीवन में 'अन्य' के माध्यम से होती है। इस 'अन्य' में परिस्थितियाँ भी शामिल हैं। अन्ध शक्तियों के बीच जूझता हुआ आदमी कभी विभाजित हो जाता है। इसी स्थिति में राकेश के पात्र कभी एक से अपने को जोड़ते हैं कभी दूसरे से। फिर कोई भी ज़िन्दगी उन्हें रास नहीं आती—न कवि की, न राज्याश्रय की; न सुन्दरी का भोग ही तुष्ट करता है और न बुद्ध का विराग। यह नियति ही उन्हें अपने विरोध में खड़ा करती है। यह विरोध ही राकेश के नाटकों का केन्द्रीय तत्त्व है। पर विरोध कहीं विद्रोह बनता नहीं लगता। इस केन्द्र के दोनों ओर अति सीमाओं पर अस्तित्व के विकल्प एक-दूसरे से पीठ फेरे दिखाई देते हैं।

इन विकल्पों के बीच असुरक्षा का भाव बहुत सक्रिय दिखाई देता है। कालिदास को असुरक्षा का भाव भटकाता है। महेन्द्रनाथ और उसके बेटे-बेटी इसी भाव के कारण उस घर से चिपके दिखाई देते हैं, जो उन्हें अपना नहीं लगता। सावित्री की पूरे पुरुष की कामना के पीछे भी यही भाव है। **लहरों के राजहंस** में सुन्दरी सबसे अधिक असुरक्षा की भावना से ग्रस्त है। गौतम और यशोधरा के प्रति उसके मन में जो विरोध का भाव है, उसका कारण भी यही है। वह नन्द को अपने रूप का दास बनाकर रखती है क्योंकि उसके अवचेतन में इसी का भय बैठा हुआ है। नन्द के मन में असुरक्षा का भाव एक भिन्न आध्यात्मिक सन्दर्भ में उभरता है। वह ठीक ही कहता है : 'आत्मरक्षा और

आत्मविनाश इन दो प्रवृत्तियों के बीच में मैं एक साथ जिया।' असुरक्षा का भय सुरक्षा के प्रयत्नों की ओर ले जाता है, किन्तु कालिदास नन्द, सावित्री, सुन्दरी, मल्लिका, महेन्द्रनाथ सभी भँवर में घुसकर महसूस करते हैं, जैसे पैर तले की ज़मीन खिसक गई है। वे एक ऐसे सीमान्त पर पहुँचते हैं जहाँ जाकर परिवेश और परिस्थिति में न उससे आगे जाना सम्भव है, न पीछे।

असुरक्षा की भाँति अलगाव राकेश के पात्रों में मुख्य है। सभी पात्र इस नियति को भोगते हैं। अलगाव और अकेलापन कालिदास के अस्तित्व के प्रमुख अंग बन जाते हैं। वह प्रारम्भ में ही अपने ग्रामीण समाज में इस स्थिति में दिखाई देता है, जब वहाँ मातुल की गौएँ चराता है और कोई उसको कवि का सम्मान नहीं देता। इस अलगाव और अकेलेपन से बचाने के लिए ही मल्लिका उसे उज्जयिनी चले जाने को बाध्य करती है। किन्तु वह एक अलगाव से बचकर दूसरे में फँस जाता है। वह राजसत्ता का अस्त्र बन जाता है। फलतः वह इस अहसास से जूझता लगता है कि उसका अस्तित्व उसका अपना नहीं, पराया है। उसके सर्जनात्मक अस्तित्व पर राज्य का अधिकार होने के नाते वह अपनी रचनाधर्मिता से विचित्र अलगाव अनुभव करता है। एक दूसरे स्तर पर अपनी सर्जन प्रेरणा मल्लिका से वह इतना विलग हो जाता है कि काश्मीर जाते हुए उससे मिलने का साहस नहीं जुटा पाता। वह अपने ग्रामीण अतीत से इतना कटा हुआ अनुभव करता है कि प्रियंगु को गाँव से कुछ वातावरण साथ ले जाने की आवश्यकता पड़ती है। वह जो कुछ है और उसे जो होना चाहिए था, कालिदास की आन्तरिक प्रवृत्ति और व्यावहारिक जीवन का यह अन्तर—उसे अतीत की स्मृतियों में लीन कर देता है। जीवन की परिस्थितियाँ जब उसे असहाय छोड़ जाती हैं तो वह मल्लिका के रूप में अतीत को प्राप्त करना चाहता है, किन्तु मल्लिका का वर्तमान उसका मोहभंग कर डालता है।

विलगाव और जुड़ाव की स्थिति का राकेश ने वैयक्तिक रूप से अनुभव किया था। उन्होंने लिखा है : 'मैं वैयक्तिक और साहित्यिक दोनों स्तरों पर अपने को ज़िन्दगी से जुड़ा पाता हूँ, पर जुड़े होने का अर्थ ज़िन्दगी की सब परिस्थितियों को स्वीकार करके चलना नहीं है। ज़िन्दगी में बहुत-कुछ है जिसके प्रति विद्रोह और आक्रोश मेरे मन में है। पर यह सब ज़िन्दगी के ऐतिहासिक उफान के अन्तर्गत आता है। वह आज जैसा है, कल वैसा नहीं रहेगा। इस विद्रोह और आक्रोश की ही कुछ परिस्थितियाँ हैं जिनमें मैं कई बार अपने को अकेला भी पाता हूँ, पर यह अकेलापन जूझने की एक स्थिति है, अलगाव नहीं। ज़िन्दगी के बीच अकेला पड़कर अपने जुड़े होने का निर्वाह करना है।'[2] सच्चाई यह है कि यह अलगाव मानसिक है, इसलिए समाज से दूर रहने का, समस्याओं को व्यक्तिवादी दृष्टि से देखने का अलगाव है। फिर भी असुरक्षा की भावना इन पात्रों में बहुत दूर तक नहीं खींची जा सकती। कई पात्रों में यह सिर्फ़ मनोवैज्ञानिक है।

लहरों के राजहंस में अलगाव और अकेलेपन की यही महत्त्वपूर्ण भूमिका है। सुन्दरी को ही लीजिए—उसका अलगाव एक आत्मकेन्द्रित और रूप-गर्विता नारी का अलगाव है। वह बुद्ध और यशोधरा के प्रति घृणा के भाव से चालित है। कामोत्सव में लोगों

के न आने से उसका अहं और भी आहत होता है। दूसरों के प्रति सहज भाव का अभाव और उन्हें मात्र उपकरण मानने की विडम्बना उसे बहुत दयनीय बना डालती है। और अन्ततः सबसे विचित्र स्थिति वह होती है जब वह अपनी अभीष्ट प्रतिमा और वास्तविक स्थिति में संगति नहीं बिठा पाती। रूप के जिस आकर्षण को वह अपने अस्तित्व का आधार मानती थी, वही जब प्रवंचना सिद्ध होता है तो अन्तिम स्थिति जो उसके पास बच रहती है, वह अलगाव की ही होती है। नन्द भी सुन्दरी की भोगों की दुनिया में जीता है; पर भोगों की वही दासता, स्कन्द की वही अवरुद्ध गति उसे आत्म-पराएपन का अहसास दिलाती है। सुन्दरी और बुद्ध के बीच बँटा वह जिस अलगाव में जीता है, उसमें वह अपने को असहाय अनुभव करता है; किन्तु फिर भी वह कहीं-न-कहीं अपने से जूझता लगता है।

अलगाव की यह व्यक्तिवादी स्थिति कई सन्दर्भों में दो विरोधी स्थितियों के बीच उत्पन्न निज की समस्या पर आधारित है। नन्द दो विरोधी मनःस्थितियों से जूझता है। जूझने की इस स्थिति के बावजूद राकेश के सभी नाटकों में ऐसी स्थिति नहीं है। उनमें पात्र जूझने के बजाय स्थितियों से जुड़े अधिक दिखाई देते हैं, जहाँ विकल्प नहीं दिखाई देते। कालिदास के सामने भी वरण का प्रश्न आता है। यद्यपि वरण मनुष्य की नियति है, पर हर स्थिति में वरण की स्वतन्त्रता विद्यमान हो, ऐसा नहीं लगता। इसीलिए राकेश के बहुत-से पात्रों के लिए वरण की सम्भावनाएँ नहीं दिखाई देतीं। महेन्द्रनाथ के सामने घर और बाहर के बीच घर को वरण करने की विवशता के अतिरिक्त और क्या है ? कालिदास के लिए मल्लिका के घर से बाहर निकलने के अतिरिक्त और कोई चारा नहीं लगता। इसी प्रकार, मल्लिका, अम्बिका, विलोम, सावित्री, सुन्दरी अनेक पात्रों के लिए वरण की सम्भावनाएँ समाप्तप्राय हो चुकी हैं। सावित्री एक के बाद एक पुरुष को चुनती है और धोखा खाती है। बेटे की बातों से तिलमिलाकर वह जगमोहन के पास जाने का निर्णय लेती है; पर वह निर्णय उसे दूर तक नहीं ले जाता; इसलिए हताश घर लौट आती है। वरण जैसे उसके लिए बचा ही नहीं। फलतः जुनेजा का यह प्रश्न उसके अस्तित्व पर कील की तरह ठुक जाता है : 'क्या सचमुच कोई चुनाव नज़र आया तुम्हें ? बोलो, आया है नज़र कहीं ?'

राकेश के पात्र चुनाव की इस विवशता में रचते दिखाई देते हैं। वे उबरने के लिए उनका आश्रय लेते हैं, पर वरण करते ही उनकी विडम्बना में फँस जाते हैं। जिनके वरण की सम्भावनाएँ चुक गई हैं, वे भी उससे मुक्त नहीं क्योंकि वरण न कर पाना भी अप्रत्यक्ष रूप से वरण करना ही है। वह भी एक प्रकार से नास्तिभाव को ही अपनाना है। सबसे बड़ी बात यह है कि सही वरण उस स्थिति में संभव नहीं जब व्यक्ति 'असत् आस्था' से ग्रस्त हो। इसीलिए कालिदास के जीवन में वरण का क्षण विचित्र संहारक शक्ति लेकर आता है। असत् आस्था के अतिरिक्त उसमें समय का तत्त्व भी विशेष महत्त्व रखता है। कालिदास उज्जयिनी के मार्ग का जब वरण करता है तो विलोम एक टिप्पणी करता है : 'घिरे हुए मेघों ने आज असमय अन्धकार कर दिया। अम्बिका, क्या

तुम्हें **समय का ज्ञान** नहीं रहा ?' समय का यही ज्ञान न होना वरण को निष्फल बना डालता है। कालिदास अन्ततः 'समय के साथ इच्छा के द्वन्द्व' में अपने को पराजित महसूस करता है और अपने वरण में काल के 'फैक्टर' को स्वीकार करता है : 'मुझे **वर्षों पहले** यहाँ लौट आना चाहिए था ताकि... ।' और सावित्री भी इसी तथ्य पर पहुँचती है : '**दस साल पहले** कहना चाहिए था मुझे...जो कहना चाहती हूँ।' और यह भी कि 'कल और आज में फ़र्क़ होता है।'

राकेश के नाटक व्यक्ति के जीवन में व्यतीत होते काल का और कालखंड में क्षण विशेष का महत्त्व प्रकट करते हैं। कालिदास जब मल्लिका की झोंपड़ी में लौटकर आता है तो उसकी अथ से आरम्भ करने की इच्छा और समय का द्वन्द्व उसे पराजित कर देते हैं। इसी प्रकार नन्द क्षण को नहीं नकार पाता और क्षण के आवेश में 'दूसरा व्यक्ति' बन जाता है। तथागत द्वारा हाथ में कमंडलु थमाने का क्षण उसके आगे के द्वन्द्व की धुरी बन जाता है। यही बात सावित्री के बारे में भी कही जा सकती है। जब क्षण आता है तो वह चुनाव नहीं कर पाती। यदि उनकी डायरी (14.12.58) का सहारा लें तो राकेश को इस बात का अहसास था कि क्षण महत्त्वपूर्ण होता है क्योंकि उसमें बरसों की सन्धियाँ टूटती हैं और जीवन की सार्थक दिशा उसमें सिमट जाती है। राकेश के नाटकों में उससे त्रासदी घटती है और वह त्रासदी किसी क्षण की होती है।

राकेश यह मानकर चलते हैं कि मनुष्य अपनी इच्छा, कर्म और वरण के माध्यम से अपने को निरन्तर बनाता जाता है। व्यक्ति को हर 'कल' बदल देता है। विलोम इसीलिए उज्जयिनी को प्रस्थान करते कालिदास से पूछता है : 'तुम अभी तक वही व्यक्ति हो जो कल थे ?' और वही कालिदास जब वर्षों बाद एक दिन मल्लिका के पास लौटता है तो अनुभव करता है, 'सब-कुछ बदल गया है।' सुन्दरी इसी बात को इन शब्दों में कहती है, 'जो बात इस क्षण नहीं सोची जा सकती, वह अगले क्षण सोची जा सकती है।' आदमी की सार्थकता इस बात में है कि वह प्रति क्षण अपना जिस रूप में निर्माण करता है, या जिस तरह अपने को दूसरे क्षण सोचने लायक़ बना लेता है, उसमें वरण का दायित्व उसका अपना ही है। इसीलिए सुन्दरी 'अपने उद्वेग का वास्तविक कारण स्वयं' अपने को मानती है। प्रश्न पूछने की इच्छा रखनेवाले नन्द को कही गई भिक्षु आनन्द की यह उक्ति भी इसी दृष्टि से सार्थक है : 'मुझसे पूछना चाहते हो ? परन्तु जो व्यक्ति तुम्हारे किसी भी प्रश्न का उत्तर दे सकता है वह **मैं नहीं** हूँ। उत्तर—तुम्हें केवल **एक ही व्यक्ति** से मिल सकता है और उस व्यक्ति का नाम है **नन्द**।' व्यक्ति को अपना वरण आप करना होता है। अपनी दिशा आप निर्धारित करनी होती है और उसका अस्तित्व भी स्वयं ही ग्रहण करना पड़ता है।

यहीं आत्मसाक्षात्कार का क्षण आता है। किन्तु आत्मसाक्षात्कार का क्षण आसानी से नहीं आता। सन्त्रास, भय, पीड़ा, असुरक्षा की भावना, अलगाव, तनाव, अकेलापन, सम्बन्धहीनता—अनेक स्थितियों से गुज़रने के बाद ही कभी वह क्षण आता है।

महेन्द्रनाथ और अम्बिका पहले से ही उस सतह पर आए लगते हैं, जिसमें वे यथार्थ को स्वीकार कर लेते हैं। मोहभंग के बाद सुन्दरी और मल्लिका भी इस स्थिति में आ जाती हैं। यह अनुभव उनके अस्तित्व को अन्दर की ओर समेट लेता है। इससे वे द्वीपों की तरह अकेले पड़ जाते हैं। यह कोई अस्वाभाविक स्थिति नहीं। सच्चे साक्षात्कार के क्षण में अकेला महसूस करना मानव का स्वभाव है। अपनी आत्यंतिक परिणति में सभी लोग अकेले होते हैं। क्षण का जड़ बर्फ की तरह जम जाना महेन्द्रनाथ को परावलम्बी बना देता है। दूसरी ओर उसकी प्रतिक्रिया नन्द को आक्रामक बना डालती है—वह जंगल में जाकर व्याघ्र से युद्ध करता है। कालिदास उज्जयिनी में निर्वासन को भोगता है और इसी में अपने स्वत्व को सुरक्षित रखना है। यह अलगाव और अकेलापन जहाँ अभिशाप है, वहाँ शक्ति का स्रोत भी है। मल्लिका, अम्बिका, कालिदास जैसे पात्र उसे शक्ति के रूप में लेते हैं।

राकेश ने अपने नाटकों में जीवन की जो अवधारणा की है, वह न एकदम रोमानी है, न आदर्शवादी। उनकी दृष्टि ने मानव की नियति के उस कटु यथार्थ को पकड़ा है जिसने उनकी बरसों पुरानी प्रतिमा को भंग कर डाला है। आज मानव अस्तित्व और अनस्तित्व के बीच जिस प्रकार प्रश्न-चिह्न बना हुआ भयंकर यन्त्रणा भोग रहा है, राकेश के नाटक उसका जीवन्त चित्र उपस्थित करते हैं। सन्त्रास, पीड़ा, टूटन, सम्बन्धहीनता, अजनबीपन, अलगाव की स्थितियों के बीच उन्होंने मानव-जीवन की तलाश की है। इस तलाश पर यत्र-तत्र अस्तित्ववादी दृष्टि की छाप है। किन्तु राकेश की महानता इस बात में है कि वे दर्शन को मनोवैज्ञानिक आयाम प्रदान कर यह अहसास नहीं होने देते कि वे कहीं दर्शन को ऊपर से आरोपित कर रहे हैं। जहाँ भी वे अस्तित्ववादी दर्शन को टिकाने का प्रयास करते हैं, वहाँ पर्याप्त मनोवैज्ञानिक पृष्ठभूमि बना लेते हैं। इसीलिए कालिदास, नन्द, सुन्दरी, सावित्री, महेन्द्रनाथ एक नपी-तुली दृष्टि से जीवन जीते और झेलते हैं, पर वह जीना या झेलना अस्वाभाविक या आरोपित नहीं लगता। उनके नाटकों में हर व्यक्ति एक घटना है। व्यक्ति के रूप में पात्र का महत्त्व उसके अन्दर और बाहर घटित होनेवाली घटनाओं, सम्बन्धों और सन्दर्भों में प्रतिपादित हुआ है। नाटक का अपना स्वरूप और संरचना उनको स्वीकारने या उन्हें निरर्थक कर देने पर निर्भर करता है। संकल्प-विकल्पों को कौन पात्र किस दृष्टि से देखेगा-जानेगा इसी पर उनकी जीवन-सम्बन्धी दिशा का निर्धारण होता दिखाई देता है।

राकेश की अपनी जीवन-दृष्टि भी कम महत्त्वपूर्ण नहीं रही है। उनके लिए अपने नाटक आत्माभिव्यक्ति के साधन भी रहे हैं। यह आत्माभिव्यक्ति भी उनके नाटकों की रचना-दृष्टि पर हावी दिखाई देती है। अगर उनके नाटकों में एक सर्वसामान्य तत्त्व की खोज की जाए तो दो बातें विशेष रूप से उभरकर आती हैं—एक, घर की तलाश; दूसरी, स्त्री-पुरुष का आपसी सम्बन्ध। ये दो अलग-अलग समस्याएँ न होकर एक ही समस्या के दो पहलू हैं। वस्तुतः घर की तलाश भी और कुछ नहीं—स्त्री-पुरुष के आपसी

सम्बन्धों की तलाश ही है।

डॉ. इन्द्रनाथ मदान ने ठीक ही लिखा है, 'यदि राकेश के सभी नाटकों पर सरसरी नज़र डाली जाए तो एक संकेत बार-बार उभरता है—नायक लौटने के लिए अभिशप्त है। **आषाढ़ का एक दिन** का नायक हो या **लहरों के राजहंस** या **आधे अधूरे** का। वह घर में लौटने को विवश है। इनके उपन्यास **अँधेरे बन्द कमरे** में भी हरबन्स-नीलिमा की स्थिति इसी तरह की है। इनकी कहानियों में भी इस तरह का संकेत बार-बार मिलता है।[3] एक ही संकेत का बार-बार अनेक रचनाओं में दुहराया जाना निश्चितः कुछ अर्थ रखता है। वस्तुतः राकेश के मुख्य पात्रों में अतीत-मोह या गृह-विरह की स्थिति का निदर्शन मिलता है। कालिदास और नन्द घर लौटते और बाहर निकल जाते हैं; महेन्द्रनाथ घर से निकलकर भी लौट आता है। कालिदास कभी दुबारा लौटेगा, इसकी सम्भावना नहीं लगती; नन्द इसकी सम्भावना लिए हुए भी है, पर ज़्यादा सम्भावना यह है कि लौटेगा ही नहीं। महेन्द्रनाथ किसी भी दिन फिर पहले की तरह घर से निकल सकता है, यह भी असम्भव नहीं। तात्पर्य यह है कि घर की तलाश आदमी की आदिम भूख है। वह उसे विगत अनुभवों से जोड़ती है जहाँ उसे राहत और आत्मीयता का अहसास होता है; किन्तु जीवन की विसंगति उसे यह सौभाग्य भोगने कहाँ देती है !

घर की तलाश व्यक्ति के स्वत्व की खोज का एक अंग है। राकेश के पात्र एक प्रकार की असत् आस्था के शिकार होते हैं जिससे वे ऐसा वरण करते मिलते हैं जिसकी ओर उनकी सहज प्रवृत्ति नहीं दिखाई देती। यह वरण उन्हें एक ऐसी भटकाववाली जीवन-डगर पर ले जाता है जहाँ उनके चरण ठीक से नहीं पड़ते। फलतः इस डगर पर वे अपने स्वत्व से दूर पड़ जाते हैं और वे ऊब, थकान, अलगाव, पीड़ा और निराशा से इतने टूट जाते हैं कि या तो फिर अपने को दूसरा आदमी महसूस करने लगते हैं या वे स्वयं दूसरे आदमी लगने लगते हैं। कालिदास जब मल्लिका के द्वार पर पहुँचता है तो इस आत्मस्वीकृति को दुहराता है, 'मैं वह व्यक्ति नहीं हूँ जिसे तुम पहले पहचानती रही हो। दूसरा व्यक्ति हूँ।...और सच कहूँ तो वह व्यक्ति हूँ जिसे मैं स्वयं नहीं पहचानता।' **आधे अधूरे** में सावित्री की नज़रों में कुछ समय बाद हर आदमी 'दूसरा' बन जाता है—जुनेजा, सिंघानिया, जगमोहन एक के बाद एक। **लहरों के राजहंस** में केश कटवाने के बाद सुन्दरी को नन्द दूसरा आदमी लगने लगता है : 'लौटकर वे नहीं आए। जो आया है वह व्यक्ति दूसरा ही है।' यह दूसरा व्यक्ति ही शायद असली व्यक्ति है। इसी व्यक्ति को वास्तव में राकेश के नाटकों में घर की तलाश है। वह बार-बार भटक जाता है और जीवन की विसंगति उसे घर लौटा लाती है। व्यक्ति की असहाय स्थिति, निरर्थकता का अहसास, परिस्थिति को अपने अनुकूल न बना सकने की विवशता, द्वन्द्व तथा आत्म-बोध भी उनके पात्र को घर लौटा लाने में सहायक होता है।

स्पष्ट है कि राकेश की नाटकीय अवधारणा में 'घर' सभी नाटकों के केन्द्र में है। वास्तव में उनके नाटकों के मूल संघर्ष का केन्द्र भी घर ही है। घर में ही वह 'कुछ' है जिसके लिए कालिदास, नन्द या महेन्द्रनाथ लौटते हैं या बाहर निकल जाते हैं। एक

अर्थ में उनके पास घर की कल्पना है, घर है ही नहीं। **आधे अधूरे** के हर पात्र को शिकायत है कि जिसमें वे रह रहे हैं, वह घर नहीं, कुछ और है। नन्द के घर की नींव भी जैसे हिल जाती है। इसीलिए भिक्षु आनन्द उससे कहता है, 'तुम्हारे पास कक्ष और उद्यान सब-कुछ है—घर नहीं **जिसमें तुम्हारी आत्मा को विश्राम मिल सके।**' घर होने के लिए अनुकूल भावात्मक वातावरण चाहिए। राकेश ठीक सोचते रहे कि रेगिस्तान में घर कैसे बनाया जा सकता है ? सच्चाई यह है कि घर रेत से भी बनाए जा सकते हैं पर जब बनानेवाला यह सोचे कि घर कोई और बनाकर दे। घर का घर न रहने का कारण घर बसानेवालों की असत् आस्था है। घर की मूल भावना का आधार वैवाहिक सम्बन्ध और दाम्पत्य प्रेम है। यह प्रेम जब व्यक्ति को वस्तु बना डालता है, तो नारी का अधिकार-भाव और पुरुष की स्वतन्त्रता का हनन घर को रेगिस्तान बना डालता है। स्त्री-पुरुष के पारस्परिक प्रेम की यह विडम्बना घर को घर और व्यक्ति को व्यक्ति नहीं रहने देती, जिससे संघर्ष अनिवार्य हो जाता है। **चन्द सतरें और** की भूमिका में इन्द्रनाथ मदान राकेश के इस सन्दर्भ में यही बात कहते हैं कि 'उनके नाटक घर के माध्यम से व्यक्तित्व की और व्यक्तित्व के माध्यम से घर की खोज करते हैं।'

इस संघर्ष का एक कारण यह भी है कि घर-बार और प्रेम का अर्थ स्त्री-पुरुष के लिए एक-सा नहीं होता। इसीलिए मल्लिका के लिए प्रेम सम्पूर्ण अस्तित्व है, एक सम्पूर्ण समर्पण और आत्मदान का भाव। वह 'भावना में भावना का वरण' करती है और कालिदास के प्रति अपने सम्बन्धों को सब सम्बन्धों से बड़ा मानती है। किन्तु इसके विपरीत कालिदास आत्मकेन्द्रित है—वह उस पर अपना अस्तित्व न्यौछावर नहीं करता—उसके साथ उसका इतना ही सम्बन्ध है कि वह एक उपादान है, सजीव व्यक्ति नहीं। इसकी जो परिणति होती है उसमें मल्लिका एक वस्तुगत अस्तित्व को भोगती दिखाई देती है और उसका 'भावना में भावना का वरण' उसकी टूटन का प्रमुख कारण बन जाता है। कालिदास से वह सीधे संघर्ष में नहीं आती, क्योंकि वह अपने को स्वतन्त्र रखती है और कालिदास को भी; किन्तु विलोम के साथ विवाह भी एक अप्रत्यक्ष संघर्ष की ही देन है।

राकेश के नाटकों में सर्वत्र स्त्री-पुरुष का प्रेम सापेक्ष्य माना गया है। स्त्री वही बनी रहे और और कुछ भी चाहे वह निर्वाह की बाधा बनता है। कोई भी सम्बन्ध अपनी शर्त पर सम्भव नहीं होता। स्त्री-पुरुष में देने और पाने की क्षमता होनी चाहिए। दूसरे से उतना ही लो जितने के अधिकारी हो, समर्थक हो। कुछ भी अनिवार्य नहीं। नन्द और सुन्दरी का प्रेम-सम्बन्ध भी इसी प्रकार का है। नन्द के प्रेम में सुन्दरी का भोग्या स्वरूप मुखर है, दूसरी ओर सुन्दरी भी भोग के स्तर पर जीती है और नन्द उसके लिए उपादान मात्र है। प्रेम की तीव्रता के लिए ऐन्द्रिय राग अनिवार्य है; किन्तु केवल देह का सुख ही काम्य नहीं, प्रेमास्पद की समस्त चेतना को अधिकृत करने की कामना भी मुख्य होती है। सुन्दरी यहीं विफल होती है और सम्भवतः सावित्री भी। सुन्दरी और सावित्री चाहती हैं कि वे अपने पति/प्रेमास्पद का एकमात्र केन्द्र बनकर जीएँ और सब

मूल्यों से ऊपर एक मूल्य बनकर रहें। इस परार्थजीविता में प्रेम के विघटनकारी तत्त्व सक्रिय रहते हैं। नन्द और सुन्दरी दोनों दो ओर से व्यक्ति को वस्तु में परिणत करने की प्रक्रिया के बीच से गुज़रते हैं। सुन्दरी 'स्व' को 'अन्य' में एकाकार कर देती है। उस रहस्यपूर्ण तृप्तिकारक एकान्विति में वह पूर्ण अस्तित्व प्राप्त करने का स्वप्न देखती है; किन्तु बदले में उसके ऊपर एक ऐसा अधिकार-भाव दिखाती है कि वह स्वतन्त्रता के लिए आकुल हो उठता है। और जब वह उसे स्वतन्त्र करती है तो वह उससे दूर चला जाता है। अकेली मल्लिका ही ऐसी है जो स्वयं वस्तु बनकर कालिदास की स्वतन्त्रता को सुरक्षित रखती है। वह जानती है कि कोई भी रेखा उसे घेर सकती है, पर स्वयं उसे वह घेरकर नहीं रखती। सुन्दरी नन्द को घेर नहीं पाती तो संघर्ष में आ जाती है। अहं की चोट खाकर वह जानती है कि शरीर केवल विषय है, विषयी नहीं। वह उपकरण मात्र है। उसे दूसरों की दृष्टि से देखना आत्मविनाश की राह पर जाना है।

आधे अधूरे और **लहरों के राजहंस** स्त्री-पुरुष के सम्बन्धों को एक ही तुला पर प्रस्तुत करते हैं। अन्तर इतना ही है कि सुन्दरी नन्द में ही अपनी भावना का पूर्ण बिम्ब देखना चाहती है और यह जानकर कि 'जितने साधारण और लोग हैं उतने ही साधारण आप भी हैं' अपने को सिकोड़ लेती है; सावित्री महेन्द्रनाथ तक ही अपनी तलाश को सीमित नहीं रखती; वह एक के बाद दूसरे को आज़माकर अन्ततः उसी तथ्य पर पहुँचती है। सुन्दरी पति में 'दूसरा व्यक्ति' देखती है और सावित्री को सब पुरुषों के चेहरों पर एक ही मुखौटा दिखाई देता है। यह इसलिए कि सावित्री और सुन्दरी दोनों के प्रेम-सम्बन्ध आत्मिक स्तर तक ऊपर नहीं उठ पाते। एक अनुकूल पुरुष की तलाश सुन्दरी को तोड़ डालती है और सावित्री को विवशता से झटका देती है।

आधे अधूरे और **लहरों के राजहंस** दोनों में पुरुष का अस्तित्व वस्तुगत है और नारी उस पर आमादा दिखाई देती है। **आषाढ़ का एक दिन** में स्त्री-पुरुष का सम्बन्ध दूसरे प्रकार का है। मल्लिका कालिदास से प्रेम करती है, विवाह नहीं। विवाह की सीमा के बाहर दोनों का भाव-प्रवण प्रेम अपने को जीवित रखता है। किन्तु दूसरी ओर जिस विलोम से उसे घृणा थी, उसी से विवाह कर मल्लिका घृणा के पात्र को चुनती है और विवाह के नाम पर विसंगति को जीती है। यह विवाह-सम्बन्ध केवल अपने अस्तित्व की सिद्धि के लिए, अपनी सत्ता को प्रकट करने, एक शून्य को भरने के प्रयास में क्षतिपूर्ति जैसा विकल्प प्रतीत होता है। मल्लिका कालिदास के सन्दर्भ में रोमानी प्रेम-कल्पना को निरूपित करती है और विलोम के सन्दर्भ में सम्प्रेषणहीन दाम्पत्य सम्बन्ध को। कुल मिलाकर मोहन राकेश के नाटक स्त्री-पुरुष के सम्बन्धों में 'कोई किसी को समूचा नहीं पा सकता' जैसी स्थिति को स्वीकार करके चलते हैं जिससे परम्परागत आदर्श घर-परिवार और पत्नी का बिम्ब ढहकर रह जाता है। कुछ लोगों ने इसी आधार पर उन्हें स्त्री-विरोधी तक कहा है।

राकेश के पक्ष में एक ही बात कही जा सकती है कि वे नाटक के साथ ही जीवन का अन्त नहीं मानते—वास्तव में जहाँ नाटक समाप्त होता है, वहाँ जीवन का वास्तविक रूप प्रारम्भ होता है। वे नाटक को उस बिन्दु पर पहुँचाकर छोड़ देते हैं जो आकुलता का महत्त्वपूर्ण क्षण होता है। जीवन की सही दिशा को खोजने के लिए भिक्षु आनन्द व्याकुलता को ही वास्तविक प्रारम्भ बताते हैं। यह 'वास्तविक प्रारम्भ' राकेश के नाटकों का विषय नहीं। इसीलिए उसको ध्वनित करने के बावजूद अपने सन्देश में वे अधूरे रह जाते हैं। पात्र अपने लिए एक नई दिशा की खोज करेगा—इस सम्भावना के बावजूद वर्तमान में वह जिस बिन्दु पर खड़ा दिखाई देता है, वह अनास्था और विफलता का बिन्दु है ! नियति को किसी तरह झेल-भर लेने की विवशता के इस बिन्दु पर वह अपने को कितना निस्तेज और पंगु अनुभव करता है ! इसमें उस तत्त्व का समावेश कहाँ है जो सही गई वेदना और अपेक्षित सहन-शक्ति को अधिक महिमामय बना देता है ? इससे मनुष्य की सम्भावना के द्वार कहाँ खुलते हैं ? जीवन में पीड़ा, अलगाव, विवशता, नास्तिभाव, निस्सारता का भाव सब-कुछ है, पर जीवन को जीने के लिए एक उच्चतर आस्था भी है जिसका आभास राकेश के नाटक दे नहीं पाते। उनमें ठहराव का क्षण इतना प्रमुख हो गया है कि पात्र न निर्णय की शक्ति रखते हैं और न विद्रोह का वरण करते हैं। यह मान भी लें कि 'जिए जाने पर जीवन धीरे-धीरे चुक जाता है' और 'कि सुख सुख नहीं, काई पर फिसलते पाँव का स्पन्दन मात्र है, मात्र रेत में डूबती बूँद की अकुलाहट'; पर यह मान लेना अत्युक्ति होगा कि 'यह अकुलाहट ही क्या जीवन का पूरा अर्थ, जी लेने का कुल पुरस्कार नहीं है ?' जी लेने का पुरस्कार आकुलता कभी नहीं हो सकती और जीवन अर्थहीन हो सकता है; पर यह मनुष्य-मनुष्य पर निर्भर करता है कि वह उसे क्या अर्थ दे।

राकेश लघु मानव की हताशा, असंख्य वासनाओं, कामनाओं और आकांक्षाओं से घिरे आधुनिक मानव के मन के पक्षधर लगते हैं। सुन्दरी, कालिदास, सावित्री सबमें चेतना की दौड़ है। ये पात्र अपनी कामनाओं के पीछे आकुल हैं और बदले में आन्तरिक शान्ति चाहते हैं। यह असम्भव की खोज है। सांसारिक विधि से आत्मिक लक्ष्य की सिद्धि दुष्कर है। निजत्व का त्याग, अहं की विकृति और जीवन को दूसरों के मानदंडों या अपेक्षाओं के अनुसार ढालना कुछ ऐसी स्थितियाँ हैं जो तनाव पैदा करती हैं। और स्वयं से भिन्न हो जाने की कामना भटकाती है। एक आन्तरिक नग्नता, एक 'स्व-भाव' की स्थिति ही वास्तविक स्थिति है। राकेश के नाटक इस तथ्य पर पहुँचते दिखाई देते हैं।

राकेश ने जीवन के सन्दर्भ में एक तथ्य को बड़ी मज़बूती से पकड़ा है। वह है मनुष्य का अधूरापन जिसको वे थोड़ा-बहुत अपने तीनों नाटकों में घसीटते रहे हैं। **आधे अधूरे** में तो सारा कथ्य ही वही है, **लहरों के राजहंस** के अन्त में नन्द के अधूरेपन की बात कही गई है। **आषाढ़ का एक दिन** में उसकी चर्चा नहीं हुई है, पर ध्वनि उसमें भी विद्यमान है। वस्तुतः किसी एक को दोषी ठहराना उचित नहीं लगता। दाम्पत्य में न कोई पूर्ण होता है न पूरक। पूर्णता और पूरकता में स्त्री-पुरुष दोनों का हिस्सा होता

है। ऐसा तभी होता है जब पुरुष और स्त्री एक दूसरे को उपादान मात्र समझते हैं। यों कह सकते हैं कि राकेश के नाटकों के मूल में अधूरेपन की विसंगति ही है जहाँ से वे शुरू होते हैं और जहाँ पर ख़त्म हो जाते हैं। अच्छा होता राकेश इसे किसी सामाजिक सन्दर्भ में उभारते, पर मनोवैज्ञानिक आधार पर वह रुग्ण मानसिकता से जुड़ गई है। इसलिए यह निरी वस्तुवादी दृष्टि लगती है, जो मानव के विघटन को तो उजागर करती है, पर उसे ऊर्ध्वमुखी विराट् चेतना के अहसास से वंचित करती है। मनुष्य को उसकी पार्थिवता और सीमाओं में देखना ही काफ़ी नहीं है, इस सम्भावना के बीच भी उसे देखना ज़रूरी है कि वह पार्थिव जीवन और अति-मानसिक सत्ता के बीच का सेतु भी बन सकता है। राकेश के नाटक मानव की उस आशा-आकांक्षा को नकारते हैं और इसीलिए उस आन्तरिक दीप्ति को नहीं जगा पाते जो तनाव, नास्तिभाव और निराशा के बीच भी बनी रहती है। राकेश के पात्र आकुलता ज़रूर दिखाते हैं, पर वे इतने पस्त, निकम्मे और अशक्त लगते हैं कि कुछ कर गुज़रने या आन्तरिक शक्ति जगाने का माद्दा उनमें कहीं दिखाई ही नहीं देता।

तब सवाल उठता है कि जो रूप राकेश प्रस्तुत करते हैं, वह वास्तव में किस आदमी का है ? क्या यही दिशाहारा, थका, हिम्मतपस्त, टूटा आदमी ही समकालीनता का प्रतिनिधि है ? क्या आदमी अपने को आक्रान्त, अभिशप्त और निर्वीर्य समझकर चुपचाप बैठ जाए? ये प्रश्न किसी समय (1967 में) स्वयं राकेश ने समय से कटी समकालीनता के सन्दर्भ में उठाए थे।[4] और तब उन्हें लगा था कि साहित्यिक अस्वीकार और जीवन के वास्तविक अस्वीकार में बहुत बड़ी खाई विद्यमान है : 'हमारा साहित्यिक अस्वीकार उतना अपने आसपास की आज की परिस्थितियों का प्रतिफलन नहीं है जितना युद्धोत्तर काल की उन देशों की परिस्थिति का, जहाँ युद्ध की वास्तविक तोड़-फोड़ ने बीस साल पहले उस पस्ती, बीमारी, नपुंसकता, निरर्थकता और बुढ़ापे की मानसिकता को जन्म दिया था।...परन्तु **पिछले दस वर्षों में उन देशों का जीवन भी उस स्थिति से आगे निकल आया है। आज के संवेदनों को न तो** युद्धोत्तर काल के हॉरर साहित्य के आधार पर आँका जा सकता है और **न युद्ध और विभाजन-ग्रस्त व्यक्ति-मानस के भीतर अतीत विरह के आधार पर।** इस बीच यदि कुछ नए सम्बन्ध-सूत्रों ने जन्म न लिया होता तो आज के अस्वीकार में वह सामूहिकता और सहभावना न होती जो आज के साहित्य को नई शक्ति का रूप दे रही है। **यह अपने अकेलेपन में आत्महत्या की बात सोचनेवाली पीढ़ी नहीं है...यह पीढ़ी मजबूरी के वक्तव्य देनेवाली पीढ़ी भी नहीं।**'[5]

राकेश की बात में सत्य है, पर उनसे कोई यह पूछ सकता है कि कालिदास, नन्द या महेन्द्रनाथ की सृष्टि करनेवाला लेखक किस पीढ़ी का है ? वह कहाँ का है ?—भारत जैसे देश का, जहाँ युद्ध का प्रभाव उस रूप में नहीं पड़ा जैसा पश्चिम में; या फिर उन पश्चिमी देशों का जहाँ युद्ध ने साहित्य में निराशावादी भूमिका निभाई ? प्रायः राकेश

सिद्धान्त रूप से एक बात करते हैं और अपनी रचनाओं में बिल्कुल दूसरी मुद्रा अख़्तियार कर लेते हैं। वक्तव्य और कृतित्त्व का यह अन्तर्विरोध उनमें कई स्तरों पर मिलता है। पश्चिम के जिस पराजयवाद को वे युद्ध की देन मानते हैं और निर्माणरत राष्ट्रों की मानसिकता में जो उन्हें अप्रासंगिक लगता है, अपने इस देश में उससे भिन्न परिस्थितियों को वे अपने नाटकों का विषय बनाते हैं। इस प्रकार पश्चिम की नकल करने के विरुद्ध जिस भारतीयता की वे अपने लेखों में हामी भरते रहे हैं, वह उनके नाटकों के लेखन में नदारद दिखाई देती है। सच्चाई यह है कि उनके नाटकों का कथ्य पश्चिम के अस्तित्ववादी चिन्तन की देन है। शुक्र इतना ही है कि उन्होंने भारतीय परिस्थितियों और मानसिकता के बारे में उसे कुछ 'डाइल्यूट' करके स्वीकार किया और उस सीमा तक नहीं गए जिस सीमा तक उसके प्रभाव में ऐब्सर्ड नाटककार पश्चिम में जा पहुँचे। मैं उनके इस कथन से सहमत हूँ कि 'मेरे विचार से पाश्चात्य लेखकों का हमारे लेखकों पर बुरा प्रभाव भी पड़ा है जो कि इस तरह से सिर्फ़ इसीलिए लिखते हैं क्योंकि वे समझते हैं कि यही एक तरीक़ा है।'[7] निस्सन्देह राकेश उन लेखकों में से नहीं हैं जो इसी तरीक़े को लिखने का एकमात्र तरीका समझते हैं। बेकेट का सन्दर्भ देकर उन्होंने अपनी बात काफ़ी स्पष्ट कर दी है : 'पाँच बरस पहले मुझे **वेटिंग फ़ॉर गोडो** जैसा नाटक पसन्द आया था। लेकिन आज उसे पढ़ने पर लगता है जैसे कोई फ़ॉर्मूला नाटक पढ़ रहा हूँ।...पाँच वर्ष पहले लगता था कि इसमें कुछ ग्रहण करने के लिए है, लेकिन अब इस नाटक के प्रति परित्याग-भावना उत्पन्न होती है...। मैं यह कहना चाहूँगा कि सामूहिक रूप से इन प्रभावों ने मेरे दिमाग़ को विकसित ज़रूर किया, लेकिन उसके बाद फिर और प्रभाव आए। मैं अपने आपको ऐसे प्रभावों से दूर नहीं रखता हूँ, उन्हें ग्रहण करता हूँ और फिर उन्हें विकसित होने देता हूँ और उसके बाद उन्हें अस्वीकृत भी कर देता हूँ।'[8]

राकेश की रचना-दृष्टि की सीमा यह भी है कि वे अपने नाटकों में समाज और युग की समस्याओं को व्यक्ति के परिप्रेक्ष्य में नहीं लेते। वे व्यक्ति की टूटन, निराशा, जड़ता, वेदना से क्षत-विक्षत अस्तित्व को तो देख लेते हैं, पर उसकी पृष्ठभूमि में वृहत्तर समाज की आर्थिक-सामाजिक समस्याओं को नहीं देख पाते। व्यक्ति के विघटन के पीछे वे जिन मुद्दों को उठाते हैं, वे कमज़ोर और नितान्त वैयक्तिक प्रतीत होते हैं। इस आक्षेप से वे परिचित थे। इसलिए निराकरण में उन्होंने कहा है : 'कथ्य जो भी हो, वह अकेले व्यक्ति का नहीं, हमारे समय का है और वह है एक आकुलता। आकुलता में गहरा असन्तोष भी है और विद्रोह भी...। आकुलता की परिणति आस्था और संघर्ष में हो सकती है, साथ ही अनास्था, सुविधा और समर्पण में भी।'[9] बात ठीक है। किन्तु राकेश के नाटकों में दूसरी परिणति ही अधिक है, उनमें अनास्था, सुविधा और समर्पण ही मुखर हुआ है।

क्या साहित्य का कोई जीवनगत मूल्य नहीं है ?—इस प्रश्न का उत्तर हर रचनाकार को देना होता है। राकेश सही उत्तर उभार पाए हों, ऐसा नहीं लगता। उन्होंने प्रश्न उठाए हैं, पर वे जो प्रश्न उठाते हैं वे व्यक्ति/पात्र के अन्तर्विरोध में खो जाते हैं। वे धुन्ध में

तैरते हैं, तीर की तरह नहीं गड़ते। इसीलिए वे तिलमिलाहट नहीं पैदा करते और न दिशा देते हैं। इस आरोप का भी मोहन राकेश के पास एक पिटा-पिटाया उत्तर है : 'मेरी दृष्टि में यदि कोई कलाकार कहे कि उसके पास कोई ऐसी विशिष्ट दृष्टि है जो दूसरों का मार्ग-दर्शन कर सकती है तो यह उसका भ्रम है। वह तो ज़्यादा से ज़्यादा इतना ही कर सकता है कि पाठक या दर्शक के मन में बैठे द्रष्टा को जगा दे जहाँ से वह चाहे तो अपना पथ आलोकित कर सके।'[10] सिद्धान्त रूप से यह उत्तर बुरा भी नहीं है। पर पूछा जा सकता है, राकेश के नाटकों में वह बिन्दु कहाँ है जो द्रष्टा को जगाता हो; एक सड़ने-गलने की प्रक्रिया है जो द्वन्द्व के बीच चलती रहती है। एक आत्महन्ता आस्था का घुन उनकी रचना-दृष्टि को खाता नज़र आता है। कह सकते हैं कि आज भौतिकता के नशे में खोए लोग गहरे प्रश्न उठा ही कहाँ पा रहे हैं ? इसीलिए शायद राकेश के प्रश्नों की नोक पैनी नहीं है। और उन पर स्थापित बिन्दु ?—बिन्दु मात्र हैं। राकेश का उत्तर होता—कोई भी उत्तर पूरा नहीं होता, न अन्तिम होता है। कोई भी बिन्दु अन्तिम नहीं; अभी तो और बिन्दु खोजने हैं। और उन बिन्दुओं में सत्य मिल ही जाए, यह भी वे नहीं कह सकते, क्योंकि लेखक स्वयं नन्द की हालत में है : 'आकाश में लटकते नीले काले बिन्दु—कोरे सिद्धान्तों के—वे अधिक स्थायी, अधिक सत्य कैसे हैं ?'

हिन्दी के लेखकों का यह कम दुर्भाग्य नहीं रहा है कि उन्होंने अन्धों की तरह हाथी को पहचाना है। अस्तित्ववाद भी ऐसा ही हाथी रहा है। और फिर कौन जाने इस बदलती दृष्टि-बिन्दुओं की दुनिया में इन विचारों का चौखटा कब चरमरा जाए और दूसरे बिन्दु उभर आएँ। तब राकेश के नाटक किस बिन्दु पर टिक पाएँगे यह शंका स्वतः उठती है। और बिन्दुओं की खोज का तरीक़ा क्या है ? दरअसल उन्हें तर्क-बुद्धि से नहीं खोजा जा सकता जैसा कि नन्द करता है। आज तो धर्म में ही नहीं, विज्ञान में भी ऐसे तत्त्व देखने को मिलते हैं जो बुद्धि की अवज्ञा करते हैं। वस्तुतः बुद्धि-तत्त्व वह आधारभूत सत्य नहीं है जिस पर मनुष्य जीवित रह सके। एक सार्थक जीवन बिताने के लिए आन्तरिक अनुभूतियों, जीवन-मूल्यों और नैतिकता का भी महत्त्व है। यह कहना कि आदमी जैसा है वैसा वह उन कारणों से है जो उसके हाथ में नहीं हैं, यह उसकी सारी उपलब्धियों को झुठलाना है। उसके कार्य-कलाप और चारित्र्य को केवल एक निरपेक्ष तथ्य के रूप में मानकर चलना अस्तित्व की विसंगति को स्वीकारना मात्र है। वस्तुतः आत्मिक शक्ति और स्वतन्त्रता का भी ज़िन्दगी में कम महत्त्व नहीं है। मानव का संकल्प भी कार्य का कारण बनता है और स्वतन्त्रता और आन्तरिक बाध्यता भी पूर्णता के द्वार खोलती है। संसार में रहकर विवशता और निरर्थकता का अहसास होता है, पर ऐसी स्थितियाँ ही तो व्यापक आध्यात्मिक अनुभव का संकेत देती हैं। राकेश के नाटकों में विशिष्ट क्षण आते हैं, पर अनुभव के दबाव में उनमें विलक्षण दृष्टि नहीं जागती। वही दृष्टि उस बिन्दु को खोजती है जो समयातीत है।

विडम्बना यह है कि नाट्य रचना करते राकेश अपने को दो हिस्सों में बाँटकर लड़ रहे हैं। पार्थिवता और अपार्थिवता जब एक ही तल पर हों, किसी विजय पर पहुँचना

सम्भव नहीं। पात्रों में द्रष्टा भाव नहीं और उनकी लड़ाई केन्द्र पर न होकर परिधि पर है। स्वयं नन्द केन्द्र पर जीवन का योद्धा नहीं बन पाता क्योंकि उसके आगे दो विपरीत बने हुए हैं। दो विपरीत एक साथ नहीं रहते—चुनाव के लिए निर्णायक शक्ति ज़रूरी होती है। और निर्णय की मनःस्थिति कभी भी विक्षिप्तता की स्थिति नहीं हो सकती—वह उसके पार ही सम्भव है। पार जाने के लिए आत्मबोध और सही चुनाव की शर्त आवश्यक है। **लहरों के राजहंस** के पात्रों की सारी शक्ति पुंसत्वहीन आन्तरिक लड़ाई में चुक जाती है। अन्तर्दृष्टि उस लड़ाई का निदान कर सकती है। यद्यपि अन्तर में लड़ी जा रहे अपने ही भावों के द्वन्द्व को सहसा ख़त्म कर देना सम्भव नहीं होता, किन्तु अचेतन को उससे मुक्त कर देना त्याग और विवेक से सम्भव है।

पुरानी सड़ी-गली मान्यताओं का ठुकराना बुरा नहीं है, पर सही नई मान्यताओं का संकेत न देना किसी भी अच्छे लेखक के लिए बुरा है। संसार में बहुत-कुछ ऐसा है जिसके बारे में वस्तुनिष्ठ रूप से विचार करके कुछ निर्णय नहीं किया जा सकता। सत्य की खोज अनुभवातीत क्षेत्र में भी सम्भव है। मनुष्य आज के जगत् में इतना गड़बड़ा गया है तो क्या अब सारे मूल्य उसके लिए पुराने पड़ गए हैं ? व्यक्ति को उसके ही केन्द्र में देखना काफ़ी नहीं। वस्तुतः व्यक्ति ही सब-कुछ नहीं है। उसे समाज में रहना होता है, वह समाज की उपज है। पर राकेश ने अपने नाटकों में व्यक्ति को समाज के प्रतिनिधि के रूप में देखने का प्रयास नहीं किया है। इसीलिए वे व्यक्ति की ह्रासोन्मुखी प्रवृत्ति को ही चित्रित करते हैं। व्यष्टिवाद उसी प्रकार आज के चिन्तन की अति है जिस प्रकार समष्टिवाद की अति है; दोनों आदर्श को नहीं समझते। तो यदि कोई बिन्दु है जो खोजा जा सकता है, तो उसे ऊर्ध्वगामी अन्तश्चेतना और सामाजिक सन्दर्भों में ही खोजा जाना चाहिए। अकेले विषाद का मसीहा बनकर खड़े रहने से यह तलाश कभी पूरी नहीं होगी।

नाट्य शिल्प

रूपबन्ध : रोयें-रेशे

आज की साहित्यिक अभिव्यक्ति में हम एक चीज़ की प्रमुखता देखते हैं और वह है—बिम्ब और विचार का सामंजस्य अर्थात् बिम्बों का ऐसा संगठन हो कि विचार उसके बीच से ही प्रस्फुटित हो, चरित्र और घटनाएँ ऐसे मूर्त्त रूप में प्रस्तुत की जाएँ कि वे ही लेखक के अभिप्राय या संकेत को स्पष्ट कर दें।

अनुभूति का एक आन्तरिक शिल्प होता है जिसकी खोज कलाकार को करनी होती है।

*—**मोहन राकेश** : साहित्यिक और सांस्कृतिक दृष्टि*

अपने एक लेख में राकेश ने घाटी में तैरते बादल के रोयें-रेशे की बात की है।[1] बादल के रोयें-रेशे होते हैं—वे ही उसे बनाते हैं, अनुभूति के भी अपने रोयें-रेशे होते हैं और अनुभूति जब आकार ग्रहण करती है तो रचना के भी रोयें-रेशे होते हैं। ये रोयें-रेशे ही रचनाकार की अनुभूति और कृतित्त्व के अविभाजित अंग बनकर उसका रूप सँवारते हैं। कृति के आविर्भाव के साथ ही उनके रोयें-रेशे और बुनावट को निर्धारित करनेवाला शिल्प भी साथ ही जन्म लेता है। किसी भी कृति के रोयें-रेशे या उनका शिल्प, वस्तुतः रचनाकार के आभ्यन्तर अनुभवों से रूपायित होता है। शिल्प की सम्भावनाएँ अनुभूति में ही सदा से निहित रही हैं। इसीलिए अनुभूति ही रचना के कथ्य से लेकर उसके बाह्य उपलक्षणों तथा उसके आभ्यन्तर संघटन तक को एक निश्चित ढाँचा प्रदान करती है। इस स्थिति में शिल्प सर्जनात्मक अनुभव से जुड़ जाता है।

राकेश को सदा इस बात की शिकायत रही है कि अनुभूति शिल्प में ढलती हुई कहीं-न-कहीं खंडित हो जाती है। फिर भी संवेदना और आन्तरिक आवश्यकताओं के दबाव में उन्होंने जो कुछ लिखा है उसमें शिल्प का एक संभार स्वयं फूट पड़ा लगता है। इसके अतिरिक्त यह भी नहीं भूला जा सकता कि वे रचना के शिल्प, हर रोयें-रेशे के प्रति बहुत जागरूक रहे। **लहरों के राजहंस** को उन्होंने कई बार लिखा था और **पैर तले की ज़मीन** चार बार लिखे जाने पर भी अपूर्ण ही रहा। वस्तुतः राकेश के नाटक निरन्तर तराशे जाने की प्रक्रिया के बीच से गुज़रे हैं। इस स्थिति में उनमें बहुत बारीक

कारीगरी के दर्शन होते हैं पर इसी श्रमसाध्य संस्कार के कारण कभी ऐसा भी लगता है जैसे शिल्प के उपादान कहीं-कहीं ऊपर से आरोपित हैं।

राकेश के नाटकों का कथ्य और शिल्प दोनों आधुनिक हैं। अपने ऐतिहासिक कहे जानेवाले नाटकों में भी उनकी दृष्टि आधुनिक ही रही है। पर इतिहास के अस्थि-पंजर से बाहर झाँकती उनके नाटकों की आधुनिकता आलोचकों के बीच संशय का विषय रही है। उन्होंने इतिहास का उपयोग एक विश्वसनीय खूँटी के रूप में किया है जिस पर वे अपने नाटकों के आधुनिक कथ्य को टाँग सकें। राकेश का विचार था : 'मेरे लिए अनुभूति का सीधा सम्बन्ध मेरे यथार्थ से है और यथार्थ है मेरा समय और परिवेश—व्यक्ति से परिवार, परिवार से राष्ट्र और राष्ट्र से मानव समाज तक का पूरा परिवेश। मैं इनसे कटकर शेष से जुड़ा नहीं रह सकता, अपने पास के सन्दर्भों से आँख हटाकर दूर के सन्दर्भों में जी नहीं सकता।'[2] इस प्रकार युग और इतिहास के बीच वैचारिक खाई की परवाह न कर राकेश ने अतीत के कथानक को अपने युग की संवेदना और ज्वलन्त प्रश्नों से जोड़ने का प्रयास किया। सवाल उठता है कि लेखक ने अपने युग के प्रश्नों की खोज अपने युग में न कर इतिहास में क्यों की ? इसका कारण सम्भवतः ऐतिहासिक कथानक का प्रभाव ही हो सकता है। इतिहास की जन-मानस पर एक विशिष्ट छाप होती है। किसी ऐसे कथानक, भाव या चरित्र को जो इतिहास-स्वीकृत हो, पाठक या प्रेक्षक के लिए स्वीकार करना भी सरल होता है। स्वयं मोहन राकेश ने इस सम्बन्ध में लिखा है—

'कभी-कभी मुझे किसी गहरी भावना, या किसी ऐसी चीज़ का जिसे लोग स्वीकार कर चुके हैं, उपयोग करना बहुत सुविधाजनक लगता है। कालिदास के नाम से लोग भली-भाँति परिचित हैं, अतः उनके नाम का प्रयोग करने की वजह से मुझे कोई अन्य प्रतीक गढ़ने की आवश्यकता नहीं पड़ी। हो सकता था कि **यदि मैं आज के किसी ऐसे सुविधाग्रस्त लेखक के नाम को गढ़ता तो मेरी रचना में दूसरे दर्जे के किसी लेखक का आभास होता अर्थात् ऐसे व्यक्ति का...जो वास्तविक लेखक दिखता ही नहीं।** कालिदास की जिन कृतियों का मैंने नाटक में उल्लेख किया है, यदि मुझे उनकी जगह छद्म नामों का प्रयोग करना पड़ता तो मैं लोगों के दिमाग में इस प्रतीक को बैठाने में समर्थ नहीं हो पाता।...मैंने इस नाटक को लिखते समय यह सोचा कि जितनी शक्ति किसी ऐसे चरित्र को गढ़ने में लगाऊँ...तो क्यों न मैं इतिहास से प्रतीक लेकर उस शक्ति और कल्पना को आज के, और आज के लिए नाटक की रचना, में लगाऊँ ? यही बात मेरे दूसरे नाटक के साथ भी हुई। इस नाटक में मैं नन्द के नाम का उपयोग नहीं करना चाहता था, मैं बुद्ध के नाम का उपयोग करता।...**मैंने इस इतिहास का उपयोग इसलिए किया क्योंकि इस कथा के माध्यम से विशेष प्रकार की** व्याख्या प्रस्तुत की जा सकती थी।'[3]

इस प्रकार राकेश ने इतिहास का उपयोग-मात्र किया है। उन्होंने ठीक ही कहा है कि 'धर्मवीर भारती, जगदीशचन्द्र माथुर और मैंने ऐतिहासिक नाटक इसलिए नहीं लिखे

हैं कि हम इतिहास की व्याख्या करना चाहते हैं या इसलिए कि हमें किसी काल-विशेष से लगाव था। न ही यह किसी प्रकार जयशंकर प्रसाद की तरह का किसी प्रकार का पुनरुत्थानवाद है, न यह किसी प्रकार का प्रतिक्रियावाद है।'[4] स्पष्ट है कि इतिहास के प्रति एक नई रुचि का परिचय देकर राकेश ने एक बद्ध परम्परा को तोड़ा है। काल की अखंड परम्परा को अतीत और वर्तमान में विभाजित करने की बजाय उन्होंने उसके निरन्तर प्रवाह को स्वीकार किया है और काल की उसी अखंडता और असीमता में उन्होंने उन क्षणों को पकड़ा है जो दिशा-सूचक हैं। इसीलिए राकेश ने अपने दो नाटकों के कथानक इतिहास की स्थूल घटनाओं से नहीं, स्थूल घटनाओं को जोड़नेवाली सूक्ष्म कल्पनाओं में से चुने हैं। उनका लक्ष्य इतिहास नहीं, इतिहास का मिथ्याभास उपस्थित करना है, जिसे वे विश्वसनीयता में मिथक की सीमा तक खींच ले जाते हैं। वस्तुतः इतिहासकार की वास्तविकता और रचनाकार की वास्तविकता में अन्तर होता है। रचनाकार की वास्तविकता आन्तरिक और विषयगत होती है। राकेश ने इतिहास में से उसी वास्तविकता को अपने नाटकों में रूपायित किया है। वर्तमान से बीते हुए काल क्षणों की यह यात्रा लेखक को उस गुहा-गह्वर में ले जाती है जहाँ वह 'अपने को नए सिरे से अनुभव करता है।' इसीलिए कालिदास, नन्द और बुद्ध के काल से जोड़नेवाला यह अहसास रहस्यमयी अनुभूतियों से जुड़कर इतिहास को भी मिथक बना डालता है। वस्तुतः मिथक रहस्यमय ही नहीं, भावना में इतिहास से भी अधिक वास्तविक और विश्वसनीय होता है। उसमें ऐसा कुछ होता है जो बहुत गहरे आत्मा को छूता है। उससे वर्तमान अतीत में अपना विस्तार पाता है और हम अतीत में नहीं वरन् अतीत को अपने में जीते अनुभव करते हैं। इस प्रकार रचनाकार और उसका पाठक अतीत से अपने को जोड़ता ही नहीं, आश्वस्त भी पाता है। वर्तमान और अतीत के इस मिलन बिन्दु पर कोई भी महानतम सर्जन सम्भव है, जहाँ दोनों कालों की सारग्रही एकता समग्र जीवन के अर्थ को उजागर करती है।

बर्नार्ड शॉ ने **सेंट जोन** लिखा तो उसकी अनैतिहासिकता को सिद्ध करने के लिए आलोचकों ने कई पन्ने रँग डाले। शॉ ने कहा : 'मैं इतिहास को अन्तःप्रज्ञा से जानता हूँ। मैं पहले लिखता हूँ और फिर इतिहास पढ़ता हूँ और फिर मैं पाता हूँ कि मैं सही हूँ।' राकेश के लिए भी नन्द या कालिदास का इतिहास के काल-प्रवाह में होना महत्त्वपूर्ण नहीं रहा—महत्त्वपूर्ण रहा है अनुभव-विशेष से गुज़रना; और साथ ही प्रेक्षक/पाठक का उसे वास्तविक महसूस करना। इस प्रकार राकेश ने इतिहास के ढाँचे को अपने विचारों, चिन्तन और दर्शन के चोले के रूप में लिया है। लॉर्ड बोलिंगब्रोक, बोल्तायर, गिब्बन आदि सभी ने कहा है कि 'इतिहास दर्शन का उर्वर क्षेत्र है और ऐतिहासिक प्रसंग दार्शनिक अभिव्यक्ति के लिए प्रायः उपयोगी होते हैं।'[5] इतिहास का एक समसामयिक जीवन्त सन्दर्भ भी होता है, जिसमें युग का तत्व मुखर होता है। इतिहास, मिथक और व्यक्ति के अस्तित्व के बीच ऐसी बहुत सारी समानताएँ विद्यमान हैं; उनसे अस्तित्व की अनेक सम्भावनाएँ उजागर की जा सकती हैं।

सवाल इतिहास के प्रति दृष्टि का है। जब भारतीय नाट्य परम्परा में ऐतिहासिक कथावस्तु और पात्र लोकप्रिय रहे हैं, तब बल तथ्यों, स्थूल घटनाओं, घटना-संकुल जीवन और आदर्शों पर था। आज का नाटककार इतिहास और मिथक में घटनाचक्र की नहीं, आन्तरिकता की तलाश करता है, जिसमें व्यक्ति का निजी अस्तित्व, मानवीय संवेदना, परिवेश और जीवन की सम्बन्धसूत्रता महत्त्वपूर्ण हो गई है। राकेश ने प्रसाद के समान इतिहास में अतीत-गौरव, राष्ट्रीय जागरण, शौर्य, त्याग ओर बलिदान जैसे तत्त्वों की तलाश नहीं की है। उनकी दृष्टि मूलतः मानव-सम्बन्धों पर केन्द्रित रही है, जिससे स्त्री-पुरुष सम्बन्ध ही उनकी कृतियों का मूल विषय बना। पहले वे इन समस्याओं को ऐतिहासिक कथानकों के माध्यम में प्रस्तुत करते रहे, बाद में उन्होंने यह अनुभव किया कि इतिहास और मिथक का इस्तेमाल प्रेक्षक का ध्यान दूसरी ओर हटा देता है।[6] और इसीलिए **आधे अधूरे** में उन्होंने समसामयिक कथावस्तु को चुना।

राकेश का नाट्य शिल्प फॉर्मूलाबद्ध नहीं है। शास्त्रीय पद्धति के अनुसार नाटक का सुनिर्धारित क्रम—आदि, मध्य और अन्त—उस प्राचीन व्यवस्था की देन था जिसमें जीवन और जगत् को एक सुनियोजित तर्कसंगत सम्पूर्ण ढाँचे के रूप में परिकल्पित किया गया था। आज की ज़िन्दगी जितनी असीमित और ऊबड़-खाबड़ है, आज के नाटक का शिल्प भी उतना ही अनिश्चित और गतिशील है जो आरम्भ, विकास और अन्त की परिभाषा में नहीं आता। प्राचीन नाट्यशास्त्र में कथानक, पताका, प्रकरी, नायक, प्रतिनायक, खलनायक सम्बन्धी दृष्टि थी—छोटे-बड़े का, सत-असत् का भेद था, आज के नाटक में यह खाई पट गई है। आज दृष्टि के कोण बदले हैं; फलतः जीवन और जगत् की एक भिन्न तस्वीर उभरी है और उसी के साथ आदमी की पुरानी प्रतिमा भंग हुई है। इसीलिए कथानक और चरित्र बदले हैं। पहले नाटक की संरचना में कथानक, चरित्र और संवाद की ही मुख्य भूमिका होती थी; अब भावों और विचारों का भी उसकी बुनावट में बहुत बडा हाथ होता है। विचार नाट्य-स्थितियों, कार्यों तथा चरित्रों के साथ इस तरह जुड़ जाते हैं कि वे बहुधा एकरूप हो जाते हैं। इसी दृष्टि से पश्चिम में नाटक के ढाँचे में भी कथ्य के साथ ही एक क्रान्तिकारी परिवर्तन हुआ है। राकेश ने भी नाटक में पुराने ढाँचे को तोड़ा है—पर उनके नाटक वस्तु और रूपबन्ध दोनों की दृष्टि से सीमा तक ही युगान्तकारी हैं। न वे कलाधर्मी हैं, न विसंगत और न प्रयोगवादी। पश्चिम का नाटक रूप में शिल्प की जितनी स्थितियों के बीच गुज़रा है, उसको देखते हुए राकेश ने काफ़ी संयम से काम लिया है। इसीलिए कहीं ऐसा लगता है कि उन्होंने अपने कथ्य को नए शिल्प में ढालने का प्रयत्न तो किया है, पर वे उसे बहुत दूर तक साथ नहीं ले गए। उनके नाटक कथ्य के स्तर पर जीवन की विसंगति, अजनबीपन और सन्त्रास को चित्रित करते हैं, पर समान्तर रूप से शिल्प के उसी स्तर को उभारने में वे लिप्त नहीं हुए। पर आधुनिक पाश्चात्य नाटककारों जैसी समसामयिक शिल्पगत चेतना कुछ-कुछ उनके बीज नाटकों में मिलती है। मृत्यु ने उन्हें समय नहीं दिया। फिर भी पाश्चात्य नाट्य शिल्प की दिशा में एक निश्चित बिन्दु तक तो उन्होंने मार्ग तय किया ही है।

राकेश की नाट्य वस्तु का मूल आधार नाट्य स्थितियाँ और उनमें उभरती हुई संवेदनाएँ हैं। वस्तुतः उनके नाटकों में स्थूल घटनाक्रम बहुत कम है—कोई एक शृंखलाबद्ध सीधा, अटूट घटनाक्रम भी उनके नाटकों में नहीं मिलता। यदि कथानक के नाम पर कुछ मिलती हैं तो कथा-स्थितियाँ जो एक-दूसरे से जुड़कर कथानक का निर्माण करती हैं। घटनाएँ अन्दर घटती हैं। इस प्रकार राकेश के नाटकों में कार्य-व्यापार के स्थान पर नाट्य स्थितियाँ और उनसे जाग्रत संवेदनाएँ मुख्य हो जाती हैं। यही जीवन का सत्य भी है—जीवन स्वयं स्थितियों और संवेदनाओं का आकलन है। मनुष्य को स्थितियों में रहना पड़ता है और उनके अहसास पर ही उसका अस्तित्व निर्भर करता है। स्थितियाँ विघटन और अनास्था के इस युग में उसे अतिवादी छोरों पर द्वन्द्व में घसीट ले जाती हैं। राकेश के नाटकों में ये स्थितियाँ ही अपनी शृंखलाबद्धता में कथानक बन जाती हैं और इन्हीं से चरित्र उभरते हैं। वस्तुतः उनका कथानक चरित्रों से, उनके क्रिया-कलाप के बीच से जन्म लेता है और चरित्र कथानक को एक विशिष्ट ढाँचा प्रदान करते हैं। फलतः कथावस्तु और चरित्र दोनों जैसे एक हो जाते हैं। कथावस्तु वस्तुतः गतिशील चरित्र का रूप धारण कर लेती है और अनेक स्थलों पर वह पात्रों के बीच सम्बन्ध-सूत्र स्थापित करने का साधन मात्र बन जाती है। स्थितियाँ कथावस्तु और चरित्र दोनों को परस्पर जोड़ती हैं। कई बार वे अवान्तर कथावस्तु की सृष्टि करते हैं जिसमें घटना की अपेक्षा राकेश आड़े-तिरछे, पूरक, सम-विषम चरित्रों का अधिक प्रयोग कर क्षति-पूर्ति करते हैं। **आषाढ़ का एक दिन** में अनुस्वार-अनुनासिक, रंगिणी-संगिनी तथा **लहरों के राजहंस** में श्वेतांग-श्यामांग, अलका आदि का क्रिया-कलाप भी इसी प्रवृत्ति का द्योतक है। **आधे अधूरे** में यह भराव पुराने भोगे हुए प्रसंगो, घर के बिखराव को व्यक्त करनेवाली वस्तुओं और व्यवहारों लटकते पायजामे, जूठी चाय की प्यालियों, कुर्सी झाड़ने, जला टोस्ट, जाँघ खुजलाने, डब्बा खोलने आदि के माध्यम से किया गया है।

राकेश कथावस्तु और चरित्र का ताना-बाना द्वन्द्व की विरोधी रेखाओं और रंगों से बुनते हैं। वे कथावस्तु और चरित्र के परस्पर विरोधी सूत्रों से अपने नाटकों का ऐसा ढाँचा खड़ा करते हैं, जो तनाव और द्वन्द्व की स्थितियों से आगे बढ़ता है। उदाहरण के लिए **आषाढ़ का एक दिन** में एक ओर पहले-पहले मेघों से मल्लिका का भीगा-भीगा महसूस करना है, दूसरी ओर अम्बिका की चुप्पी और आँसू; एक ओर ग्रामीण जीवन का निरीह अस्तित्व और दूसरी ओर राजसैनिकों द्वारा आहत मृग-शावक; एक ओर मल्लिका का 'भावना में भावना का वरण' और दूसरी ओर कालिदास का उपेक्षा भाव जैसी विरोधी स्थितियाँ कथापट को बुनती हैं। विरोधी रेखाओं और रंगों में कहीं-कहीं ऐसी निरीहता और करुणा व्यंजित होती है और उनके बीच विडम्बना के इतने तीखे 'शेड' या अर्थ-छवियाँ उभरती हैं कि ये हृदय पर अपना विलक्षण प्रभाव छोड़ जाती हैं। इसी प्रकार **लहरों के राजहंस** में भी नाटककार जीवन के द्वन्द्वों, अन्तर्विरोधों और उनसे निरन्तर उलझी हुई स्थितियों को अनुभव में ढालता है और यह अनुभव सर्वत्र द्विरूप अर्थात् दो रूपों में विभक्त हो उठता है, जिससे द्वन्द्व रचना-प्रक्रिया का ही आधार बन

जाता है। 'क्यों आज ही वर्षों बाद कामोत्सव का आयोजन किया जा रहा है ?'—इस प्रश्न के दोनों ओर चेतना जैसे दो खेमों में बँट जाती है। जो पात्र 'सोचते' हैं, वे इस आयोजन से कट जाते हैं, उसमें भाग लेने नहीं आते। वे बुद्ध का समर्थन करते हैं। इस प्रकार सुन्दरी का काम-तत्त्व, यशोधरा तथा बुद्ध के विराग-तत्त्व के आमने-सामने खड़ा कर दिया गया है। सारी कथा-स्थितियों में द्वन्द्व के दो स्तर उभरते हैं। एक के केन्द्र में सुन्दरी है और दूसरे के केन्द्र में नन्द। सुन्दरी का द्वन्द्व बाह्य स्थिति का द्वन्द्व है—वह बुद्ध के आगमन और यशोधरा के प्रव्रज्या ग्रहण का सामना कामोत्सव के आयोजन से करती है; किन्तु उसके इसी प्रेरक हेतु से नन्द के मन में एक आन्तरिक स्थिति जागती है, जो उसे बहुत गहराई से घेर लेती है। द्वन्द्व का यह दुहरा स्तर **लहरों के राजहंस** के शिल्प को कई रूपों में प्रभावित करता है।

आधे अधूरे का सारा ढाँचा बाह्य द्वन्द्व पर ही निर्भर है। **आषाढ़ का एक दिन** की बुनावट में गति-तत्त्व की एकतारता के कारण द्वन्द्व पैना नहीं है। **लहरों के राजहंस** में द्वन्द्व का स्तर भावात्मक और वैचारिक है; किन्तु **आधे अधूरे** में न भावात्मकता है, न वैचारिकता और न काव्यात्मकता—उसमें ऐसी स्थूल किन्तु यथार्थ स्थितियों का अंकन है, जिनके मूल में दैनंदिन घरेलू संघर्ष है। ये पूरा नाटक छोटी-छोटी घटनाओं और नाटकीय स्थितियों से परिपूर्ण है। यह नाट्य स्थितियाँ भिन्न-भिन्न पात्रों की हैं, पर बाह्य द्वन्द्व या क्राइसिस सबको परस्पर जोड़ देता है, जिससे खंड-चित्र एक सम्पूर्ण चित्र में समाहित हो जाते हैं और एक ही अर्थ को उजागर करने लगते हैं। आधे अधूरे संक्रान्ति (क्राइसिस) या द्वन्द्व से प्रारम्भ होकर उसी में समाप्त होता है। बाह्य द्वन्द्व की व्याप्ति के बावजूद नाटक आगे बढ़ने के बजाय परिधि में फैल जाता है। इसीलिए नाटक में गति की अपेक्षा फैलाव ज़्यादा है, जो जड़ स्थिति के दुहराव या एकांगी स्वरूप के कारण है। नाटक में जो द्वन्द्व है वह क्रिया-व्यापार के स्तर पर उतना नहीं जितना मनःस्थितियों और उनसे उद्‌भूत झल्लाहटपूर्ण संवादों के स्तर पर है। इस तरह यह नाटक स्थितियों, क्रियाओं, प्रतिक्रियाओं और तनावों का नाटक बनकर रह जाता है जिसमें आरम्भ से लेकर अन्त तक एक एकतानता दिखाई देती है। इसीलिए उसमें गति, आरोह, अवरोह, चरम आदि स्थितियों को खोजना उचित नहीं। वस्तुतः यह प्रवृत्ति अब नए नाटकों की विशेषता बनती जा रही है।

इसमें कोई सन्देह नहीं कि राकेश के नाटकों में द्वन्द्व को उभारनेवाला बहुत-सा क्रिया-व्यापार नेपथ्य में घटित होता है और मंच पर इसकी सूचना भर दी जाती है; तीनों नाटकों के अन्त में आए लम्बे संवाद इसके प्रमाण हैं। इसे बहुत से आलोचकों ने दोष माना है, किन्तु इसे विधा की सीमा भी कहा जा सकता है। इसी प्रकार गतिशीलता की अवधारणा में भी आज परिवर्तन आया है। आधुनिक नाटककार क्रिया-व्यापार की गतिशीलता के अभाव की पूर्ति टकराव से पूरी करता है और उसे बहुत अधिक महत्त्व भी नहीं देता। क्योंकि वह जानता है कि आज का जीवन घटना-संकुल न होकर मन्थर है। आदमी के जीवन में आज कुछ भी महत्त्वपूर्ण नहीं घटता, घटने को रह जाता है !

ऐसी स्थिति में नाटक में भी यदि ऐसा हो तो कोई आश्चर्य नहीं। इसके अतिरिक्त राकेश के नाटकों की संरचना पर एक दोष यह भी लगाया जाता है कि उनमें एक जाने-माने कोण के हिसाब से आरोह-अवरोह आदि नहीं हैं। पुष्पा बंसल ने रेखाकृतियों के आधार पर इस दोष को उजागर करने का अच्छा प्रयास किया है और लिखा है कि राकेश 'चरम सीमा से शरमाते हैं' और उनके नाटकों में कार्य की वह अवस्था नहीं है 'जिसके आर-पार लोगों के जीवन बदल जाएँ, जिस बिन्दु या शिखर पर खड़े होकर सम्पूर्ण नाटकीय वस्तु को एक नज़र से देख लिया जाए।'[7] हमारा विचार है कि राकेश के नाटकीय पात्रों के जीवन में एक स्थिति ऐसी आती अवश्य है जब वे एक बिन्दु पर पहुँचकर जीवन को एक नज़र देखते हैं, पर वह बिन्दु शिखर का ही हो, ऐसा नहीं लगता। सच्चाई यह है कि आज आदमी की ज़िन्दगी में शिखर आता ही कहाँ है ? वस्तुतः आज ज़िन्दगी एक ऐसी पहाड़ी यात्रा है जिसमें ऊँचाइयों के साथ उतराइयाँ भी तय करनी होती हैं और फिर ज़िन्दगी में ऐसी स्थितियाँ आती ही कितनी हैं जो उसे बदल डालें ? ज़िन्दगी आज पुरानी मान्यताओं के आधार पर इस रूप में बदलती ही कहाँ हैं कि उसे हम बहुत बड़े नाटकीय बदलाव का नाम दे सकें ? आदमी सिर्फ़ उसे झेलता है, दुहराता है। इसीलिए राकेश के नाटकों में यदि ऊपर उठनेवाले सोपान ऊँचे नहीं हैं या उनके नीचे सीधी रेखाओं का-सा ठहराव है, नियति की स्थितियाँ हैं, वर्तुल प्रत्यावर्तन है तो वह जीवन की माँग है। जीवन को नाटक के बँधे-बँधाए शिल्प पर न्यौछावर करने को नहीं कहा जा सकता। जिसको हम युगों से नाटक मानते रहे हैं, वह तो ज़िन्दगी में कभी घटित होता ही नहीं। **पैर तले की ज़मीन** के अन्तिम संवाद इस तथ्य को सही रूप में व्यक्त करते हैं—

नीरा : क्या हुआ ? क्या हुआ दीदी...मुझे बताओ तो क्या हुआ ?

रीता : कुछ नहीं। **एक नाटक पूरा होते-होते रह गया। एक इतिहास होते-होते रुक गया।**

वस्तुतः राकेश इसी अहसास से जीवन जीते रहे। अपनी डायरी में उन्होंने लिखा है : 'आगे इठलाता समुद्र है। उस पानी पर लगातार चलते रहना है जो इस साहिल से उस साहिल तक अनेकानेक चुनौतियाँ लिए फैला है और **अपनी सारी बदलती मुद्राओं के बावजूद सब जगह एक-सा गहरा, खारा और तूफ़ानी है।**'[8]

द्वन्द्व राकेश के नाट्य शिल्प का प्रेरक तत्त्व है; किन्तु साठोत्तरी कहानी का शिल्प भी कहीं-कहीं उनके नाटकों पर हावी होता दिखाई देता है। कहानी के समान ही चेतना-प्रवाह उनके नाटकों में शिल्प के तन्तुवाय बिछाता जाता है। उनमें कहीं पुराकाल की स्मृतियाँ उद्दीपन पाकर उद्दीप्त होती चलती हैं, कहीं मूलभूत वस्तु या विचार की चेतना नाटकीय अर्थ की प्रतीति के लिए प्रतीकों और बिम्बों को जुटाती प्रतीत होती हैं। वस्तु-विन्यास और चरित्र-सृष्टि में भी कहीं उसका शिल्पगत उपयोग करने का प्रयास

मिलता है। विशेषतः जहाँ शिल्प नें अनुभूति का धारा-प्रवाह दिखाई देता है। **आषाढ़ का एक दिन** का शिल्प इस दृष्टि से विचारणीय है। उसमें नाटक की सारी अवधारणा मेघ पर निर्भर है। चेतना का बिन्दु उसी पर केन्द्रित है। कालिदास उन्हीं मेघों की छाया में जीता है और मल्लिका भी। आषाढ़ के पहले-पहले मेघ और यौवन-तृप्त देह का उनमें भीगने का अनुभव प्रेम की प्रारम्भिक तल्लत, वर्षण और तृप्ति की अनुभूति कराते हैं। यह नाटक अपने में ही अवरुद्ध किशोर अनुभवों की वर्षण चेतना से प्रारम्भ होता है। किन्तु तभी मेघमय आकाश को धुँधला कर देनेवाली स्थितियाँ आती हैं। अम्बिका की यथार्थ भावना और असंपृक्त चेतना के साथ मेघों के इस बिम्ब को काटनेवाला एक और बिम्ब उभरता है—राज्य के सैनिकों और उनके द्वारा आहत हरिण शावक का। ये दोनों बिम्ब आदि से अन्त तक राकेश की नाट्य चेतना पर छाए दीखते हैं। राजशक्ति के प्रतीक सैनिक निरीह सौन्दर्य और स्वतन्त्र-जीवन के प्रतीक हरिण शावक और मेघमयी रोमानी ज़िन्दगी के भीगे अनुभवों के शत्रु बनकर सामने आते हैं। राजशक्ति हरिण को ही नहीं, अप्रत्यक्ष रूप से कालिदास और मल्लिका, यहाँ तक कि मातुल तक को घायल करती है। प्रियंगु आकर जब मल्लिका को उसके किसी सामन्त से ब्याह करने अथवा उसकी झोंपड़ी का संस्कार करने को कहती है—यह राजसत्ता का दूसरा आघात है, जो उसे गहराई से बेधता है। फिर भी मेघ की चेतना कहीं नहीं मिटती—आरम्भ के मेघ आह्लाद के मेघ थे; कालिदास के उज्जयिनी चले जाने पर वे मेघ उसके स्मृति के मेघ बन जाते हैं : कालिदास की पीड़ा के मेघ, उसकी अपनी पीड़ा के स्पर्श से घिरे मेघ—'जब भी तुम्हारे निकट होना चाहूँगी, पर्वत शिखर पर चली जाऊँगी और उड़कर आते मेघों में घिर जाऊँगी।' किन्तु कालिदास द्वारा प्रियंगु से विवाह कर लेने और ग्राम प्रांतर में आकर भी उससे मिलने से कतराने पर वे मेघ उसे कभी भिगोते नहीं, फिर भी वह उन्हें भूलती नहीं : 'सोचती थी, तुम आओगे तो इसी तरह मेघ घिरे होंगे, वैसा ही अँधेरा दिन होगा, वैसे ही एक बार वर्षा में भीगूँगी।' वस्तुतः भीगने के बजाय उसकी भावना के स्रोत सूख जाते हैं। अन्ततः जब कालिदास राजसत्ता से मुक्त होकर मल्लिका के द्वार लौटता है, तो एक बार फिर मेघ बरसते हैं और इसके साथ ही नाटक जिस तरह मेघों में शुरू हुआ था, उन्हीं में उसका अन्त हो जाता है। अन्तर इतना ही है कि मल्लिका की नई स्थिति के अहसास के बीच मेघ उतने बरसते नहीं, जितने गरजते और गाज गिराते हैं।

चेतना-प्रवाह इस नाटक के शिल्प को एक अद्भुत वातावरण और सम्बद्धता प्रदान करता है। एक अंक और दूसरे अंक के बीच काल का अन्तराल समाप्त करने में भी यह युक्ति काम देती है। मेघ के माध्यम से समय की खाई पट जाती है। बरसों बाद कालिदास के उसी स्थान और उसी वातावरण में मिलने से अन्तराल का अनुभव नहीं होता। वस्तुतः राकेश नाटक में जिस क्षण पर बल देते हैं वह अतीत होकर भी बार-बार वर्तमान बनकर जीवित हो जाता है। स्मृति के कोमल रेशों में अतीत और वर्तमान जैसे एक साथ बुन जाता है। सब-कुछ जैसे एक ही चेतना में सिमट आता है।

लहरों के राजहंस में चेतना की इस प्रवहमान अन्तर्धारा के दर्शन कुछ प्रतीकों में होते हैं। उसमें एक ओर सुन्दरी की आदिम चेतना लहरों पर तैरते राजहंसों की है, जिन्हें काँपती लहरें जिधर ले जाती हैं, उधर तैर जाते हैं। दूसरी ओर नन्द की खंडित चेतना की अभिव्यक्ति श्यामांग और अलका के द्वारा हुई है। श्यामांग के हाथ में पत्तियाँ उलझ जाती हैं। उस पर कमल ताल में हंसों पर पत्थर फेंकने का आरोप लगाया जाता है और अलका उन्हीं से सम्बद्ध सपने देखती है : 'सूखा सरोवर, पत्रहीन वृक्ष और धूल-भरा आकाश !' सपने की यह चेतना सुन्दरी की विरोधी चेतना है और यशोधरा के भिक्षुणी के वेश से मेल खाती है। नाटक के अन्तर्भाग में इन चेतनाओं का ताना-बाना बुना गया है और उससे अतिप्राकृतिक वातावरण की सृष्टि भी की गई है, जो स्थिति की विसंगति को उभारने में सहायक होती है। अलका का स्वप्न और ताल में दिखाई देनेवाली निरन्तर लम्बी होती डरावनी छाया, वह स्थिति है, जो राजहंसों को सहज नहीं रहने देती और वे चीत्कार कर उठते हैं। और यह 'छाया' आतंक की तरह सुन्दरी के मन पर छा जाती है, जिसका प्रतिकार वह श्यामांग को दंड देकर करती है। किन्तु श्यामांग और अलका की यह अतिप्राकृत मनःस्थिति केवल आक्रामक और भ्रामक नहीं है। सुन्दरी उस विसंगति के बीच उभरते सत्य को पहचानती है : 'अलका उससे प्रेम करती है।...स्वयं कुमार के मन में भी उसके प्रति अनुराग है।' यहीं श्यामांग की चेतना नन्द की खंडित चेतना से जुड़कर एक हो जाती है और श्यामांग सुन्दरी की नज़रों में 'दो आँखों का एक अनचाहा भाव' बन जाता है। अब तक जो चेतना श्यामांग को आन्दोलित किए थी, वह नन्द में उभरने लगती है। आखेट में अपनी ही क्लान्ति से मरते मृग को देखकर उसके मन का विराग उसे बुद्ध की ओर ले जाता है और सुन्दरी की चेतना आहत होने लगती है। आहत अहं की पीड़ा सुन्दरी को घर की चीज़ों की तोड़-फोड़ की ओर ले जाती है, तो नन्द और श्यामांग को अनिद्रा और प्रलाप की ओर। लेखक की शिल्पगत योजना यहाँ सफल हुई है। उस उन्माद के सन्दर्भ में नन्द की चेतना बार-बार आहत मृग पर और सुन्दरी की चेतना आहत हंसों पर टिक जाती है। गौतम बुद्ध का जो अहसास श्यामांग को कमल ताल पर हंसों के बीच में पड़ी छाया के रूप में होता है, वही नन्द को किवाड़ पर होती आहट के रूप में बेधता है : 'जैसे एक खुटकबढ़ैया सन्नाटे में लगातार चोंच से लकड़ी पर प्रहार कर रही हो।' और नन्द के गौतम बुद्ध के पास चले जाने पर स्थिति की जो कसक उभरती है, उसमें राजहंस फिर से सुन्दरी की चेतना पर तैरने लगते हैं : 'परन्तु राजहंस आहत थे...कम-से-कम उनमें से एक अवश्य आहत था। क्या उनके पंखों में इतनी शक्ति रही होगी कि वे अपनी इच्छा से उड़कर कहीं चले जाते ? फिर जिस ताल में इतने दिनों से थे, उसका अभ्यास, उसका आकर्षण क्या इतनी आसानी से छूट सकता था ?' यह प्रश्न क्या इस नाटक की मूल चेतना में नहीं है ?

आधे अधूरे कथ्य और शिल्प के स्तर पर एक भिन्न कोटि का नाटक है। इसमें मुख्य बल खंड-खंड सामान्य घटनाओं पर है, जो परिवार के बिखराव, टकराहट और टूटने को ध्वनित करती हैं। कथावस्तु की इकाइयों को नाटक की सम्पूर्णता के साथ

जोड़ने का कार्य चेतना की एकतान्ता के द्वारा सम्पन्न हुआ है। इसीलिए नाटक में घटनाएँ मनःस्थितियों में बदल जाती हैं, और मनःस्थितियाँ उसके व्यापक कथ्य के साथ जुड़ जाती हैं, जिससे सारा अनुभव बड़ा त्रासद हो उठता है। अलग-अलग ये प्रसंग बड़े साधारण, रोमानी या कृत्रिम लग सकते हैं, किन्तु नाटक की समग्र चेतना से गूँथकर केन्द्रीय अर्थ के साथ तदाकार हो जाते हैं।

वस्तुतः चेतना प्रवाह राकेश के नाटकों में अन्तरंग यथार्थ को उजागर करता है और आत्मपरक चेतना के गह्वरों में ले जाता है। इसके माध्यम से वे कथावस्तु और चरित्रों को परस्पर अनुस्यूत करके उनकी आन्तरिकता को अभिव्यक्त करते हैं। सचेतन विचार के तल में निहित अचेतन की प्रक्रियाओं को उभारकर राकेश वस्तु और चरित्र की योजना कुछ इस तरह करते हैं कि चेतना पर पड़े बिम्ब मुखर हो उठते हैं। इसीलिए उनके नाटकों में बहिरंग क्रिया-व्यापार कम है और अन्तरंग चेतना का विस्तार ही अधिक है। यही कारण है कि कथावस्तु के विभिन्न प्रसंग और चरित्र एक पारम्परिक चेतना से गुँथे हैं। एक ही पात्र अपने को कई पात्रों में दुहराता है और संवादों में भी पात्र आत्महीन चेतना के क्षणों में अपने को परिभाषित करते दिखाई देते हैं।

राकेश के नाट्य शिल्प का एक और आधारभूत तत्त्व सांकेतिकता भी है। वे सामान्य घटनाओं से असाधारण संकेत उभारते हैं। कथानक के रोयें-रेशों में वे प्रायः संवेदनीयता का मार्मिक प्रयोग करते हैं और घटना तथा चरित्र को साभिप्रायता देने के लिए वे उसे प्रतीकों, बिम्बों और विचारों से जोड़ते हैं। फलतः कथानक एक सारगर्भी सांकेतिकता की गूँज देने लगता है। **आधे अधूरे** में छोटी-छोटी घटनाओं में ऐसी ही सांकेतिकता है। बिना उठाए चाय के बर्तनों, गन्दे पाजामे, कैंची, तस्वीर की कतरनों, गर्द से 'सब रूपों में इस्तेमाल होनेवाले कमरे' की विसंगति को व्यंजित किया गया है। अर्थ की व्यंजना के लिए राकेश घटना और पात्र के स्थूल उपादानों का तो प्रयोग करते ही हैं, वातावरण की सूक्ष्म सृष्टि भी करते हैं। **आषाढ़ का एक दिन** मे मेघ का बिम्ब और **लहरों के राजहंस** में राजहंस, मृग, दर्पण आदि का प्रतीक तत्त्व नाटक के कथ्य को उभारता है। वस्तुतः राकेश बिम्बों और प्रतीकों का प्रयोग विभिन्न कोणों पर रखे आइनों की तरह करते हैं, जो गतिशील कथावस्तु और चरित्र के कई पहलुओं को प्रतिबिम्बित करते हैं। वे ऐसे मुख्य प्रतीकों व बिम्बों को बार-बार दुहराते हैं। वे पात्रों की चेतना को भी रंग-सामग्री के माध्यम से व्यक्त करते हैं। उनके माध्यम से पात्र की मनःस्थिति और कथानक की अर्थवत्ता दोनों उद्घाटित होती हैं। साथ ही नाटक की मूल समस्या और क्रिया-व्यापार को भी एक दिशा देते हैं। शिल्प के स्तर पर बार-बार उनकी पुनरावृत्ति से उनके द्वारा कथानक का एक 'डिज़ाइन' रूपायित होता है। वैचारिक पृष्ठभूमि के निर्माण में भी वे सहायक होते हैं। इसके अतिरिक्त अपनी अर्थवत्ता में पाठक/प्रेक्षक के मन में इस प्रकार बैठ जाते हैं कि आगे के घटना-क्रम के लिए उसे तैयार रखते हैं।

आषाढ़ का एक दिन में एक ही बिम्ब सारी रचना का डिज़ाइन बुनता है। उसकी

संवेदना और तरलता आदि से अन्त तक हर नई स्थिति में वह नए भावात्मक रूप को ग्रहण करता जाता है। **लहरों के राजहंस** में बिम्बों का एक पूरा क्रम है जिसमें एक बिम्ब दूसरे पर आरोपित हो जाता है। सरोवर के सूखने का स्वप्न, राजहंसों को पत्थर मारना, उड़ना, चील आदि के बिम्ब कथावस्तु, चरित्र और वातावरण में विसंगति की सृष्टि करते हैं। राजहंस का केन्द्रीय बिम्ब अन्य बिम्बों के साहचर्य में अर्थ का विस्तार करता है। ये गौण बिम्ब भी नाटक के अर्थ की ओर संकेत करते हैं और नाटक की बुनावट को एक विशेष रंग प्रदान करते हैं। इनकी एक विशेषता यह भी है कि वे प्रेक्षक के मन में अपेक्षित भाव जगाते हैं। दृश्यों को परस्पर जोड़ने में भी वे सहायक होते हैं।

राकेश के नाटकों में सांकेतिक अर्थ कभी-कभी अचानक ही पूर्व-स्थितियों से उभरता नज़र आता है। ऐसे स्थलों पर 'आइरनी' का कौशल देखा जा सकता है। 'नारी का आकर्षण पुरुष को पुरुष बनाता है, तो उसका अपकर्षण उसे गौतम बुद्ध बना देता है' सुन्दरी की यशोधरा के सन्दर्भ में कही गई यह दर्पोक्ति अन्ततः स्वयं उसकी अपनी निजी ज़िन्दगी का सत्य बन जाता है, तो विचित्र चोट करता है। विलोम से घृणा करनेवाली मल्लिका का अन्त में उसी से विवाह कर लेना, स्वयं अपने में एक विडम्बना है और इससे बड़ी विडम्बना है, थके-हारे, टूटे कालिदास का उसके द्वारा लौट आना। प्रेम में एक व्यक्ति दूसरे का स्थान नहीं ले सकता, पर प्रियंगु की यह उक्ति कि 'मल्लिका, तुम उनमें से जिसे भी अपने योग्य समझो उसी के साथ तुम्हारे विवाह का प्रबन्ध किया जा सकता है'—राजसत्ता पर तीखा व्यंग्य है विलोम व्यंग्य करता ही नहीं बल्कि कालिदास के व्यंग्य पर स्वयं एक व्यंग्य बन जाता है। निश्चयतः अर्थ की ये सब सम्भावनाएँ शिल्प की क्षमताओं को व्यक्त करती हैं।

राकेश के नाटकों का प्रारम्भ वहाँ से होता है, जहाँ अस्तित्व का द्वन्द्व मुखर होने लगता है। यह बिन्दु संक्रान्ति (क्राइसिस) से कुछ ही पहले का होता है। विरोधी स्थितियों का टकराव वहीं नज़र आने लगता है। बाद में जो कुछ घटता है, वह एकदम अप्रत्याशित नहीं होता—उसके लिए राकेश पहले ही ज़मीन बना चुकते हैं। संक्रान्ति और द्वन्द्व की विकसित स्थिति से ही प्रायः वे कथावस्तु को त्वरा से मध्य भाग की ओर ले जाते हैं, जहाँ से नाट्य स्थिति परिणति की ओर अग्रसर होती है। राकेश के नाटक परम्परागत नाटक की आदि, मध्य, अन्त की स्थिति का निर्वाह नहीं करते। इसीलिए उनमें नाटकीय वस्तु चरमसीमा की ओर अग्रसर होने के बजाय नियति की दिशा में आगे बढ़ती है। अन्तर्भाग में न कोई चौंकानेवाले मोड़ आते हैं और न पात्रों की हृदय-परिवर्तन की स्थितियाँ। उदाहरण के लिए **आषाढ़ के एक दिन** तीन अंकों का नाटक है, किन्तु उसमें आदि, मध्य, अन्त की शास्त्रीय व्यवस्थाओं से बद्ध कथा-विकास की स्थितियाँ नहीं हैं। पहला और तीसरा अंक कालिदास से सम्बद्ध है, पर दूसरा अंक उसके परोक्ष में घटता है। पहले दो अंकों में मल्लिका महत्त्वपूर्ण हो जाती है और तीसरे में कालिदास

की त्रासदी मन-प्राणों पर छा जाती है। तीसरे अंक को पहले दो अंकों का सहज विकास नहीं कहा जा सकता है। **लहरों के राजहंस** में भी तीन अंक हैं। दूसरा अंक अन्तर्भाग के रूप में द्वन्द्व का विकास करता है। किन्तु यहाँ भी पहले दो अंकों में सुन्दरी का प्रसंग महत्त्व पा जाता है जबकि तीसरे में नन्द की अस्तित्व-वेदना उसका स्थान ले लेती है। **आधे अधूरे** की कथावस्तु एक सम पर आरम्भ होती है। वस्तुतः उससे पहले जो घटित हुआ वह नाटक में कम आया है। सम्भवतः महेन्द्र और सावित्री कभी प्रेम भी करते रहे होंगे। यह सब नाटक में नहीं है। कथावस्तु में सपाटपन है, अनेक कथाखंडों में एक ही स्थिति की पुनरावृत्ति से उसमें कथावस्तु बहुत आगे नहीं बढ़ती और वह सावित्री और महेन्द्रनाथ की वापसी के साथ जहाँ को तहाँ लौट जाती है। इस प्रकार राकेश के नाटकों में चरम के क्षण लुप्त हैं और नाटक ज़िन्दगी की तरह ही सीधा-सपाट है क्योंकि उनकी नाटकीयता की धारणा घटनाओं के क्रम-विकास पर आधारित न होकर जीवन के द्वन्द्वों पर आधारित है।

राकेश के नाटकों का एक अन्त होता है; किन्तु जीवन का नहीं—उनके पात्र उसके बाद भी जीते रहते हैं। जो उनके नाटकों में अन्त की स्थिति है वह इति नहीं, वास्तविक जीवन का अथ है। वे पात्रों को जिस आकुलता के साथ दिशा की खोज में झोंक देते हैं, वही वास्तविक आरम्भ है, जिसका संकेत **आषाढ़ का एक दिन** और **लहरों के राजहंस** में किया गया है। फलतः उनमें अन्त वह प्रस्थान बिन्दु है जहाँ अनेक सवाल उभरते हैं और एक गहरे अहसास के साथ एक नई प्रयाण यात्रा शून्य में शुरू होती है। कालिदास और नन्द की यात्रा ऐसी ही है। पर सावित्री और महेन्द्रनाथ के लिए जैसे कहीं जाने के लिए कुछ बचा ही नहीं है। फिर भी कालिदास और नन्द उससे बढ़कर नहीं हैं। जीवन का युद्ध वस्तु या व्यक्ति से ही लड़ा जा सकता है, किन्तु एक बड़े शून्य या नास्तिभाव के साथ नहीं। जहाँ नन्द चौराहे पर खड़ा नंगा अनुभव करता है, जैसे दिशाएँ उसे लील जाना चाहती हैं, वहाँ लड़ने के लिए रह ही क्या जाता है ? कालिदास की अथ से ज़िन्दगी शुरू करने की कल्पना मल्लिका की बच्ची के रुदन के साथ ही मिट्टी में मिल जाती है। विवशता के इस क्षण को उभारते हुए राकेश दुखती रग पर हाथ मात्र रख देते हैं, जिससे नियति का अहसास मनुष्य में घाव बनाकर रिसता रहता है। बेचारा उस मोड़ पर कराहता नज़र आता है, जहाँ से उसे एक और यात्रा शुरू करनी है। राकेश अन्ततः वह अनुभव दे जाते हैं जो जलते अंगारे जैसा दाह दे जाता है।

अधिकांश आधुनिक नाटकों की भाँति राकेश के नाटक अन्त में नायक के ऐसे क्रिया-व्यापार या उक्ति को रेखांकित नहीं करते, जिससे सकारात्मक रूप में कोई विचार ग्रहण किया जाए। संकट-बोध मात्र उनका लक्ष्य होता है और संकट-बोध में यदि किसी सत्य के दर्शन होते हैं तो मानव की असहाय अवस्था और कष्ट को शक्तिपूर्वक सहने की आस्था के। उनके नाटक त्रासद दृष्टि के वाहक अवश्य दिखाई देते हैं, किन्तु उनका अन्त त्रासदी की आदर्शवादी प्रवृत्ति से सर्वथा मुक्त दिखाई देता है। उसमें निराशावादी जीवन दृष्टि बहुत पैना प्रभाव छोड़ जाती है।

राकेश के वस्तुशिल्प की सबसे बड़ी विशेषता यह है कि उनके नाटकों में रूपबन्ध की अद्‌भुत कसावट है। इसका मुख्य कारण क्रिया-व्यापार, विचार और अनुभूति की सघनता है, जिसमें कथ्य और रूप एकात्म हो जाते हैं। राकेश के नाटकों में अधिक रूपात्मक वैविध्य नहीं है। सामान्य ब्यौरों की उपेक्षा करते हुए वे केवल नाट्य स्थितियों, नाटकीय वातावरण और दृश्यों को मनःस्थितियों के साथ जोड़ने का प्रयास करते हैं, जिससे प्रभाव ही मुख्य भूमिका निभाने लगता है। उनके नाटकों में स्थल और काल की अन्विति मिलती है और इसके साथ ही दृश्य-बन्ध की एकता भी। इसके कारण कुछ लोग ऐसा महसूस करते हैं कि वे अपने नाटकों के अन्त को सँभाल नहीं पाते। वस्तुतः **लहरों के राजहंस** का अन्त लिखने में वे कभी सन्तुष्ट नहीं रहे, **आधे अधूरे** का अन्त विस्तार से ग्रस्त है। उसमें पुरुष चार बहुत मुखर हो गया है।

इस सन्दर्भ में चरित्र-चित्रण को लेकर कुछ कहना उपयुक्त होगा। राकेश पात्रों की सृष्टि भावना और विचार के मिले-जुले तत्त्वों से करते हैं। **आषाढ़ का एक दिन** से उन्होंने जो एक वैचारिक यात्रा प्रारम्भ की थी, वह **पैर तले की ज़मीन** में आ उभरती है। **आषाढ़ का एक दिन** में किशोर सम्बन्धों की अभिव्यक्ति है। **लहरों के राजहंस** में सम्बन्धों की उलझन है और सुलझाव की तलाश है; किन्तु **आधे अधूरे** में वह एक स्वीकृत तथ्य है। **पैर तले की ज़मीन** में सम्बन्धों का ऊपरीपन चित्रित हुआ है। इन्हीं सम्बन्धों के परिप्रेक्ष्य में राकेश ने स्त्री-पुरुष पात्रों के बिम्ब अपनी मानसिक अवधारणा के आधार पर उभारे हैं।

आयनेस्को ने एक स्थान पर कहा है : 'थिएटर को मनोविज्ञान की उपेक्षा कर उसे आधितात्त्विक (मेटाफिजिकल) आयाम देने चाहिए।'[9] राकेश ने यही किया है। उन्होंने अपने पात्रों को अस्तित्व की पीड़ा, सन्त्रास, संघर्ष, स्वप्न आदि की स्थितियों में आधितात्त्विक आयाम दिया है। वे मनोविज्ञान की उपेक्षा नहीं करते किन्तु वे चरित्र-चित्रण के क्षेत्र में एक प्रकार से मनुष्य की पुनर्सृष्टि जैसा प्रयास करते हैं। उन्होंने अपने पात्रों का जो रूप सँवारा है, वह आधुनिक मानव की वैचारिक अवधारणा के ढाँचे में ढला हुआ है। उनके नाटकों में वैचारिकता को एक आन्तरिक सत्य के रूप में स्वीकार किया गया है। यह वैचारिकता एक ओर राकेश के अपने अर्जित अनुभवों की देन है,[10] दूसरी ओर उनका अपना आधुनिकता-बोध भी पात्रों के जीवन-सन्दर्भों में एकाकार हुआ लगता है। इसके अनुसार मानव के गरिमामय शौर्यपूर्ण व्यक्तित्व की कल्पना आज ध्वस्त हो चली है। वह आज असम्बद्ध, विकृत और विसंगत जीवन को जीने के लिए बाध्य है और अपने से बाहर की कुछ ऐसी शक्तियों की दया पर वह अपने अस्तित्व को झेल रहा है, जिसमें उसकी भूमिका निष्क्रिय और कर्तृत्वहीन है। एक अनिच्छित परिवेश से सम्बद्ध होने के कारण मनुष्य अपने आत्मरूप को उपलब्ध नहीं कर पाता। फलतः उसके सामने लगातार टूटा हुआ जीवन है। टूटने का यह अहसास उसकी आन्तरिक पीड़ा, भय और सन्त्रास को जगाता है और नग्नता और निरर्थकता का आभास दिलाता है। बन्द ज़िन्दगी की दीवालों में बन्दी आज का मानव इस जगत् में निर्वासित और अपने प्रति

अजनबी हो गया है। वह असंगत के आमने-सामने विवश खड़ा है।

जीवन की यह जकड़, यह विसंगत स्थिति स्वयं आज नाटक का पात्र बन गई है। असंगत मानव का अतीत विरह और इन दोनों के संघर्ष से उत्पन्न शून्य और नास्तिभाव का अहसास ही आज वास्तविक पात्रों का रूप धारण कर चुका है। विसंगति, जैसा कि कामू ने कहा है, 'आदमी में नहीं, विश्व में भी नहीं, वह तो इन दोनों की एक-साथ उपस्थिति में है।'[11] आदमी जीवन और जगत् से इस प्रकार बद्ध है कि उससे भी अधिक महत्त्वपूर्ण भूमिका परिस्थितियों और परिवेश की हो गई है। आदमी को तो केवल उन्हें जीना है और जीना विसंगति के साथ निर्वाह करना मात्र है।

राकेश के पात्र अस्तित्व के इन्हीं संकट के प्रति जागरूक हैं। इसीलिए उनमें एक ख़ास तरह का रुख़ नज़र आता है। वे परिस्थिति के बीच भँवर में खड़े दिखाई देते हैं—कभी यथास्थिति में और कभी उससे बाहर निकलने की छटपटाहट में। स्थितियों के बाहर वे कुछ नहीं; जो कुछ हैं स्थितियों के बीच, अपने जाने-पहचाने दायरों के बीच। पहले से उनकी कोई प्रकृति नहीं—स्थितियाँ ही उनकी प्रकृति को निर्धारित करती हैं। इसीलिए राकेश के नाटक पात्र-प्रधान नाटक नहीं, स्थिति-प्रधान नाटक हैं। हर पात्र की अपनी दुनिया है, पर इस दुनिया में उसका अस्तित्व रोज़मर्रा की ज़िन्दगी जैसा नहीं है। नन्द, कालिदास, सावित्री, सुन्दरी—सबको अपनी दुनिया है, अपनी दृष्टि और जीवन स्थितियाँ हैं, जिनके अलग-अलग आयाम हैं। आत्यन्तिक स्थितियाँ इन पात्रों को वहाँ ले जाती हैं जहाँ ज़िन्दगी उलट-पुलट जाती है और आदमी स्वयं अपने ही ख़िलाफ़ खड़ा दिखाई देता है। नन्द, कालिदास, सावित्री, महेन्द्रनाथ—सब इसी आत्महन्ता स्थिति के कगार पर जीवन-यापन करते हैं—केवल एक उद्दीप्त अतिरंजित आक्रोशपूर्ण बिन्दु पर वे अपनी सारी संवेदना को टिका देते हैं। जीने के लिए कालिदास मल्लिका का, सावित्री जगमोहन का, नन्द सुन्दरी का सहारा लेना चाहता है, जैसे वे अन्तिम सत्य हों। इस प्रकार जीवन की धुरी से हटकर ये पात्र अन्ततः किसी एक छोर को अपने लिए चुन लेते हैं; पर वहाँ भी ज़िन्दगी उन्हें टिकने नहीं देती !

वस्तुतः राकेश के पात्र जीवन और जगत् से एक विशेष सन्दर्भ से सम्बन्धित लगते हैं। उनकी सृष्टि त्रासद इसीलिए है क्योंकि वह अनुक्रियाओं और भावनाओं पर अवलम्बित हैं। उनका जीवन एक ऐसी कविता है जो मानवीय नियति के कठोर बिम्ब को उभारती है। सम्भवतः मानव नियति को अलग रखकर राकेश के पात्रों की कल्पना ही नहीं की जा सकती। हम में से कोई भी उनकी जगह हो सकता है—इस भाव के बावजूद वे अपनी नियति में अलग हैं। इसीलिए वे जीवन का प्रतिनिधित्व नहीं, बल्कि उसकी व्याख्या करते हैं। वे जीवन का पूरा चित्र भी प्रस्तुत नहीं करते; नियति से जूझते मानव की एक झलक, एक मुद्रा प्रस्तुत करते हैं जिस पर त्रासद अनुभवों की विडम्बनाएँ हावी दिखाई देती हैं।

राकेश के पात्र प्रायः प्राकृतिक स्थिति से एक भिन्न अजनबी स्थिति में अपने को पाते हैं—एक अज्ञात देश में, जहाँ वे धकेल दिए गए हैं। ऐसे में निर्वासन की उस

ज़िन्दगी को भोगते हैं जिसमें वे विभाजित महसूस करते हैं। कालिदास ग्राम के स्नेहिल और नगर के ऐश्वर्यपूर्ण, सुख-सुविधापूर्ण जीवन के द्वन्द्व में बँट जाता है; सावित्री घर और बाहर के बीच बँट जाती है। नन्द दो वास्तविकताओं, दो ज़रूरतों के बीच वरण की दुविधा से ग्रस्त हो जाता है। महेन्द्रनाथ की भी दो जिन्दगियाँ हैं—एक वह जो घर में पत्नी के साथ जीता है और दूसरी राहत की जो उसे जुनेजा के पास जाकर मिलती है। इस प्रकार ये पात्र दो प्रकार की ज़िन्दगी, दो प्रकार की दुनिया में बँटे हुए हैं, चुनाव और अन्तिम सत्य की तलाश में जुलाहे की फिरकी की तरह इधर से उधर और उधर से इधर चालित होते दिखते हैं। यहीं उनके सामने चुनाव की समस्या आती है।

राकेश की पात्र-सृष्टि में अवधारणा का आधार मानव त्रासदी है जिसे वे अच्छी तरह पहचानते थे। उन्होंने लिखा है : 'ज़िन्दगी में व्यक्ति कुछ भी चुने, उसमें एक विशेष प्रकार की आइरनी होती है क्योंकि परिस्थितियाँ फिर-फिर वही बन जाती हैं।'[12] परिस्थितियों का फिर-फिर वही बन जाना इच्छाशक्ति और संकल्प के मार्ग में बाधक सिद्ध होता है। वस्तुतः वरण की स्वतन्त्रता मानव की वास्तविक त्रासदी है। राकेश के नाटकों में वरण एक आत्मोपलब्धि न होकर कारुणिक स्थिति है, क्योंकि वरण करने का अर्थ है—दूसरी सम्भावनाओं से हाथ धोना। इस प्रकार हर वरण में पात्र के अस्तित्व का त्रासद अंश विद्यमान लगता है। कालिदास, नन्द, सावित्री कहीं-न-कहीं अपने चुनाव में ऐसी स्थितियों में फेंक दिए जाते हैं, जो उन्हें एक स्थिति से काटकर दूसरी से जोड़ देती हैं। यहीं वे पीड़ा, सन्त्रास और त्रासद अनुभव के बीच से गुज़रते हैं। राकेश के पात्रों का त्रासद अनुभव उनके वरण, भ्रम और आत्मान्वेषण की असमर्थता पर निर्भर करता है। नन्द इसलिए पीड़ित होता है कि वह अपने को पहचान नहीं पाता और वरण से पलायन करता है। कालिदास की त्रासदी दूसरे प्रकार की है। उसकी अत्यधिक संवेदना उसे बाह्य जगत् से सामंजस्य नहीं करने देती; और सामंजस्य का थोथा समझौता ही उसके जीवन की करुणा उभारता है। ओ' नील और टेनेसी विलियम्स ने कलाकार-कोटि के ऐसे पात्रों का चित्रण किया है जिनमें अति संवेदनशील पात्र संवेदनशून्य जगत् में कष्ट झेलते दिखाए गए हैं। ऐसी स्थिति में या तो वे विनष्ट होते हैं या पाशविक बन जाते हैं—कालिदास कुछ-कुछ दोनों सीमाओं को छूता है। **आधे अधूरे** में त्रासदी का एक भिन्न आयाम है। उसमें पारिवारिक विघटन और उनमें बन्दी व्यक्ति की त्रासदी के साथ राकेश ने इससे वंशक्रम की त्रासदी को भी ध्वनित किया है। भ्रष्ट वंशक्रम और भ्रष्ट वातावरण माता-पिता पर ही नहीं उनकी सन्तान पर भी हावी हो जाता है और वे उन्हीं के जीवन को दुहराते हैं। उनके माता-पिता जैसी नियति उनकी भी दिखाई देती है। वे केवल दर्पण बनकर रह गए हैं।

एक और त्रासद अनुभव राकेश के पात्रों को घेरे हुए दिखाई देता है और वह है उनका दो विरोधों के बीच जीवन-यापन करना; दो नावों में सवार ज़िन्दगी उन्हें अपनी निजता से अलग कर देती है। अपनी अस्मिता और पूर्णता की खोज में कभी वे आत्मरूप दिखाई देते हैं और कभी नहीं। यही द्वैत उनकी अस्मिता का विभाजन कर देता है।

नन्द, कालिदास, महेन्द्रनाथ ऐसे अन्तर्विरोध को जीवन में जीते हैं। वे एक निष्क्रिय तत्त्व के रूप में सामने आते हैं, जो मानव की नियति को स्वीकार करते हैं, छटपटाते हैं, पर उससे लड़ते नहीं। अगर वे जूझते भी हैं तो अपने मन से, उन विरोधों से जो उन्हें निगल जाते हैं। (यह जूझना परम्परावादी नाटकों के पात्रों-सा नहीं है जो सत्-असत् या आदर्श के लिए जूझते थे।) परन्तु राकेश के हर नाटक में ऐसे पात्र हैं जो अन्तर्द्वन्द्व को खो चुके हैं। विलोम, अम्बिका, मल्लिका, सुन्दरी, ऐसे पात्र हैं और वे अपेक्षाकृत सुखी दिखाई देते हैं। किन्तु फिर भी अपने-अपने ढंग से सभी दुखों को उठाते हैं। यही कारण है कि राकेश के सभी पात्रों के प्रति सहानुभूति होती है—अम्बिका, मल्लिका, सावित्री, महेन्द्रनाथ, नन्द, सुन्दरी, कालिदास, मातुल, विलोम—सबके प्रति। किन्तु इस सहानुभूति को नाटकीय विडम्बना नियन्त्रित करती है जिससे उसकी तीव्रता कम हो जाती है।

राकेश पात्रों को विभिन्न कोणों से प्रस्तुत करते हैं। पात्र के प्रति पाठक की दुर्भावना को कम करने और सहानुभूति जगाने के लिए वे उसकी पूरी वकालत करते दिखाई देते हैं। **आधे अधूरे** में महेन्द्रनाथ के लिए जुनेजा ऐसा ही करता है। नन्द अपने बचाव में सुन्दरी के और कालिदास मल्लिका के सामने एक लम्बा वक्तव्य देता है। इससे पाठक की सहानुभूति मल्लिका और कालिदास, सुन्दरी और नन्द, महेन्द्रनाथ और सावित्री दोनों पक्षों से होने लगती है। इसके अतिरिक्त राकेश के पात्र एक निराशा में जीते हैं जो जीवन का नकार है और कुछ लोगों का कहना है कि निराशा की स्थिति में कोई त्रासदी सम्भव नहीं। यदि जीवन व्यर्थ है तो उसके संकट में त्रास की अनुभूति अपनी तीव्रता खो देती है। उस हालत में पाठक/प्रेक्षक की सारी सहानुभूति का कोष व्यर्थ ही जाता है। राकेश पात्र को त्रासदी के भँवर में छोड़ देते हैं और अपने को उससे अलग कर देते हैं।

राकेश ने पात्रों के अन्तर्विरोध को गहराई से पकड़ा है। किन्तु यह अन्तर्विरोध व्यक्ति का अन्तर्विरोध है—इसको युग के अन्तर्विरोध से जोड़ने में वे असमर्थ रहे हैं। व्यक्ति का जो चित्र वे उभार पाए हैं, वह न समाज की पृष्ठभूमि में है और न वह उसका प्रतिनिधित्व ही करता है। व्यक्ति जैसा भी है, उसकी अपनी अस्मिता नहीं है—वह आधा- अधूरा है, खंडित प्रतिबिम्ब की तरह, परिस्थितियों से घिरा और विरोधाभासों और अन्तर्विरोधों से परिपूर्ण। उनके पात्र युग्म बनाते हैं। कई बार वे एक ही सिक्के के दो पहलू के रूप में चित्रित हैं। एक-दूसरे के विरोध में खड़े पात्र अपनी चेतना में एक-दूसरे से जुड़े लगते हैं। पर उसके द्वारा सबको एक सतह पर लाने का जो प्रयत्न हुआ है, उससे मानव की प्रतिष्ठा घटी है और उसका बौनापन ही उभर पाया है।

अधूरेपन को लेकर राकेश के सब पात्र एक हो जाते हैं, चाहे वह कालिदास हो, नन्द हो या महेन्द्रनाथ। मल्लिका जानती है कि कालिदास को कोई भी रेखा घेर सकती है, स्वयं उसको विलोम घेर लेता है। सुन्दरी का व्यक्तित्व एकांगी है। नन्द 'सब जगह अपने को एक-सा अधूरा अनुभव करता है' और सुन्दरी यह जानती है कि 'जितने साधारण और लोग हैं, उतने ही साधारण आप भी हैं।' महेन्द्रनाथ अधूरा आदमी है।

अधूरापन स्त्री-पुरुष दोनों में है, किन्तु राकेश के नाटकों में नारी अपनी धुरी पर पुरुष को घुमा न पाने की असमर्थता में पुरुष का अधूरापन देखती है, जबकि जिसे वह उसका अधूरापन मानती है, पुरुष-पात्र उसे आन्तरिक आवश्यकता, अपेक्षा मानते हैं और अधूरेपन को नारी में संकेतित करते हैं। अधूरेपन की यह तलाश एक 'मिथ' है क्योंकि आज आदमी एक ऐसे समय में रह रहा है, जब उसके पूरे होने की गुंजाइश ही नहीं है। पूर्णता एक लक्ष्य पर पहुँचने और वहाँ टिके रहने की स्थिति है, जो कि आज के जीवन में सम्भव नहीं। राकेश के नाटकों में व्यक्ति इन्हीं मुद्राओं में विद्यमान है।

इसीलिए राकेश के नाटक नायक की परम्परागत अवधारणा को नकारते हैं। बहुत-से लोगों ने आक्षेप किया है कि नन्द, कालिदास, बुद्ध आदि का व्यक्तित्व इतिहास की गरिमा के अनुकूल चित्रित नहीं हुआ है। प्रसाद के गौरवयुक्त पात्रों से अभ्यस्त हिन्दी का संस्कारी पाठक इस तरह से सोचे तो कोई आश्चर्य की बात नहीं। किन्तु इस बात को नहीं भूला जाना चाहिए कि प्रसाद का युग राष्ट्रीय आन्दोलन और सांस्कृतिक पुनर्जागरण का युग था। फलतः इतिहास की कथावस्तु के माध्यम से महान् नायकों का चित्रण अनिवार्य था। उस समय सारा ज़ोर आदर्श पर था या कर्म पर। अतः धीरोदात्त नायक युग की आवश्यकता के लिए अनिवार्य था। स्वतन्त्रता के बाद नायक के ऊपर से सारा प्रभामंडल हट गया, जिससे आरोपित दिव्यता कपूर हो गई। इतिहास की तलाश धीरोदात्त नायकों के लिए नहीं, वरन् मानव की निजी सत्ता, मानवीय संवेदना और उसकी अन्तरंग झाँकी प्रस्तुत करने के लिए की जाने लगी। आज यह अहसास निरन्तर गहराता जा रहा है कि मानव जागतिक स्थितियों के पाटों के बीच पिसता जा रहा है। आधुनिक नाटककार इस धारणा से गहराई से प्रभावित हुआ है। इसीलिए आज वह नायक की पुरानी अवधारणा से विहीन हो गया है। कहने के लिए नायक अब भी है, पर जहाँ भी है वहाँ अनाम, अज्ञात और अति साधारण है, वह चाहे बुद्ध हो, नन्द हो या कालिदास। सब लघु मानव हैं। उनकी भावनाएँ, कामनाएँ स्वप्न और पीड़ाएँ सब महान् हैं, पर वे स्वयं महान् नहीं हैं। वे महान् नहीं हैं, क्योंकि महानता स्वभावज नहीं। वस्तुतः राकेश के पात्र निर्व्यक्तिकरण की प्रक्रिया को अपने में समाहित किए हैं। व्यक्ति अव्यक्ति होकर रह गया है। मनःस्थितियाँ ही व्यक्तित्व अर्जित कर लेती हैं और पात्र में जो व्यक्तित्व दिखाई देता है, वह उसका नहीं, मनःस्थिति, संवेदना या विचार का है अथवा युग का है। पात्र युग और समूह से जुड़े हैं और उन्हीं की नियति को जीते हैं। इसके अतिरिक्त वे विचारों या मानवीय भावों की रूपरेखा के प्रतीक बन जाते हैं। वे किसी भाव या विचार को उभारते हैं और चरित्र अमूर्त-सा होकर रह जाता है। **आषाढ़ का एक दिन** और **लहरों के राजहंस** में पात्र रेखाओं में अंकित हैं। **आधे अधूरे** में जहाँ पात्रों का अत्यधिक सामान्यीकरण है, वहाँ बल उन आकांक्षाओं और पीड़ाओं पर ही है जिन्हें आम आदमी जीवन में भोगता है। सब पात्र महानता में नहीं, पर भावना में एक भरपूर जीवन जीते हैं। इसीलिए चाहे राकेश के नाटकों में नायक न हों, या बहुत साधारण हों, पर इतना तो निश्चित है कि वे एक 'हीरोइक' ज़िन्दगी जीते लगते हैं।

यह बात दूसरी है कि स्थितियों की बागडोर उनके हाथ में नहीं होती और उनके निर्णय दुनिया को बहुत दूर तक प्रभावित नहीं करते। राकेश के पात्र कहीं निरपेक्ष, एकाकी, निर्वासित या निरीह लग सकते हैं, पर उनमें एक अद्भुत जीवन्तता भी है। इसका एक कारण यह भी है कि उन्होंने उन्हें अपने अनुभवों से ढाला है और उनके साथ लेखक का एक सीधा रिश्ता दिखाई देता है। इसके अतिरिक्त राकेश के चरित्र-चित्रण की एक विशेषता यह भी है कि नायक द्वन्द्व के केन्द्र में है, पर नाटक के केन्द्र में नारी है। नारी में द्वन्द्व नहीं, पर नाटक पर वही छाई दिखाई देती है। इसीलिए राकेश के नाटक नायिका-प्रधान बन गए हैं।

चरित्र-सृष्टि राकेश का आत्यंतिक उद्देश्य प्रतीत नहीं होता। उनके नाटकों में चरित्र को सुनिश्चित रेखाओं में उभारने का प्रयत्न नहीं है। चरित्र जैसी स्थिर कोई वस्तु है भी नहीं; जैसा कि उन्होंने **आधे अधूरे** में का.सू.वा. से बुलवाया है, व्यक्ति की कोई निश्चित भूमिका नहीं। दूसरों के साथ टकराते हुए किसी नाट्य स्थिति में जो उभरकर आता है, वही चरित्र है। फलतः उनके चरित्र-चित्रण का मुख्य उपादान है—स्थिति पर आरोपित बहुमुखी चेतना। ऐसी चेतना जिसमें शक्ति की हीनता है और जिसमें समस्या के समाधान की स्थिति नहीं दिखाई देती। स्थिति और पात्र जैसे एक हो जाते हैं, स्थिति पात्र की अनुभूति बन जाती है और पात्र उसका भोक्ता। पर यह भोग चेतना के संकट का भोग मात्र है। राकेश के नाटकों में चेतना का अर्थ मनोवैज्ञानिक ऊहापोह नहीं। बल थोथे मनोविज्ञान पर नहीं, वरन् घटनाक्रम कार्य और संवाद पर दिखाई देता है। यद्यपि राकेश में मनोविज्ञान की गहरी पैठ दिखाई देती है; फिर भी मानव के स्वतन्त्र विवेक और वैयक्तिक संकल्प पर प्रहार कर उसे अचेतन ग्रंथियों के सहारे चालित पशु के रूप में प्रतिष्ठित करनेवाली भोंडी मनोवैज्ञानिकता उनमें नहीं। इसीलिए उनके कई पात्र अन्तर्द्वन्द्व को खो देते हैं और वे फिर स्थिर आस्था की तलाश करते हैं। जहाँ अन्तर्द्वंद्व पात्रों में है भी, वह छटपटाहट में व्यक्त हुआ है, भावनाओं की ऊहापोह में उलझकर नहीं रह गया है। सारा अन्तर्द्वन्द्व मंच के पीछे घटित होता है। कालिदास, नन्द आदि का अन्तर्द्वन्द्व ऐसा ही है। अलका नन्द के सन्दर्भ में कहती है, 'ओह कैसे अन्तर्द्वन्द्व के क्षण रहे होंगे कुमार के लिए ! पर इस अन्तर्द्वन्द्व की केवल कल्पना ही की जा सकती है। निश्चयतः अन्तर्द्वन्द्व चाहे कालिदास का हो, नन्द या महेन्द्रनाथ का—उसका पता बाद में उल्लेख से चलता है। फिर उनके अन्तर्द्वन्द्व के बीच लेखक कहीं खड़ा नहीं हुआ—दो विरोधी मूल्यों में कभी पात्र झूलते दिखाई देते हैं तो लेखक ग़ायब दिखाई देता है। उसी प्रकार अन्तर्द्वन्द्व के लिए विषय के दोनों पक्षों को प्रस्तुत करने की जो अनिवार्यता होती है, वह भी राकेश के पात्रों में नहीं। वास्तव में वे मानते थे कि नाटक के लिए किसी भी चीज़ का संकेत काफ़ी है, उसका पूरा विकास दिखाना आवश्यक नहीं। जिस आधारभूत अन्तर्द्वन्द्व को लेकर उन्होंने अपने पात्रों की रचना की, उसमें बहुत आगे बढ़ जाना उन्हें स्वीकार्य नहीं था। यह उनकी दृष्टि में 'समाज कल्याण है, नाटक नहीं।' राकेश के नाटकों में ऐसे पात्रों का भी अभाव नहीं जो अन्तर्द्वन्द्व को

बहुत पहले पचा चुके हैं। इसका एक उदाहरण महेन्द्रनाथ है जो नाटक में जड़ता की स्थिति भोगता है। वह भी एक तरह से सावित्री और जुनेजा में बँटा है, पर सब-कुछ अन्तर्निहित है। सुन्दरी, अम्बिका, मल्लिका, विलोम आदि कई पात्र उसकी तरह जड़ नहीं, पर अन्तर्द्वन्द्व के अभाव में भी अधिक व्यावहारिक, अधिक स्थिर और जीवन्त हैं।

वस्तुतः राकेश के नाट्य-शिल्प की एक बहुत बड़ी विशेषता चुनाव की प्रक्रिया है। उन्होंने अपनी नाट्यवस्तु को सदा एक विशेष कोण से चुना है। इसीलिए **लहरों के राजहंस** में बुद्ध पात्र बनने से रह गए या **आषाढ़ का एक दिन** में कहीं-कहीं विलोम कालिदास की समकक्षता प्राप्त कर लेता है। राकेश कथावस्तु और चरित्र की उन सार्थक इकाइयों को ही चुनते हैं, जो सर्जनात्मक लक्ष्य की पूर्ति में सहायक होती हैं। इसीलिए उनके नाटकों में पात्र क्रिया-व्यापारों की अन्तःशक्तियों के संश्लिष्ट रूप में सामने आते हैं और क्रिया-व्यापार चरित्र की विशिष्ट क्षमताओं को ध्वनित करता है। चरित्र और क्रिया-व्यापार का अन्तःसम्बन्ध यादृच्छिक नहीं, बल्कि नाटककार दोनों की परिस्थिति और मानवीय सम्बन्धों के बदलते स्वरूपों पर छोड़ देता है। यही कारण है कि पात्र अपने को परिभाषित करता है, रूढ़ नहीं बन पाता। राकेश पात्र और उसके क्रिया-व्यापार को किसी निर्णायक बिन्दु पर नहीं ले जाते, वरन् उसे प्रश्नों और परिस्थितियों के भँवर में डालकर स्वयं अलग हो जाते हैं। इसका मुख्य कारण यह है कि कथावस्तु और चरित्र को वे कथ्य तक पहुँचने के लिए साधन की तरह प्रयुक्त करते हैं और उसी में नाटक अपना रूप ग्रहण करता है।

परिस्थितियों के आग्रह से बद्ध होने के कारण राकेश के नाटकीय पात्र मिलकर जीवन की विडम्बना को भोगते हैं। इस प्रकार एक सामान्य अनुभव के बीच वे एक-दूसरे से कटा अस्तित्व नहीं रखते। वे एक-दूसरे के सन्दर्भ में ही अपने को परिभाषित करते दिखते हैं और पारस्परिक सम्बन्धों पर आधारित लगते हैं। इस प्रकार वे अलग-अलग होते हुए भी एक-दूसरे की संवेदना से जुड़े दिखाई देते हैं। यही बँटी हुई संवेदनशीलता कहीं पात्र को खंड-खंड में तोड़ती है, तो कहीं विरोधी खंडों के रूप में आपस में जोड़ती है। कहीं पात्र बाहर से बिल्कुल ठोस और इकहरा लगता है, तो कहीं दुहरे-तिहरे आवरण में लिपटा दिखाई देता है। यह विधि पात्र के समरूप व्यक्तित्व को नहीं गढ़ती, विविधता के उपकरण जुटाने का प्रयत्न करती है। बिन्नी-किन्नी और सावित्री अनेकता में एकता और एकता में अनेकता का परिचय देते हैं। इसी प्रकार महेन्द्रनाथ, जुनेजा, जगमोहन आदि अनेक पात्र भी हैं और एक भी हैं। कालिदास और विलोम विपरीत होकर भी एक ठहरते हैं। अम्बिका, मल्लिका, नन्द, श्यामांग सब क्षैतिज विस्तार के सूचक हैं। अतः पात्रों की अवधारणा कहीं युग्मों और कहीं एकल विस्तार के रूप में हुई है। इस प्रकार कहीं एक पात्र का अन्य पात्र के रूप में विस्तार है तो कहीं किसी एक की किसी प्रवृत्ति का पात्रीकरण।

इस प्रकार राकेश के पात्र परम्परागत पात्रों की तरह ठोस नहीं—वे तरल हैं। अहं उन्हें ठोस बनाता है और राग तरल। पर दोनों स्थितियाँ सापेक्ष भी हैं। अम्बिका, सुन्दरी,

विलोम जैसे कुछ पात्र बाहर से ठोस अवश्य दिखाई देते हैं, पर उनका यह ठोसपन किन्हीं अन्य पात्रों की तरलता के विरुद्ध उभारा गया है; अन्यथा उनमें भी सारी तरलता विद्यमान है, जो बाहर से जम गई है। ये पात्र अपने में बन्द दिखाई देते हैं, जो दूसरों की ओर अतिक्रमण नहीं करते। ये भीड़ के बीच जीनेवाले पात्र नहीं हैं, भीड़ में खो जानेवाले पात्र हैं। अम्बिका परिवेश की पीड़ा से जुड़ी है और सुन्दरी शरीर के साथ एकत्व बोध की मानसिकता की शिकार है। वे अहं के घेरे में संवेदनाओं में जीती हैं, सत्यों में नहीं। ये ठोस पात्र ऊपर से बहुत दुराग्रही और आक्रामक लगते हैं। दूसरी ओर आस्था और आशा के बल पर जीवन को झेलनेवाले पात्र कोमल और दुर्बल हैं। ये दो प्रकार के पात्र अलग-अलग से लगते हैं, पर अलग-अलग हैं नहीं। उनमें परस्पर गहरा सम्बन्ध है—इस रूप में कालिदास और विलोम, मल्लिका और अम्बिका, नन्द और श्यामांग, महेन्द्रनाथ और अशोक, सावित्री और बिन्नी-किन्नी आपस में सम्बद्ध हैं और सम्बन्धों की पारस्परिकता में ही विकसित होते हैं। **आधे अधूरे** के माता-पिता और सन्तान में अद्‌भुत समीकरण है। मल्लिका, निक्षेप और अम्बिका कहीं संवेदना में एक हो जाते हैं, तो कहीं विलोम और अम्बिका। अलका श्यामांग से जुड़ जाती है और श्यामांग नन्द से। एक ही संवेदना से घिरे ये पात्र एक ही सामान्य भावना से मुखर हो उठते हैं, जैसे दो तार एक ही झंकार से झंकृत हो जाते हैं और एक-सी गूँज पैदा करते हैं। पात्रों की आन्तरिक चेतना की यह समानता पात्रों को एक प्रकार की समानांतरता प्रदान करती है।

राकेश के चरित्र-शिल्प का एक दूसरा आयाम भी है। उनके लिए पात्र की चेतना-परिधि मुख्य है। इसीलिए पात्र न किसी ढाँचे में ढले हैं और न उनमें व्यक्तित्व का कोई सायास सामंजस्य है। पात्र निर्धारित सीमाओं में निर्मित नहीं है जैसा कि **आधे अधूरे** का का.सू.वा. कहता है : 'आप शायद यह सोचते हों कि मैं कोई निश्चित इकाई हूँ (पर) नाटक के बाहर या अन्दर मेरी कोई निश्चित भूमिका नहीं है।' वस्तुतः राकेश एक ही पात्र को व्यक्तित्व की अनन्य विशेषताओं से गुंफित करने के बजाय उसके खंड व्यक्तित्व के आधार पर एक-एक स्वतन्त्र पात्र की परिकल्पना करते हैं। फलतः उनके नाटकों में खंड चरित्र मिलते हैं। पात्र के अवचेतन में स्थित अप्रकाशित व्यक्तित्व को खंड रूप में दूसरे पात्र में दिखाने की विधि का राकेश ने अपने नाटकों में सफल प्रयोग किया है। कालिदास और विलोम, नन्द और श्यामांग (अलका और सुन्दरी ?) इसी कोटि के पात्र हैं, जो पूरक भी हैं और विरोधी भी। श्यामांग नन्द का ही विचारक रूप है जो सुन्दरी से उलझ जाता है। जो कुछ वह महसूस करता है वह नन्द के अन्दर होनेवाली प्रच्छन्न प्रक्रिया है। आज के आदमी के जगत् के साथ जो सम्बन्ध हैं वे बौद्धिक हैं, सोचने-विचारने तक रह गए हैं। सोचना आदमी का काम ज़रूर है, पर वह उसका सम्पूर्ण अस्तित्व नहीं, एक सामर्थ्य मात्र है। कर्म के साथ न जुड़नेवाली ज़िन्दगी कमल ताल की विकृत छाया की अयथार्थ कल्पना से जुड़ जाती है। श्यामांग की, हर सोचनेवाले की विसंगति यही है। नन्द इसी विसंगति को भोगता है, पर प्रारम्भ में उसका माध्यम

श्यामांग बनता है—एक ऐसा पात्र अपने अस्तित्व में जो पराई पीड़ा भोगता है। पर वह जैसा भी है, उसका खंडित, विकृत, विक्षिप्त और अतिरंजित व्यक्तित्व नाटक की अर्थगर्भिता में बहु-आयाम धारण कर लेता है। वस्तुतः वह सत्य का वाहक पात्र है। इस विराट् शून्य जगत् में वह अपना स्थान चाहता है और भरे-पूरे अन्धकार में एक किरण की खोज में है। जगत् उसके साथ नहीं, इसीलिए वह भी जगत् के साथ नहीं प्रतीत होता है। फिर भी वह यथार्थ है और जिन प्रश्नों को वह उठाता है वे भी यथार्थ हैं।

इसी प्रकार विलोम कालिदास का आलोचक व्यक्तित्व है। वह अपनी पैनी दृष्टि से कालिदास की आन्तरिक दुर्बलताओं को उघाड़कर रख देता है। कालिदास के सन्दर्भ में वह अम्बिका के समान ही कटु है। पर विरोध के बावजूद वह अपने और कालिदास के बीच एक रिश्ता निकाल ही लेता है : 'विलोम क्या है; एक असफल कालिदास। और कालिदास ?—एक सफल विलोम। हम कहीं एक-दूसरे के बहुत निकट पड़ते हैं।' अपने नाम के अनुरूप ही वह कालिदास के विपरीत लगता है, पर 'सभी विपरीत एक-दूसरे के बहुत निकट पड़ते हैं'—इस दृष्टि से विलोम को विपरीत कहना मुश्किल है। कालिदास महान् कवि बनता है, राज्याधिकार प्राप्त करता है; किन्तु विलोम कुछ भी न कर पाने पर भी मल्लिका को पा जाता है जिसे कालिदास न पा सका। कालिदास और मल्लिका में भावना का सम्बन्ध है। पर क्या विलोम और मल्लिका में ऐसा कुछ नहीं जो दोनों को जोड़ता है ? यह उलझाव राकेश की चरित्र-सृष्टि के कौशल को व्यक्त करता है जिसकी सबसे बड़ी उपलब्धि तो यह है कि विलोम प्रेम के त्रिकोण का एक भद्दा पात्र बनकर नहीं रह जाता। मल्लिका के लिए उसके मन में कालिदास से भी अधिक आत्मभाव दिखाई देता है—अधिक आस्था और उदारता भी। यहीं वह खल पात्र की स्थिति से ऊपर उठ जाता है। पर अपनी टूटन में वह कालिदास के बहुत निकट पड़ता है। इस प्रकार वह चाहे कालिदास हो या विलोम या फिर नन्द और श्यामांग, ये पात्र एक दूसरे-के अन्तरंग विरोधी प्रतिबिम्ब हैं, जिनमें प्रक्षेपण (प्रोजेक्शन) की विधि का प्रयोग हुआ है। इन छाया-पात्रों का विशेष मूल्य है, क्योंकि एक-दूसरे के माध्यम से ही उनके दूसरे पहलू व्यक्त होते हैं।

अन्य पात्रों में भी इसी प्रकार की एक अन्तर्निहित चेतना का सक्रिय सूत्र नज़र आता है। राजसत्ता के द्वारा प्रदत्त अनुभव के सन्दर्भ में मातुल और कालिदास अन्ततः एक हो जाते हैं। राजशक्ति मातुल को पंगु बना देती है और कालिदास को सर्वहारा। जगत् का कटु यथार्थ मल्लिका को अम्बिका बना डालता है। इसी प्रकार मल्लिका-निक्षेप, श्वेतांग-श्यामांग, अलका-सुन्दरी, अलका-श्यामांग, अलका-नन्द के बीच सम्बन्धों के पूरक कोमल तन्तु दिखाई देते हैं। **आधे अधूरे** में पुरुष एक, दो, तीन, चार शिल्प के ही नहीं, चेतना के स्तर पर भी एक-दूसरे से एक सूत्र में बँधे हैं। यही बात सावित्री और बिन्नी-किन्नी के बारे में कही जा सकती है। सावित्री और महेन्द्रनाथ में परस्पर कितना वैमनस्य है किन्तु कोई तन्तु है जो उन्हें जोड़े हुए है, जो महेन्द्रनाथ को हर मंगल-शनीचर को घर लौटा लाता है और सावित्री को घर से बाहर भटकाता है।

राकेश ने उन्हें प्रतीकों और बिम्बों का आधार बनाया है। अपने पहले दो नाटकों में वे कालिदास, नन्द, बुद्ध आदि का चारित्र्य अतीत के बिम्बों के माध्यम से उभारते हैं। इसी प्रकार चरित्रों का प्रतीक तत्त्व भी राकेश को प्रिय रहा है। कालिदास को उन्होंने स्वयं 'सृजन का प्रतीक' कहा है। विलोम उन्हीं के शब्दों में 'दुराग्रह की आक्रामक शक्तियों को संकेतिक करता है।' नन्द और सुन्दरी धरती और आकाश के, पुरुष और प्रकृति के प्रतीक हैं। इसी प्रकार **आषाढ़ का एक दिन** में रंगिणी-संगिनी, अनुस्वार और अनुनासिक भी किंचित् प्रतीक तत्त्व लिए हुए हैं। मल्लिका को आस्था के प्रतीक रूप में अंकित किया गया है। कुछ आलोचकों ने **आधे अधूरे** के चरित्रों में भी प्रतीक तत्त्व ढूँढ़ निकाला है। चरित्रों में निहित प्रतीकार्थ निश्चयतः इन नाटकों को नई क्षमता प्रदान करता है। श्यामांग को ही लीजिए, एक प्रतीक चरित्र के रूप में वह नाटक की अर्थवत्ता को बढ़ाता है।

राकेश के पात्रों के सम्बन्ध में एक धारणा यह भी है कि कई अर्थों में उनके नारी पात्र विशिष्ट ठहरते हैं। **आषाढ़ का एक दिन** में मल्लिका और प्रियंगु भारतीय नारी की परम्परागत भूमिका निभाते हैं। दोनों पुरुष को समर्पित हैं और आत्मदान के भाव को व्यक्त करती हैं। **लहरों के राजहंस** की सुन्दरी दाम्पत्य सम्बन्धों को भोग और कामना के बीच निभाती है; किन्तु इस नाटक में आकर स्त्री-पुरुष सम्बन्धों में एक नई दृष्टि का प्रवेश हो जाता है। दोनों दो भिन्न बिम्ब को लेकर लेखक की कल्पना में उभरते हैं—पुरुष मूर्ति आकाश की ओर बाँहें फैलाए हुए और नारी मूर्ति आँखें धरती की ओर किए हुए। जैसे यही नारी और पुरुष के बीच की विषमता है। दोनों दो जीवन दृष्टियाँ हैं जो उलझ जाती हैं, उद्घटित नहीं होतीं। पर नारी और पुरुष का सम्बन्ध मात्र आकर्षण का ही नहीं, विकर्षण का भी है। राकेश के नाटक इस तथ्य को भी उभारते हैं। **आधे अधूरे** में यह स्थिति आगे बढ़ जाती है। पर आकर्षण और विकर्षण को एक साथ जीना स्त्री-पुरुष की नियति है। एक ओर सावित्री ने महेन्द्रनाथ को 'शिकंजे में जकड़ रखा है' दूसरी ओर वह 'उस औरत' को बहुत चाहता है। और वह औरत इतने गिले-शिकवे के बावजूद उस अधूरे आदमी को छोड़ने का इरादा ठीक समय पर नहीं कर पाती—जैसे एक यथास्थिति में जीना और लड़ते हुए जीना दोनों की नियति है। राकेश ने पुरुष को नारी पर आश्रित रखा है पर पुरुष नारी को एक आन्तरिक आवश्यकता के रूप में कभी प्राप्त नहीं कर पाता। नारी आमादा दिखाई देती है और पुरुष उसे झेलता है, मुक्ति चाहकर भी उससे मुक्त नहीं हो पाता।

राकेश की चरित्र-योजना में कहीं कुछ दोष भी नज़र आते हैं। एक सबसे बड़ा आरोप उन पर यह है कि उनके चरित्र कई नाटकों में अपने को दुहराते हैं। तीनों नाटकों के पात्र एक-दूसरे के बहुत निकट जान पड़ते हैं। उदाहरण के लिए मल्लिका, सुन्दरी और सावित्री में तथा कालिदास, नन्द और महेन्द्रनाथ में काफ़ी साम्य है। सभी जीवन के सपनों में पराजित अनुभव करते हैं और अधूरे अस्तित्व को निभाते हैं। कालिदास थका-हरा है, नन्द अपने ही द्वन्द्व का शिकार है और महेन्द्रनाथ इससे भी आगे जड़ हो

गया है। तीनों एक ही दिशा की ओर उन्मुख हैं और किसी बिन्दु पर एक हो जाते हैं। कुछ पात्र बाहर से टूटते हैं, अन्दर से ठोस बने रहते हैं, कुछ अन्दर से टूटते हैं। वे किसी आन्तरिक आवश्यकता से विवश होने के कारण 'नक्षत्र की तरह एक वृत्त में घूमते हैं और टूटे नक्षत्र की तरह उनका वृत्त भी नहीं रह जाता।' ब्रजमोहन शाह ने तीनों नाटकों की चरित्र-योजना में एक भिन्न कोटि का साम्य भी देखा है, 'राकेश के **आषाढ़ का एक दिन** के कालिदास, मल्लिका और विलोम, **लहरों के राजहंस** के नन्द, सुन्दरी और भिक्षु और **आधे अधूरे** के पुरुष, एक स्त्री और पुरुष चार—एक दूसरे के प्रतिरूप हैं। उनका कार्य-व्यापार एक-सा है। कालिदास मल्लिका से भागना चाहता है, नन्द सुन्दरी से, पुरुष एक स्त्री से—पर भाग कोई नहीं पाता। विलोम, भिक्षु और पुरुष चार अभिन्न हैं—अलग करने और मिलाने की कड़ी हैं। दोष और भी गिनाए गए हैं, जैसे—श्यामांग एक अस्वाभाविक पात्र है, उसे नाटक में नहीं होना चाहिए था; कालिदास का चित्रण ऐतिहासिकता का निर्वाह नहीं करता; मल्लिका के साथ विलोम का सम्बन्ध भारतीय परम्परा के प्रतिकूल है अथवा उसका 'अभाव की सन्तान' को जन्म देना उसकी प्रतिमा के विरुद्ध है; **आधे अधूरे** के पात्र एकस्तरीय और रूढ़ हैं। या फिर यह कि राकेश के नाटकों में वे पात्र कहाँ हैं जो बेबसी में आत्मसमर्पण नहीं करते और दर्शक बनने के बजाय भोक्ता बनकर जीते हैं ? ये सारे आरोप नगण्य हैं; किन्तु राकेश के पात्र अपने को जिस तरह दुहराते हैं, वह इस बात का परिचायक बनकर दोष बन जाता है कि लेखक के पास कुछ नया अनुभव नहीं है।

फिर भी सन्तोष की बात है कि राकेश के वस्तु और चरित्र शिल्प सुविचारित और सुगठित हैं। उनका पूरा लेखन एक मानसिक तैयारी पर आधारित है। इसीलिए वे अपने नाटकों को कई बार लिखते थे। जो इस तरह नहीं भी लिखे गए उनमें भी एक-एक शब्द, एक-एक नाट्य-स्थिति, एक-एक चरित्र, एक-एक प्रतीक चुनकर रखा गया है। इसीलिए उनमें कहीं बारीक कारीगरी और नपी-तुली पच्चीकारी का आभास मिलता है। इस आभास की रूढ़ स्थिति का परिणाम यह हुआ है कि राकेश के नाटक एक जैसे शिल्प में ढल गए हैं। इससे उनका बहुत-सा नैसर्गिक सौन्दर्य और सामर्थ्य लुप्त हुआ हो तो कोई आश्चर्य नहीं। फिर भी कला का निखार तो आया ही है।

भाषा : शब्द की बुनावट

अभिव्यक्ति के लिए भाषा से उलझने की समस्या हर लेखक के सामने आती है—अर्थात हर सचेत और अनुभूति प्रवण लेखक के सामने। समस्या नहीं आती तो उन लोगों के सामने, जो अनुभूति और चिन्तन की दृष्टि से अन्दर से बिल्कुल साफ़ हैं या केवल कुछ एक बेसिक अनुभूतियों और विचारों के दायरे में सोचते रहते हैं।...उन्हें अपने मन की हर बात के लिए उपयुक्त भाषा ज़रूर मिल जाती है क्योंकि उनके मन में की हर बात पहले से ही किसी-न-किसी की कही और सोची हुई बात होती है। हज़ारों बार कही और सोची हुई बातों की कार्बन कापी तैयार करने में समस्या पैदा ही कहाँ होनी है? जब एकाध विचार और दो-एक अनुभूतियों से आगे व्यक्ति का मन यात्रा ही न कर पाता हो तो उलझे हुए रास्ते पार करने की समस्या ही नहीं होगी, उनके लिए साधन ढूँढ़ने की बात तो बाद में आती है।

*—**मोहन राकेश** : बक़लम .ख़ुद, पृ. 73-74*

सवाल अनुभूतियों और विचारों की जटिलता का है। निश्चयतः आज के लेखक को अभिव्यक्ति के स्तर पर उसके कई आयामों का सामना करना पड़ता है। जीवन और अनुभूतियों की जटिलता ने लेखक का काम पहले से काफ़ी मुश्किल कर दिया है। राकेश ने जब लिखना शुरू किया था, उससे पहले छायावाद के कवियों ने इस समस्या को महसूस किया था और उसके समाधान के रूप में उन्होंने हिन्दी को एक ऐसा साहित्यिक ढाँचा दे डाला था, जो जन-जीवन की जीवन्त धारा से अलग छिटक गया था। इस भाषा का संस्कार रोमानी स्पर्श, कोमलकान्त पदावली, संस्कृत के चिकने-चुपड़े शब्दों, अलंकरण और चमत्कार की प्रवृत्ति से किया गया था। तब हिन्दी का वैभव संस्कृत की पुनर्जीवित परम्परा पर आधारित था; बाद में जैनेन्द्र, प्रेमचन्द और अज्ञेय ने इस भाषा को एक नई दिशा में आगे बढ़ाने का प्रयत्न किया, पर प्रेमचन्द की भाषा में संवेदनशीलता न थी, प्रसाद, अज्ञेय और जैनेन्द्र की भाषा संवेदनशील थी, पर वह न सहज थी न अन्तरंग। फलतः अब भी हिन्दी की मौलिक सर्जनात्मक क्षमताओं का अन्वेषण बहुत-कुछ होने को था। प्रयोगबाद, नई कविता और नई कहानी के उदय के

साथ रूढ़ साहित्यिक किन्तु जन-चेतना से विरहित भाषा की व्यर्थता का अहसास बड़ी तेज़ी से प्रारम्भ हुआ। इसके फलस्वरूप जो तीव्र प्रतिक्रिया हुई, उसने इस विचार को पुष्ट किया कि पूर्ववर्ती युग की भाषा 'भाषा नहीं शब्द-योजना की शैली-मात्र' थी।[1] राकेश के अनुसार यह एक ऐसी ऐंद्रजालिक भाषा थी जिसमें जीवन की जटिल और साहसिक अभिव्यक्ति सम्भव न थी। किशोर वय की भावुकता में तत्सम शब्दों के माध्यम से प्रौढ़ता का आभास देने का प्रयास कराने अथवा किसी मनःस्थिति के इर्द-गिर्द हेत्वाभासी दार्शनिकता का जाल बुनकर चमत्कार उत्पन्न करने तक ही इस भाषा की उपलब्धि मानी जा सकती है।[2]

यह मानना एक हद तक ग़लत भी न था। प्रसाद की नाट्य-भाषा जिस प्रकार हिन्दी नाट्य में रूढ़ हो गई थी, वह भाषा की भावी सम्भावना के विकास की दृष्टि से एक बहुत बड़ा ख़तरा बन गई थी। वस्तुतः यह भाषा युग और जीवन की नई अपेक्षाओं की पूर्ति नहीं कर सकती थी। अतः आगे चलकर बदलती हुई परिस्थितियों में उसका सन्दर्भ से कटा हुआ लगना स्वाभाविक था। राकेश और उनके साथियों ने जब लिखना शुरू किया तो अपने को उस युगसन्धि पर खड़ा पाया, जहाँ पुराने जीवन-मूल्यों और शुद्धिवादी भाषा-परम्परा के विरुद्ध एक सजग चेतना सक्रिय हो उठी थी। ज़िन्दगी बदल रही थी—एक अतीत अपनी मौत मर रहा था और अस्तित्व और अनास्था के संकट से ग्रस्त एक नया वर्तमान उठ खड़ा हो रहा था। बहुत कुछ था जो अतीत में खो चुका था और बहुत-कुछ था जो नई परिस्थितियों में जीवन से जुड़ गया था। कमलेश्वर के शब्दों में, 'चारों तरफ विचारों, प्रतिक्रियाओं, वादों-प्रतिवादों, आन्दोलनों-नारों, शोषण, अत्याचार, असुरक्षा वगैरह की इतनी उलझी हुई आवाज़ें थीं कि आदमी अपनी पुरानी भाषा की आवाज़ें सुन ही नहीं पा रहा था।'[3] इस प्रकार एक नए आन्तरिक संकट के कगार पर खड़े आदमी के लिए पुरानी भाषा अर्थहीन और छलछंदपूर्ण हो गई। अनुभव के बदलते ही भाषिक अभिव्यक्ति की स्थितियाँ बदल गई थीं, और बदली हुई सम्वेदनाओं के साथ एक नई भाषा की खोज होने लगी थी। खोज—एक ऐसी भाषा की जिसमें युग की उन मनःस्थितियों और सम्वेदनाओं को व्यक्त किया जा सके, जो पूर्ववर्ती भाषा में ठीक से नहीं उतर पा रही थीं। यह खोज नाटक के क्षेत्र में बाद में हुई है, कविता और कहानी के क्षेत्र में पहले हुई। चूँकि नाटक की विधा को इनकी भाँति विकसित होने का मौक़ा नहीं मिला और प्रसाद बहुत समय तक नाट्य-रचना की शैली पर हावी हो रहे, इसलिए मोहन राकेश को भी अपने पहले दो नाटकों में भाषा के साथ कुछ समझौता करना पड़ा, जिससे वे अपने को **आधे अधूरे** में जाकर मुक्त कर सके।

पूर्ववर्ती परम्परा में नाटक साहित्य था; इसीलिए नाटक की भाषा की इयत्ता साहित्यिकता तक मानी जाती थी। नाटककार साहित्यिक प्रभाव के लिए लिखते थे। जब यह भाषा रंगमंच पर प्रयुक्त होती थी तो प्रेक्षक और मंच के बीच एक बड़ी दीवार खड़ी हो जाती थी। राकेश को ऐसी मंचीय भाषा की तलाश थी जो समसामयिक प्रेक्षक से सम्बन्ध स्थापित कर सके। ऐसी सिद्धि बोलचाल की भाषा के सर्जनात्मक स्वरूप को

ही मिल सकती है। इस समय बोलचाल की भाषा का सर्जनात्मक प्रयोग हो रहा था, पर उसमें कविता की भाषा अत्यधिक वैयक्तिक और कथा-साहित्य की भाषा कुछ-कुछ वस्तुपरक बन बैठी थी। नाटक के लिए जिस भाषा की ज़रूरत थी उसकी संभावनाओं को संवाद के दायरे में ही देखा-परखा जा सकता है। संवाद के दायरे में नाटक की भाषा जीवन की भाषा ही हो सकती है। मोहन राकेश इस समस्या से पूरी तरह अवगत थे। वे कहते थे, 'मैं जानने की भाषा के बजाय निरन्तर जीने की भाषा की ओर जाना चाहता हूँ।'[4] वे मानते थे कि नाट्य लेखन में 'साहित्यिकता के अतिरेक को जितनी जल्दी झाड़ा जा सके उतना अच्छा।' उनका बल साहित्यिकता पर उतना नहीं जितना शब्द के सही प्रयोग पर था। अतः उन्होंने तत्सम या तद्भव को व्यर्थ का महत्त्व न देकर शब्दों के सही चयन पर बल दिया।

जीने की भाषा समसामयिक जीवन के अनुभवों से जुड़ी होती है। इसीलिए राकेश अपने पहले दो नाटकों में अपनी भाषा से संतुष्ट नहीं थे। उनका विचार था कि वे अपनी साहित्यिकता के कारण सराहे गए; किन्तु इसी तत्त्व के कारण सम्भवतः वे हमारी समसामयिक ज़िन्दगी के परिवेश, उसकी लय और मनःस्थिति को नहीं पकड़ पाए।[5] वस्तुतः नया कथ्य नई भाषा भी माँग करता है। पश्चिम में अनेक बार भाषा की नई चेतना और सर्जनात्मकता से कथ्य की जड़ता को खंडित करने का प्रयास हुआ है। हिन्दी में बहुत दूर तक प्रभावित करनेवाली ऐसी कोई बात नहीं हो सकी है; किन्तु राकेश और उनके सहयोगियों ने भाषा को नए कथ्य के अनुरूप ढालने का प्रयास अवश्य किया है। पश्चिम में अभिव्यक्तिवादियों से लेकर विसंगतवादियों तक ने भाषा की नई भूमि तैयार की है, और इस दिशा में बहुत-से सार्थक और चौंकानेवाले प्रयोग किए हैं। राकेश उस दिशा में बहुत दूर तक नहीं गए; किन्तु इधर वे नाटक की भाषा के सम्बन्ध में बहुत जागरूक थे और इस सम्बन्ध में शोधरत भी थे। उनका विचार था कि दृश्यात्मकता में नाटक सिनेमा और टेलीविज़न का मुक़ाबला नहीं कर सकता। इसलिए नाटक की समग्र दृश्यात्मकता स्वीकार करते हुए भी वे शब्द को आधार-भूमि प्रदान करने के पक्ष में थे। उनका विचार था कि 'रंगमंच में दृश्य की आपेक्षिक स्थिरता के बावजूद जो एक आन्तरिक गति रहती है, वह शब्दों और ध्वनियों की निरन्तरता से उपजती है क्योंकि यहाँ जो 'देखा' जाता है वह 'सुने जा रहे' का ही रूपान्तर होता है।'[6] इसी आधार पर वे नाटक को दृश्य के बजाय श्रव्य माध्यम मानते थे।[7] राकेश नाटक को शाब्दिक या श्रव्य माध्यम बताकर यह कदापि नहीं कहना चाहते थे कि शब्द का जादू खड़ा करना नाटक का एकमात्र लक्ष्य है। साहित्यिक भाषा के भी वे आग्रही न थे, वरन् वे 'अनुकूल नाटकीय शब्द' की अनिवार्यता की तलाश में थे। नाटक में भाषा और शब्द के सूक्ष्म अन्तर्संबन्धों पर उनकी दृष्टि थी। इसीलिए वे शब्द का रंगमंच के सन्दर्भ में विभिन्न दृष्टियों से अध्ययन कर रहे थे। शब्द और ध्वनि, ध्वनि और मौन, ध्वनि और बिम्ब, नृत्य, संगीत और माइम अथवा किसी अन्य विश्वजनीन माध्यम की उपादेयता के प्रति वे जागरूक थे। स्पष्ट है कि प्रयोग के स्तर पर ही नहीं, बौद्धिक स्तर पर भी वे भाषा

के आग्रही चिंतक थे।

प्रयोग के स्तर पर मोहन राकेश ने अपने नाटकों में नवीन संवेदना के अनुरूप भाषा का सर्जनात्मक संस्कार किया है। यह समकालीन संवेदना से सीधे साक्षात्कार करनेवाली भाषा है, जो पूर्ववर्ती भाषा के बने-बनाए ढाँचे को तोड़कर उभरी है। यह भाषा मूलतः द्वन्द्व और तनाव की भाषा है जिसमें रोमानी स्थिर स्थितियाँ नहीं, गतिशील जीवन की टकराहट है। वह सदा विरोध से टकराती है और इस टकराहट में कथन या वक्तव्य की तरह कम और औज़ार की तरह ज़्यादा काम देती है। **आषाढ़ का एक दिन** और **लहरों के राजहंस** जैसे नाटकों में ऐतिहासिक कथानक के बावजूद इसीलिए यह भाषा आधुनकिता के भाव-बोध से मंडित है। दूसरी सबसे बड़ी उपलब्धि है इसकी अर्थ-गर्भिता। राकेश की भाषा अर्थ के विभिन्न आयामों को उद्घाटित करती है, बिम्बों का विधान करती जाती है और प्रतीकों के माध्यम से मार्मिक अन्तरंग क्षणों का बोध कराती दीखती है। वह आन्तरिक अनुभवों से बहुत दूर तक जुड़ी है, इसीलिए एकदम जीवन्त और प्रभावशाली है।

नाट्य भाषा के सम्बन्ध में मोहन राकेश की स्पष्ट धारणा थी। इस सम्बन्ध में उनका विचार था : 'भाषा केवल शब्द योजना नहीं।...भाषा की सजीव अर्थवत्ता जो कि शब्दों की रूढ़ अर्थयुक्तता से अलग है, कुछ अंश तक ध्वनि, स्वराघात और शब्दों के विशिष्ट विन्यास पर निर्भर करती है। परन्तु मुख्य रूप से इस अर्थवत्ता की उपलब्धि प्रयोग की ऐतिहासिकता, विचार और अनुभूति की आन्तरिक लय तथा बिम्बों के संयोजन में लेखक की आन्तरिक अपेक्षा से होती है। भाषा का अर्थ साथ-साथ रखे गए शब्दों का सम्मिलित अर्थ ही नहीं, उस संयोजन से प्राप्त होनेवाला एक और अर्थ भी है।'[8] राकेश की नाट्य भाषा सही रूप में इन अपेक्षाओं को पूरा करती है। उसमें ध्वनि, लय और व्यंजना का सुन्दर सामंजस्य मिलता है।

राकेश की भाषा के कई स्तर हैं। इसका कारण उनका विकसित होता हुआ लेखन और कथानक दोनों हैं। **आषाढ़ का एक दिन** और **लहरों के राजहंस** ऐतिहासिक कथानक के कारण अपनी भाषा का स्वरूप एक विशिष्ट प्रकार से निर्धारित करते हैं। मोहन राकेश की एक विशेषता यह है कि वे शब्दों का एक तन्तुजाल बुनते हैं जिससे नाटक में एक भावपूर्ण वातावरण निर्मित होता है। उनके दो नाटकों में मेघ, वर्षा, भीगना, भावना, अधिकार, आकृति, दूध, हरिण, भूमि, अतिथि तथा हंस, ताल, लहर, छाया, मृग, झूला, जंगल, हवा, दर्पण आदि शब्द भाषिक संरचना में ही नहीं, संवेदना में भी सहायक होते हैं। उन्होंने प्रचलित संस्कृत शब्दावली—क्षण, दूसर व्यक्ति, अपरिचित, व्यक्तित्व बोध, अकेलापन, समय, अयाचित अतिथि, विशेषण, अनधिकार प्रवेश आदि का प्रयोग भी अंग्रेज़ी मुहावरे में आई नई अवधारणा के साथ कर दिखाया है। **आधे अधूरे** की भाषा कथ्य के अनुरूप नई संवेदना से जुड़ी है; किन्तु भाषा के सन्दर्भ में नवीनतम प्रयोग उन्होंने अपने बीज नाटकों में किए थे। इसी आधार पर डॉ. गिरीश रस्तोगी ने लिखा है : नाट्य-भाषा के तीन नमूने उन्होंने दिए हैं—साहित्यिक होते हुए भी गहरी

नाट्यानुभूति से जन्मी भाषा ('आषाढ़ का एक दिन' और 'लहरों के राजहंस'), जीवन्त बोलचाल की भाषा ('आधे अधूरे'), और बेतरतीब, बेतुकी, टूटती-फूटती भाषा– कुछ-कुछ ऐब्सर्ड नाटकों का आभास देती हुई (बीज नाटक)।'[9] **आषाढ़ का एक दिन** से लेकर बीजनाटकों तक भाषा के अन्वेषण का जो रास्ता वे तय कर पाए, उसमें फिर भी एक समानता है और वह है–एक प्रकार की प्रचलित किताबी भाषा से मुक्ति और रंगमंच के लिए उसकी उपयुक्तता।

आषाढ़ का एक दिन से ही यह प्रक्रिया प्रारम्भ हो जाती है। इस नाटक की भाषा अपने कथानक के अनुरूप कहीं भावना से भीगी लगती है; किन्तु वह रोमान की सीमा तक पहुँचे, इससे पूर्व ही राकेश उसे तनाव में झोंक देते हैं। चरित्र की माँग पर मल्लिका की भाषा में जहाँ रोमान उभरने लगता है, वहीं अम्बिका और विलोम के संवाद उस पर हथौड़े जैसा प्रहार करते हैं। इसीलिए कोमल भावनाओं और कठोर यथार्थ के जैसे तन्तु **आषाढ़ का एक दिन** की भाषा बुनती है, वैसे अन्यत्र दुर्लभ हैं। इसमें कहीं-कहीं लेखक संस्कृत शब्दावली का प्रयोग करने के लिए बाध्य हुआ है; पर यह बाध्यता दुराग्रह की नहीं, नाटक की आन्तरिक अपेक्षा की स्वीकृति की रही है। उन लोगों को क्या कहा जाए जो इसकी भाषा को क्लिष्ट और बोझिल मानते हैं, किन्तु जो प्रसाद के नाटकों से परिचित हैं, उन पाठकों को राकेश की यह भाषा बड़ी राहत देती है। इसमें कोई सन्देह नहीं कि इस भाषा का स्तर साहित्यिक है किन्तु इसकी साहित्यिकता बहुत गहरे कहीं नाट्यानुभूति से जुड़ी है और वास्तविक रंगानुभव प्रदान करती है। राकेश इससे भी सरल भाषा लिख सकते थे। सम्भवतः इस नाटक को रेडियो के लिए लिखते हुए उन्होंने ऐसा किया भी; किन्तु बाद में उन्होंने सरल शब्दों (जो कोष्ठक में दिए गए हैं) के स्थान पर उपयुक्त तत्सम शब्दों का प्रयोग ही उचित समझा, जैसे तल्प (सामने), उपत्यका (घाटी), शर्करा (शक्कर), आस्तरण (शैया), उपादान (साधन), उद्घोष (घोषणा), अभिस्तुति (हाथ जोड़कर), क्रीत (बिका हुआ), अभ्यागत (अतिथि), प्रांतर (प्रदेश), उर्वर (उपजाऊ), स्थपति (निर्माता) श्लक्षण (चिकनी), विक्षुब्ध (अस्थिर) आदि।[10] यह परिवर्तन बोलचाल के शब्दों के स्थान पर संस्कृत के शब्द रखने जैसा प्रयोग नहीं है। इसके पीछे सांस्कृतिक और सर्जनात्मक दृष्टि स्पष्ट है। राकेश ने संस्कृत शब्दों की कलात्मक संगति बिठाई है। इसीलिए वे भाषा के शरीर पर ग्राफ्ट की गई त्वचा की तरह नहीं लगते। उसी में समाहित हो जाते हैं। राकेश के शब्दों में, 'एक काल विशेष की पृष्ठभूमि रखने के कारण अपेक्षित यथार्थ भ्रम भी सृष्टि के लिए मुझे तब भी यह आवश्यक लगा था और अब भी मेरी यह धारणा बदली नहीं है। साधारण दर्शक तथा नाटक के सम्प्रेषण में इससे कोई बाधा आती हो, ऐसा अब तक के अनुभव से सिद्ध नहीं होता।[11] वास्तव में भाषा बोझिल और क्लिष्ट तब होती है जब वह नाट्य-स्थिति और रंगानुभव से विलग हो। **आषाढ़ का एक दिन** की भाषा दुधारी तलवार की तरह काम करती है। एक ओर से वह कथावस्तु को आज के वातावरण से काटकर अतीत से जोड़ती है; दूसरी ओर नई संवेदनाओं और कथ्य के आधार पर उसे अतीत

से विलग कर वर्तमान से मिलाती है। पुराने कथानक को यह भाषा नए कथ्य और नई संवेदना से युक्त करती है, साथ ही देश, काल, पात्र और स्थिति के अनुरूप अपना आभिजात्य भी बनाए रखती है। इसका मुख्य कारण लेखक की सर्जनात्मक क्षमता है। इसीलिए यह भाषा आरोपित नहीं लगती क्योंकि उसमें शब्दों का घटाटोप, लफ़्फ़ाज़ी, अलंकरण, हेत्वाभासी दार्शनिकता कुछ भी नहीं। यह भाषा वक्तव्य बनकर सामने नहीं आती—नाट्य-स्थितियों के बीच से पैदा होती है और द्वन्द्व के बीच पलती है। इसीलिए वह एकदम मर्म को छूती है और काव्यमय प्रभाव छोड़ जाती है। काव्यमय से तात्पर्य उस अभिव्यक्ति के स्वरूप से है जो नाटकीय क्रिया-व्यापार और कहे गए शब्दों के समग्र अर्थ और वातावरण तथा अनुभूति-बिम्ब को उजागर करता है। यह काव्यमयता शब्दों से नहीं, शब्दों के बीच से उभरती है और एक निश्चित लक्ष्य की ओर संक्रमण करती है।

लहरों के राजहंस में अनुभूति की तीव्रता की अपेक्षा वैचारिक तत्त्व की प्रधानता है। सारे नाटक की अवधारणा विचार पर आधारित है। ऐसा प्रतीत होता है जैसे नाटक की कथावस्तु लेखक के ज़ेहन में बहुत गहराई से नहीं उतर पाई। लगता है नन्द की भाँति लेखक भी अपने को राग-विराग किसी से जोड़ नहीं पाया। नाट्य-स्थितियों की कमी भी शायद उसके लिए उत्तरदायी हो। इसीलिए किसी स्थिति विशेष को बहुत दूर तक खींचा गया है। इससे कहीं-कहीं भाषा में पैनापन नहीं रह गया है और सबसे विचित्र स्थिति नाटक के अन्त में है, जहाँ नन्द और सुन्दरी आमने-सामने हैं, पर नन्द के पास वह भाषा ही नहीं रह जाती जिससे सम्प्रेषण हो सके। कुल मिलाकर इस नाटक की भाषा में एक प्रकार का आरोपित संयम, आभिजात्य और सौष्ठव विद्यमान है पर नाटकीयता भी है। विभिन्न नाट्य-स्थितियों से भाषा को बचाने का आग्रह भाषा को एक नपा-तुला स्वरूप प्रदान करता है। उद्वेग की स्थिति में वे शब्द की अपेक्षा अन्य नाटकीय माध्यमों का प्रयोग करते दिखाई देते हैं, जिससे भावना की तरलता तल में ही छँट जाती है। नाटक की भाषा सीधे द्वन्द्व से पैदा नहीं होती। प्रारम्भ को ही लीजिए, तनाव की स्थितियाँ तो हैं, पर सीधा संघर्ष नहीं है। बाद में बुद्ध के आगमन के बाद कुछ त्वरा आती है और भाषा तब अपनी पूरी नाटकीयता निभाती है, किन्तु अन्त तक पहुँचते-पहुँचते कुछ फैल-सी जाती है। फिर भी तीव्रता न सही, **लहरों के राजहंस** की भाषा अपनी अनुगूँज छोड़ जाती है। यह इन्द्रिय संवेदनों को जगाती है और साथ ही विचारों को भी। भाषा की इस शक्ति के कारण ही यह नाटक एक ओर यथार्थ को छूता है, दूसरी ओर अतिकल्पना को। यह भाषा यथार्थ और स्वप्न दोनों को जगाती है और चेतना की ज्ञात और अज्ञात गुहाओं को उद्‌घटित करती है। समस्त सीमाओं के बावजूद **आषाढ़ का एक दिन** और **लहरों के राजहंस** की भाषा प्रसाद-युग की रूढ़ियों को पूरी शक्ति से तोड़ती है और उसका अतिक्रमण करती है। इसमें कोई सन्देह नहीं कि उसका संस्कार मूलतः साहित्यिक है और संवेगात्मक है। फिर भी विचार की ताजगी और सर्जनात्मक रूप से नएपन में ढली है।

आधे अधूरे भाषा की दृष्टि से राकेश का अन्यतम प्रयोग है। इस नाटक की भाषा

नाटक के क्षेत्र में वर्षों से व्याप्त जड़ता को भंग करने में सफल हुई है। इसमें जो सहजता, ताजगी, लोच और चालूपन है, वह नाट्य भाषा की सम्पूर्ण संभावनाओं और आन्तरिक शक्तियों का उपयोग करता दीखता है। सबसे बड़ी बात यह है कि इस आभिजात्य का आरोप नहीं है और वह मूलतः बोलचाल की भाषा है। अपने कथ्य के अनुरूप यह भाषा जीवन के सामान्य अनुभवों को उन्हीं के शब्दों में बख़ूबी व्यक्त करती है। एक-सी स्थितियों की व्यापकता के कारण पूरे नाटक की भाषा में एकरसता ज़रूर आ गई है, पर समकालीन जीवन की पकड़ में यह भाषा बड़ी तेज है। **आधे अधूरे** में पहुँचकर राकेश की नाट्य भाषा की एक तलाश पूरी होती है। काश, इस तलाश में वे इस नाटक की भाषा को विविध आयाम दे पाते, जिससे यह महसूस नहीं होता कि वह भावना के एक क्षेत्र को ही सींचती है, इन्द्रियों को समग्रता में तृप्त नहीं करती।

नाटक में भाषा की पहचान संवाद के दायरे में होती है क्योंकि उसमें भाषा संवाद में ढलकर आती है। इसीलिए नाटक में संवाद से अलग भाषा का सवाल नहीं उठता। संवाद नाटक के कथ्य, कथानक, चरित्र और उसकी स्थितियों से पैदा होता है, भाषा उसका माध्यम मात्र बनती है। फिर भी संवाद के स्तर पर भी भाषा का एक निश्चित दायित्व होता है। वह जब संवाद में ढलती है तो नाट्य-स्थिति, चरित्र, सम्बोधक और सम्बोधित व्यक्ति और प्रेक्षक की मनःस्थिति के अनुरूप अपना सहज स्वरूप निर्धारित करती है। इसमें भावना, प्रतिक्रिया और अनुक्रिया जगाने में शब्द महत्त्वपूर्ण भूमिका निभाते हैं। राकेश अपने संवादों में निर्दिष्ट प्रभाव के लिए शब्दों का इन्द्रजाल खड़ा करते हैं। ऐसे शब्द उनके नाटकों में कहीं-कहीं पर 'की-वर्ड' बन जाते हैं, जिनकी पुनरावृत्ति से वे नाटकीय संवेदना का अद्‌भुत ताना-बाना बुनते हैं। उदाहरण के लिए, **आषाढ़ का एक दिन** में मेघ, वर्षा, भीगना, गीले वस्त्र, अँधेरा, भावना, अतिथि, आकृतियाँ और वातावरण; **लहरों के राजहंस** में हंस, छाया मृग, श्यामांग, बुद्ध, यशोधरा, अँधेरा, दीपक, किरण, उलझन, नींद, रात, दर्पण, कामना और मदिरा; तथा **आधे अधूरे** में घर, हूँ, हवा, अधूरापन आदि इसी प्रकार के शब्द हैं। बहुत दूर तक नाटकों की संवेदना इन शब्दों के इर्द-गिर्द घूमती है। अकेले **आधे अधूरे** में 'घर' शब्द इसका प्रमाण है। **आषाढ़ का एक दिन** में सम्पत्ति, अधिकार, भूमि जैसे शब्दों को लेकर संवादों को बड़ी कुशलता से चिना गया है। एक उदाहरण लीजिए–

कालिदास : मैं अनुभव करता हूँ कि ग्राम प्रान्तर मेरी वास्तविक **भूमि** है। मैं कई सूत्रों से इस **भूमि** से जुड़ा हूँ।...यहाँ से जाकर मैं अपनी **भूमि** से उखड़ जाऊँगा।

मल्लिका : यह क्यों नहीं सोचते कि नई **भूमि** तुम्हें यहाँ से अधिक सम्पन्न और उर्वर मिलेगी। इस **भूमि** से तुम जो कुछ ग्रहण कर सकते थे, कर चुके हो। तुम्हें आज नई **भूमि** की आवश्यकता है जो तुम्हारे व्यक्तित्व को

अधिक पूर्ण बना दे।

कालिदास : नई **भूमि** दुखा भी तो सकती है।

मल्लिका : कोई **भूमि** ऐसी नहीं होती जिसके अन्तर मे कोमलता न हो।

स्पष्ट है कि जिन शब्दों की अर्थ-छवियों से राकेश अपने नाटकों का ढाँचा खड़ा करते हैं उनकी संवादों में महत्त्वपूर्ण भूमिका होती है। यहाँ केवल पुनरावृत्ति नहीं है बल्कि एक ही शब्द-तन्तु संवेदना और अर्थ को बुनता जाता है। इसी प्रकार के समानांतरता के भी कई प्रयोग संवादों में मिलते हैं : मल्लिका मेघ के सौन्दर्य को लेकर कहती है : 'मैं उसे छू सकती थी, देख सकती थी, पी सकती थी।'

शब्द के नाटकीय प्रयोग से उसकी अर्थवत्ता का विस्तार राकेश की नाट्य-भाषा और संवादों का एक बहुत बड़ा गुण है। **आषाढ़ का एक दिन** में अम्बिका कालिदास के प्रति मल्लिका का राग देखते हुए उससे कटुता रखती है (जिससे कालिदास के आहत होने की कल्पना की जा सकती है)। कालिदास ऐसी स्थिति में आहत हरिण शावक को गोद में लिए हुए सब-कुछ अनदेखा/अनसुना करके जो प्रतिक्रिया व्यक्त करता है उसकी व्यंजना केवल एक शब्द पर निर्भर करती है : '**हम** जिएँगे हरिण शावक ! जिएँगे न ? एक बाण से आहत होकर **हम** प्राण नहीं देंगे। हमारा शरीर कोमल है तो क्या हुआ ? **हम** पीड़ा सह सकते हैं...।' इस संवाद में एक साधारण शब्द **हम** आहत हरिण शावक और कालिदास को एक ही भावभूमि पर खड़ा कर देता है। इससे कालिदास का स्वयं आहत होना व्यंजित होता है। हरिण शावक राजसैनिक के बाणों से आहत है, तो कालिदास अम्बिका और अन्ततः राजसत्ता दोनों से। **आधे अधूरे** में इसी प्रकार सामान्य से सामान्य शब्द को लेकर कई अर्थ-स्तर खोले गए हैं। उदाहरण के लिए **खड़ा होना** साधारण-सी क्रिया है, किन्तु अर्थ सन्दर्भ से जुड़कर अभिधा तक ही सीमित नहीं रहता—

पुरुष एक : (गुस्से से उठता) तुम तो ऐसे बात करती हो जैसे...।

स्त्री : **खड़े** क्यों हो गए ?

पुरुष एक : क्यों मैं **खड़ा नहीं हो सकता** ?

स्त्री : (हलका कहकहा लेकर तिरस्कारपूर्ण स्वर में) **हो तो सकते हो** पर **घर** के अन्दर ही।

इसी प्रकार **नया** कोई, बाहर एक सामान्य शब्द होने पर भी सावित्री को जैसे नंगा करके रख देते हैं—

स्त्री : अभी **कोई** आनेवाला है बाहर से और...।

बड़ी लड़की : कौन आनेवाला है ?

पुरुष एक : सिंघानिया। इसका बॉस। यह **नया** आना शुरू हुआ है आजकल।

राकेश संज्ञा, सर्वनाम, विशेषण शब्द का ऐसा प्रयोग करते हैं जिससे उसका अनुपस्थित अर्थतत्त्व भी प्रकट हुए बिना नहीं रहता। वस्तुतः कई ऐसे शब्द हैं, लगता है जैसे वे पात्रों के अन्तर्मानस में कई सन्दर्भों के आगार बने हैं। एक-एक शब्द मनःस्थितियों का पुंज-सा बना हुआ अवचेतन में पड़ा है; ज़बान पर आते ही वह अर्थ

के कई स्तरों को खोलने लगता है। **आधे अधूरे** में ऐसे भावगर्भित शब्दों की प्रचुरता है। सामान्य-से शब्द जैसे **वह** (वह आज फेर आएगा), **कहीं** (तुम कहीं जा रहे हो ?...क्या पता कहीं और से आ रही हो), **नहीं** (नहीं जाता, नहीं ढूँढ़ता), **आदमी** (अच्छा...वह आदमी), **और** (कोई और भी आनेवाला है), हूँ आदि ऐसे ही कुछ शब्द हैं जो सम्बोधित और सम्बोधनकर्त्ता के अन्तर्निहित व्यंग्य को उभारते हैं। इस प्रकार शब्द मानवीय सम्बन्धों की भाषा का निर्माण करते हैं।

आज का मानव जिस अन्तर्विरोध की स्थिति से गुज़र रहा है उसमें मानवीय भावना के सूक्ष्म तन्तु बुरी तरह क्षत-विक्षत हो चुके हैं। फलतः शब्द या तो निरर्थक हो गए हैं या धोखा देते हैं। राकेश ने इसी स्थिति का संकेत करते हुए लिखा है : 'हर चीज़ पर सन्देह होता है।...उन सब शब्दों पर सन्देह होता है जिनके अर्थ नकारात्मक नहीं हैं। लगता है कि व्यक्तियों की भाँति शब्द बेईमान हो उठे हैं।'[12] यह बेईमानी कहीं-कहीं महेन्द्रनाथ के संवादों में बुरी तरह झलकती है। वह शब्द का अर्थ ही नकार जाता है—

पुरुष एक : तो लोगों को पता है, वह आता है यहाँ ?

स्त्री : क्यों, **बुरी** बात है ?

पुरुष एक : मैंने कहा बुरी बात है ? मैं तो बल्कि कहता हूँ **अच्छी** बात है।

स्त्री : तुम जो **कहते हो,** उसका मतलब समझ में आता है मेरी।

पुरुष एक : तो अच्छा यही है कि मैं **चुप रहा** करूँ...।

स्त्री : तुम **चुप रहते** हो ! और न कोई।

यहाँ अर्थ शब्दों से नहीं, शब्दों के बीच से उभरता है। अर्थ की इस विवृत्ति में संवादों का शाब्दिक गठन महत्त्वपूर्ण हो जाता है। राकेश एक शब्द से दूसरे शब्द तक, एक अर्थ से दूसरे अर्थ तक पहुँचने के लिए उपयुक्त सेतु का निर्माण करते हैं। इससे संवाद के चरण निरन्तर आगे बढ़ते जाते हैं और अर्थ के नए आवरण स्वयं खुलते जाते हैं—

पुरुष एक : यह सब **कहता** है **वह** ? और क्या-क्या कहता है ?

स्त्री : वह इस वक़्त **तुम से बात नहीं** कर रही।

पुरुष एक : पर **बात** तो मेरे **घर** की हो रही है।

स्त्री : तुम्हारा **घर** ? हंह !

पुरुष एक : तो मेरा **घर** नहीं। **कह दो नहीं** है।

स्त्री : सचमुच, **तुम** अपना **घर** समझते हो इसे ?

पुरुष एक : **कह** दो, **कह** दो जो **कहना** चाहती हो।

स्त्री : **दस साल** पहले **कहना** चाहिए था मुझे...जो **कहना** चाहती हूँ।

पुरुष एक : **कह** दो अब भी...इससे पहले कि **दस साल ग्यारह साल** हो जाएँ।

स्त्री : **नहीं होने** पाएँगे **ग्यारह साल**...इसी तरह चलता रहेगा सब-कुछ तो...।

पुरुष एक : **नहीं होने** पाएँगे सचमुच ? काफ़ी अच्छा **आदमी** है **जगमोहन**।... मिला था उस दिन कनाट प्लेस में। **कह** रहा था **आएगा** किसी दिन **मिलने**।

बड़ी लड़की : (धीरज खोकर) **डैडी !**

पुरुष एक : ऐसी क्या **बात कही** है मैंने ? **तारीफ़** ही की है **उस आदमी** की।

स्त्री : **ख़ूब** करो **तारीफ़**...और भी **जिस-जिस** की हो सके **तुम** से।

इस संवाद में वह, तुम, बात, घर, दस साल, कहना, जगमोहन और तारीफ़ जैसे शब्द अर्थ के सोपानों का निर्माण करते चलते हैं।

शब्द और उसके अभिधेय अर्थ के अलगाव के चित्रण में मोहन राकेश ने अपूर्व कौशल दिखाया है। उनके नाटकों में ऐसे संवादों की कमी नहीं जो शब्द और अर्थ के अलगाव के बीच से केवल स्थिति का बेहूदापन या बेतुकापन व्यंजित करते हैं। **आषाढ़ का एक दिन** में रंगिणी-संगिनी के संवाद इसी कोटि के हैं; अनुस्वार-अनुनासिक के संवादों की बानगी भी कुछ इसी प्रकार की है—

अनुस्वार : मैं इससे सहमत हूँ।

अनुनासिक : **तो।**

अनुस्वार : **तो** ?

अनुनासिक : **तो** इसे हटा देना चाहिए।

अनुस्वार : हाँ, अवश्य हटा देना चहिए।

अनुनासिक : **तो** ?

अनुस्वार : **तो** ?

अनुनासिक : हटा दो।

अनुस्वार : **मैं** ?

अनुनासिक : हाँ।

अनुस्वार : **तुम नहीं !**

अनुनासिक : **नहीं।**

अनुस्वार : **क्यों** ?

अनुनासिक : क्यों का कोई अर्थ नहीं।

अनुस्वार : फिर भी।

अनुनासिक : पहले मैंने तुमसे कहा था।

अनुस्वार : परन्तु चौकी पहले देखी तुमने है ?

अनुनासिक : **तो** ?

अनुस्वार : **तो** ?

अनुनासिक : तुम हटा दो

अनुस्वार : तो रहने दो।

अनुनासिक : रहने दो।

ये संवाद वस्तुतः किसी शाब्दिक अर्थ की अभिव्यक्ति के लिए प्रयुक्त नहीं हुए हैं। जो अर्थ है वह अलग-अलग संवाद से नहीं बल्कि उनके एक पूरे समुदाय से उभरता है; और अर्थ है राजसत्ता और उसके अधिकारियों की विसंगति को उभारना। **आधे अधूरे**

में सिंघानिया के संवाद इसी प्रकार शाब्दिक अर्थ का अतिक्रमण कर स्थिति और सम्बन्धों के बेहूदेपन को प्रकट करते हैं। ऐसे संवादों में राकेश ने पश्चिम के विसंगत (ऐब्सर्ड) नाटकों के भाषा-प्रयोग का यथेष्ट लाभ उठाया है। बीज नाटकों और **छतरियाँ** में भी ऐसा ही प्रयोग है।

ध्वन्यार्थ की ऐसी ही विवृत्ति उन संवादों में भी होती है जो नाटकीय विडम्बना से युक्त दिखाई देते हैं। नाटकीय विडम्बना एक ऐसी युक्ति है जो सारी कृति के अर्थ को समझने में सहायता देती है। नाटकीय विडम्बना से युक्त संवाद सारी कृति के सन्दर्भों, प्रसंगों और पात्रों के कथनों को परस्पर जोड़ते हैं। राकेश के नाटकों में वह जीवन की क्रूर नियति को व्यंजित करती है। **आषाढ़ का एक दिन** में विलोम के संवादों में नाटकीय विडम्बना का अंश बहुत अधिक है। विलोम की मल्लिका के प्रति यह उक्ति कि 'अनचाहा अतिथि सम्भवतः फिर भी कभी आ पहुँचे' इसका एक सामान्य उदाहरण है। प्रियंगु का मल्लिका से यह प्रस्ताव कि 'तुम इनमें से जिसे भी योग्य समझो उसी के साथ तुम्हारा विवाह किया जा सकता है' उसके जीवन की सबसे बड़ी विडम्बना बन जाती है।

प्रियंगु : तुम भी **हमारे साथ** क्यों नहीं चलती ?

मल्लिका : **मैं** ?

प्रियंगु : बाधा क्या है ? यहाँ तुम किसी **सूत्र से जुड़ी** तो नहीं हो कि...। तुम्हारे मन में **कल्पना** नहीं कि तुम्हारा **अपना** घर-परिवार हो ?

अम्बिका : **इसके** मन में यह कल्पना नहीं, भावना के स्तर पर जीती है।

स्पष्ट है कि दोनों के सम्वादों में यहाँ अर्थग्रहण की समस्या अभिधा पर निर्भर नहीं। प्रियंगु मल्लिका की आवश्यकता को मात्र ऊपरी धरातल पर महसूस करती है, पर जो बात वह नहीं समझती, प्रेक्षक समझता है और कथन की चुभन को महसूस करता है। यही बड़ी मार्मिक नाट्य विडम्बना बनकर उभरती है। ऐसे स्थलों पर अनकही बात के अर्थ की गूँज बहुत दूर तक जाती है। **लहरों के राजहंस** में भी कामोत्सव सम्बन्धी आयोजन के सन्दर्भ में बुद्ध और यशोधरा के सम्बन्ध में कहे गए संवाद बाद में स्वयं सुन्दरी की नियति का उपहास करते हैं और उसे अनुभूति का कटु दंश दे जाते हैं। वह कामोत्सव में न आनेवाले अतिथियों के लिए कहती है : 'उन सबसे कह दें मेरे यहाँ आने के लिए वे **कल** की प्रतीक्षा में न रहें वह **कल** उनके लिए कभी नहीं आएगा ! सचमुच वह उन्हीं के लिए ही नहीं, नन्द और सुन्दरी के लिए भी कभी नहीं आता। इसी प्रकार यशोधरा के प्रति कही उसकी यह उक्ति कि 'नारी का आकर्षण पुरुष को पुरुष बनाता है तो उसका अपकर्षण उसे गौतम बुद्ध बना देता है' क्या स्वयं उसे नहीं छलती ? क्या अन्ततः उसका आकर्षण नन्द को उससे बाँध पाता है ? क्या उसकी अपनी स्थिति यशोधरा के अनुरूप नहीं हो जाती ? **आधे अधूरे** में सावित्री जगमोहन के पास जाते हुए अपनी लड़की से जो कुछ कहती है उसमें नाटकीय व्यंग्य की मार्मिक अभिव्यक्ति हुई है : 'तुझसे एक बात कहना चाहती थी।...अगली बार मैं तुझे यहाँ न

मिलूँ शायद। मैं जानती थी कि एक दिन आना ही था ऐसा।' पर जगमोहन की उपेक्षा से वह दिन कभी आता ही नहीं ! इसी प्रकार सिंघानिया और सावित्री, सावित्री और जगमोहन के बीच के संवाद नाटकीय विडम्बना के सुन्दर उदाहरण हैं।

सिंघानिया और सावित्री के संवादों में कोई तार्किक संगति नहीं दिखाई देती। यह विसंगति ही वास्तविक नाटकीय विडम्बना को जन्म देती है। सावित्री सिंघानिया से लड़के को नौकरी देने के लिए कहती है; पर वह कहती कुछ है, वह समझता कुछ और है—असल में दोनों में से हर एक के पास दूसरे से भिन्न अपनी एक बात कहने को है—

स्त्री : **उस विषय** में सोचा आपने कुछ ?

पुरुष दो : **किस विषय** में ? (मुँह चलाता)

स्त्री : वह जो मैंने बात की थी आपसे कि कोई ठीक-सी जगह हो आपकी नज़र में तो...।

पुरुष दो : **बहुत स्वादिष्ट है।**

स्त्री : **याद है न** आपको

पुरुष दो : याद है। कुछ बात की थी तुमने एक बार। अपने किसी **कजिन** के लिए कहा था...**नहीं,** वह तो **मिसेज मल्होत्रा** ने कहा था। **तुमने किसके लिए** कहा था। ?

स्त्री : (लड़के की तरफ़ देखती हुई) **इसके लिए**।

...

स्त्री : ...देखिए, जैसे भी हो, **इसके लिए** आपको कुछ-न-कुछ करना है।

पुरुष दो : **ज़रूर। किसके लिए** क्या करना है ?

स्त्री : (लड़के की तरफ़ देखकर) **इसके लिए**...कुछ-न-कुछ।

पुरुष दो : **...तुम घर पर आना किसी दिन।** बहुत दिनों से नहीं आईं।

स्त्री : मैं **भी** सोच रही थी **बेबी से मिलने...।**

पुरुष दो : वह पूछती रहती है आंटी इतने दिनों से नहीं आईं। **वह** बहुत **प्यार** करती है अपनी **आंटियों से। माँ के न होने** से बेचारी...।

स्त्री : बहुत ही प्यारी बच्ची है। मैं पूछ लूँगी किसी दिन आपसे ? **इससे** भी कह दूँ कि आकर **मिल** ले आपसे एक बार।

पुरुष दो : (**बड़ी लड़की** को देखता) **किससे** ? **इससे** ?

स्त्री : अशोक से।

पुरुष दो : (बाहर दरवाज़े की तरफ़ बढ़ता **बड़ी लड़की** से) **तुम** नहीं करतीं कहीं नौकरी ?

बड़ी लड़की : हाँ...नहीं...ऐसा है कि...।

स्त्री : **डरती है** ?

पुरुष दो : **डरती है** ?

स्त्री : अपने **पति** से।
पुरुष दो : पति से ?
स्त्री : हाँ, उसे **पसन्द नहीं** है।
पुरुष दो : **यह लड़की** ?
स्त्री : नहीं, इसका नौकरी करना।

ये संवाद स्थिति की विसंगति से ही क्रूर नाटकीय व्यंग्य को भी उभारते हैं। इस कोटि के संवादों में पात्रों के बीच संवाद की कोई सहज स्वाभाविक स्थिति नहीं दिखाई देती। पात्रों के बीच स्वार्थों, आग्रहों और विकारों की ऐसी अदृश्य दीवारें दिखाई देती हैं जो मानवीय सम्बन्धों के आड़े आती हैं। पात्र परस्पर बातें करते हैं, पर जैसे बात होती ही नहीं। संवाद अचानक शुरू होते हैं, बीच में ही कट जाते हैं। इसके कारण या तो पात्र का अवचेतन बोल उठता है या चेतन के रूप में वह मन की बात कहने से कतराता है। सावित्री घर से सम्बन्ध तोडने का निर्णय लेकर जब जगमोहन के पास जाती है तो वह सारी मुलाक़ात को दैहिक स्तर पर लेता है, जिससे उसका सही मन्तव्य उसके ज़ेहन तक पहुँचता ही नहीं—

स्त्री : (उसके हाथ पर हाथ रखकर) जोग !
पुरुष तीन : (हाथ सहलाता हुआ) **क्या बात** है कुकू!
स्त्री : मैं **वहाँ पहुँच गई हूँ** जहाँ पहुँचने से डरती रही हूँ ज़िन्दगी-भर। मुझे **आज लगता है** कि...।
पुरुष तीन : **परेशान** नहीं हो इस तरह।
स्त्री : मैं **सच** कह रही हूँ। **आज मुझसे कहो कि...।**
पुरुष तीन : (अन्दर की तरफ़ देखता) **घर में कोई नहीं** है ?

...

स्त्री : मैंने कल **एक फैसला** कर लिया है।
पुरुष तीन : **हाँ हाँ** ?
स्त्री : वैसे उन दिनों भी **सुनी होगी** तुमने ऐसी बात मेरे मुँह से...पर इस बार सचमुच फ़ैसला कर लिया है।
पुरुष तीन : (**जैसे बात को आत्मसात् करता**) हूँ !
स्त्री : उधर क्यों चले गए ?
पुरुष तीन : (**शेल्फ से किताब निकालना**) ऐसे ही। यह किताब देखना चाहता था ज़रा।
स्त्री : तुम्हें शायद विश्वास नहीं आया मेरी बात पर ?
पुरुषी तीन : **सुन रहा हूँ** मैं।
स्त्री : मेरे लिए असम्भव था यह सहना। पर अब आकर बिल्कुल **असम्भव हो** गया है।
पुरुष तीन : (**पन्ने पलटता**) तो मतलब यह कि...?

स्त्री : **ठीक सोच रहे हो तुम।**

पुरुष तीन : (**किताब वापस रखता है**) हूँ।

स्त्री : मुझे अफ़सोस है इस बात का कि मेरी वजह से तुम्हें भी इतनी तकलीफ़ उठानी पड़ी ज़िन्दगी में।

पुरुष तीन : देखो...सच पूछो तो **मैं अब ज़्यादा सोचता ही नहीं इस बारे में।**

जगमोहन और सावित्री के ये संवाद नाटककार की क्षमताओं को प्रकट करते हैं। दोनों पात्र दो विभिन्न दिशाओं में सोच रहे हैं फिर भी एक को भ्रम है कि दूसरा भी उसी दिशा में सोच रहा है ! **आधे अधूरे** की भाषा में जहाँ तीखापन, तुर्शी, क्षिप्रता, आक्रोश, दैन्य के विभिन्न स्तर हैं वहाँ शब्दों का छलावा भी है। यह आड़े-तिरछे सम्वादों का छलावा ही शब्दों के नए अर्थ गढ़ता जाता है—

बड़ी लड़की : क्या **बात** है डैडी ?

पुरुष एक : **बात** ? कुछ **भी नहीं।**

बड़ी लड़की : है तो सही **कुछ न कुछ** बात।

पुरुष एक : **ऐसे** ही। **तेरी मम्मी** कह रही थी।

बड़ी लड़की : क्या कह रही थी ?

पुरुष एक : मतलब **वह** नहीं **मैं** कह रहा था।

बड़ी लड़की : क्या कह रहे थे ?

पुरुष एक : तेरे बारे में बात कर रहा था।

स्पष्ट है कि इस संवाद में एक सुनिर्धारित शब्द-योजना के आधार पर नाटककार ने शब्द प्रयोग की विसंगति को उजागर करते हुए शब्दों का नहीं, शब्दों के बीच छिपे दुराव को ध्वनित करने का प्रयास किया है।

वस्तुतः राकेश अपनी सम्वाद योजना में कथ्य की विभिन्न भंगिमाओं का प्रयोग करते हैं। तर्क, कुतर्क, प्रश्न, अनुत्तरता, अवज्ञा, तर्जना, जैसी मनस्थितियों का वे भिभिन्न शैल्पिक विधियों—साम्य, प्रतिसाम्य, व्यंग्य, विडम्बना आवृत्ति, रिक्ति आदि कई माध्यमों से कर दिखते हैं।

मन की कोठरी में अपने को बन्द कर जीनेवाले पात्र भाषा का प्रयोग अभिव्यक्ति के बजाय गोपन के लिए करते हैं। पारस्परिकता और अन्तरंगता के अभाव में पात्र जिस प्रकार संवादों का आश्रय लेते हैं, उसमें अभिधा अर्थहीन हो जाती है और अनकही बातें ही अर्थवान हो जाती हैं। राकेश के नाटकों में ऐसे पात्र कम नहीं जो एक-सी समानांतर स्थितियों में नहीं जीते और अपनी प्रतिक्रियाओं में एक-दूसरे से काफ़ी अलग हो जाते हैं। फलतः उनके मन का भाव खुले संवाद का रूप धारण नहीं कर पाता। **लहरों के राजहंस** में नन्द के मुंडित-शिर हो जाने के बाद सुन्दरी इसीलिए अधिक मुखर नहीं है—यह एक प्रकार का अस्वीकार ही है, जो आघात करता है। ऐसे ही क्षण कभी-कभी कालिदास के कारण मल्लिका और अम्बिका के बीच आते हैं। नाटक के प्रारम्भ में मल्लिका आषाढ़ की धारासार वर्षा और उसमें भीगने के सुख का वर्णन करती है; उतनी

देर अम्बिका सारी प्रतिक्रियाएँ दबाए रहती है। मल्लिका कहती है : 'तुमने सुना माँ, राज्य उन्हें राजकवि का आसन देना चाहता है।' तो अम्बिका उसकी बात को अनसुना कर देती है : 'गीले वस्त्र सूखने को रख दिए हैं। थोड़ा-सा दूध शेष है, इसमें शर्करा मिला दी है।' अम्बिका की उपेक्षा उसे खलती है और वह अपनी बात फिर दुहराती है : 'तुमने सुना नहीं माँ, राजपुरुष क्या कह रहा था ?' पर अम्बिका कहती है : 'दूध पी लो।' और इसी तरह बहुत-सारी उपेक्षा के बाद उस मन का बादल फूटता है। संवाद जैसे शून्य में फेंक दिए जाते हैं; ठीक जगह पर चोट भी करते हैं; पर साथ ही एक विचित्र अस्वीकार की भावना के कारण टकराते नहीं, तिलमिलाहट मात्र पैदा करते हैं। ये संवाद माँ और बेटी के बीच की सम्बन्धहीनता को बाहर-बाहर से संकेतित करते हैं; किन्तु दोनों के बीच जो अप्रकट आत्मीयता है वह शब्दों में नहीं, आँसुओं में प्रकट होती है। बेटी मेघ से तन-मन से भीगने की बात माँ को सुनाती है। पर उमंग ही उसकी पीड़ा का कारण बनती है—वह उसके भविष्य को जानती है, इसलिए उसे बढ़ावा न देने के लिए उसकी बातों की उपेक्षा करती है। पर कौन माँ है जो अपनी बेटी का सुख नहीं चाहती ? उसे जो आशंका होती है वह उसे आँसुओं से भिगो देती है। मल्लिका उसे समझ नहीं पाती : 'क्या हुआ माँ, तुम रो क्यों रही हो ?' यह वह स्थिति है जहाँ शब्द संवाद में नहीं, भावना में ढल जाते हैं और फिर अपना सही अर्थ ग्रहण करते हैं।

वास्तव में अर्थ-विस्तार की दृष्टि से संवाद में शब्द ही पर्याप्त नहीं होते। शब्दों के कई अन्तर्सम्बन्ध होते हैं जिन्हें वे वाक्यों और वाक्य-समूहों में धारण किए रहते हैं। ऐसी स्थिति में जब भाषा अवचेतन की गहराइयों से प्रवाहित होती है, तब वह अर्थ के दुहरे स्तरों पर क्रियाशील होती है। कालिदास लौटकर आता है, तो आसन पर रखे उन ग्रथित कोरे पृष्ठों को पाता है जिनके विषय में मल्लिका कहती है : 'ये पन्ने मैंने अपने हाथों से बनाकर सिए थे। सोचा था तुम राजधानी आओगे तो मैं तुम्हें भेंट दूँगी। कहूँगी कि **इन पृष्ठों पर महाकाव्य की रचना करना। अब तो ये पन्ने टूटने भी लगे हैं और मुझे कहते संकोच होता है कि ये तुम्हारी रचना के लिए है।** इसके उत्तर में कालिदास जो उत्तर देता है उसमें भी अर्थ के कई स्तर उद्घाटित होते हैं : 'स्थान-स्थान पर इन पर पानी की बूँदें पड़ी हैं जो निःसन्देह पानी की बूँदें नहीं हैं। लगता है **तुमने अपनी आँखों से इन कोरे पृष्ठों पर बहुत कुछ लिखा है।** और आँखों से नहीं, स्थान-स्थान पर ये पृष्ठ **स्वेदकणों से मैले** हुए हैं, स्थान-स्थान पर **फूलों की सूखी पत्तियों ने अपने रंग इन पर छोड़ दिए हैं।** कई स्थानों पर तुम्हारे **नखों ने इन्हें छीला है,** तुम्हारे **दाँतों ने इन्हें काटा है।** और इसके अतिरिक्त ये ग्रीष्म की धूप के **इनके गहरे रंग, हेमन्त की पत्र धूलि** और इस घर की **सीलन... ये पृष्ठ अब कोरे कहाँ हैं मल्लिका ? इन पर एक महाकाव्य की रचना हो चुकी है...अनन्त सर्गों के एक महाकाव्य की।** इन पृष्ठों पर **अब नया कुछ क्या लिखा जा सकता है ?'**

यहाँ जो ध्वनित है वह प्रतीयमान अर्थ से कहीं अधिक मार्मिक है। बड़ी लालसा से सँजोए गए कोरे पृष्ठ स्वयं मल्लिका के अपने अस्तित्व के थे, जिन्हें कालिदास द्वारा

अपने ऊपर जीवन का महाकाव्य लिखे जाने की कामना थी। किन्तु कालिदास के लौट न आने के कारण एक भिन्न स्तर पर इस काव्य की रचना आँसू, स्वेद, घर की सीलन, कड़ी गर्मी और वर्षा कर देती है। और अन्ततः बचा-खुचा काव्य विलोम लिख जाता है। जीवन भी एक रचना है, चाहे उसे वह या दूसरा रचे या व्यक्ति परिस्थितियों को उसे रचने दे। रचना हर स्थिति में होती है। मल्लिका के जीवन की एक बहुत बड़ी विडम्बना को एक सामान्य तथ्य के माध्यम से व्यक्त करने में राकेश का यह कौशल प्रशंसनीय है।

वस्तुतः राकेश के संवाद सीधे और सपाट नहीं हैं। कोई विशेष मनःस्थिति जब पात्रों के मन को आ घेरती है तो वह अतिशय सघनता और सांकेतिकता प्राप्त कर लेती है। तब शब्द जैसे संकेतों में बदल जाते हैं और वाक्य-विन्यास भी कहीं सघन और कहीं खंडित हो जाता है। पात्र जिस तनाव और विकृति को झेलते हैं उसमें एक के शब्द दूसरे से उलझ जाते हैं। कोई कुछ कहता है, तो कुछ उत्तर मिलता है और फिर होती है बेतुकी बातें, टोका-टोकी, सुनी-अनसुनी करना, टूटे-फूटे संवाद और चुप्पी, जो समस्त विसंगति में अर्थ की संगति बिठाती है। कुछ संवाद शुरू होते हैं और ख़त्म होने से पहले टूट जाते हैं या पीछे छूट जाते हैं। और कहीं पात्र कुछ कहकर स्वयं ही बात को अनकहा छोड़ देते हैं—

सुन्दरी : क्या सोच रहे हैं ?

नन्द : (अव्यवस्थित)...नहीं, सोच कुछ नहीं रहा।

सुन्दरी : (टोकरी उसकी ओर बढ़ाकर) मेरे माथे पर विशेषक नहीं बनाएँगे ?

नन्द : (अपने में खोया-सा) विशेषक ! अ...हाँ ! (हाथ आगे बढ़ाकर) लाओ दो मुझे।

सुन्दरी : नहीं। इस मन से नहीं। पहले बताएँ क्या सोच रहे हैं ?

नन्द : सोच रहा था कि...।

यह दुराव, यह खंडित अभिव्यक्ति तब और भी सार्थक हो जाती है जब राकेश आगे चलकर उन खंडित सूत्रों को फिर से संवादों में समेट लेते हैं या पिछले रहस्यों को आगे व्यंजित कर डालते हैं। इस प्रकार संवादों का सारा उलझाव एक रचनात्मक रूप धारण कर लेता है।

इसी प्रकार एक स्थिति में चलते हुए संवाद को एक चरम पर पहुँचाने के क्षण से पहले ही राकेश अप्रत्याशित रूप से विषयान्तर कर डालते हैं, जिससे बात कहने को रह जाती है। ऐसा आकस्मिक परिवर्तन संवाद को एक विलक्षण चमत्कार प्रदान करता है। तारतम्य को तोड़कर वे एक ऐसा नाटकीय झटका पाठक/प्रेक्षक को देते हैं कि संवाद एक स्थिति से दूसरी स्थिति को फलाँगते जाते हैं। उदाहरण के लिए **लहरों के राजहंस** में कामोत्सव की तैयारी के बीच श्यामांग के संवाद और अलका का स्वप्न-विवरण नाटक को एक स्थिति से दूसरी स्थिति में ले जाते हैं। उसके बाद तुरन्त अपनी ही क्लान्ति से मर जानेवाले मृग से चलकर बात कामोत्सव में अतिथियों के

न आने की शंका पर पहुँच जाती है और वहाँ से श्यामांग पर और फिर अतिथियों के कामोत्सव में न आ पाने की वास्तविक स्थिति पर। इससे तारतम्य खंडित होता है, किन्तु विच्छिन्नता की विभिन्न स्थितियों में चेतना के सूत्र बने रहते हैं। यही कारण है कि वे अन्ततः सब मिलकर एक आवेगपूर्ण क्षण का निर्माण करती हैं जहाँ उनका समग्र प्रभाव देखने को मिलता है।

चेतना की यह एकसूत्रता राकेश के नाटकों में मेघ, राजहंस, छाया, मृग आदि के बिखरे सन्दर्भों में भी दिखाई देती है। किन्तु इसका सबसे सुन्दर उदाहरण **लहरों के राजहंस** में नन्द और श्यामांग के 'संवादों' में दिखाई देता है। श्यामांग ज्वर में प्रलाप करता है और नन्द अनिद्रित अवस्था में उसका प्रलाप सुनकर आत्मगत कुछ कह उठता है। दोनों के संवादों की समानान्तर स्थिति पात्रों की समान मानसिक स्थिति को ध्वनित करती है—

नेपथ्य (श्यामांग) : कहाँ हूँ **मैं** ? क्यों हूँ यहाँ...मेरा स्वर, पानी की लहरों का स्वर, सब-कुछ एक आवर्त्त में घूम रहा है।...एक चील...एक चील सब-कुछ झपटकर लिए जा रही है। इसे रोको ! इसे रोको !

नन्द : आधी रात में अब तक यह स्वर नहीं रुका। सोचता था यह रुके, तो कुछ देर सोने का प्रयत्न करूँ। परन्तु यह स्वर जैसे रात पर ही नहीं, मेरी चेतना पर भी पहरा दे रहा है। यही मुझे **भी** सोने नहीं देता।

नेपथ्य (श्यामांग) : यह चील मुझे लिए जा रही है...जाने कहाँ ? इसे रोको।...नहीं, चील नहीं है यह। एक छाया है। काले अँधेरे कूप में भटकती छाया, अकेली।

रंगमंच पर नेपथ्य का बड़ा योग होता है। राकेश ने **लहरों के राजहंस** में इसका बहुत ही सार्थक प्रयोग किया है। नेपथ्य से श्यामांग का कथ्य रहस्यपूर्ण भयावह बिम्बों की सर्जना करता है। अदृश्य के उपयोग के माध्यम से यह पद्धति दृश्य और दिक् का विस्तार करती है। श्यामांग के शब्दों को नेपथ्य में ले जाकर उन्हें संवाद की स्थिति से मुक्त किया गया है, पर वे संवाद से भी अधिक प्रभावकारी लगने लगते हैं। इस प्रकार नेपथ्य के प्रयोग से राकेश प्रेक्षक के मानस को कई दिशाओं से आक्रांत करते हैं। नेपथ्य की ध्वनि अथवा शब्दों का सबसे अधिक प्रभावशाली प्रयोग इस नाटक में उस अवसर पर हुआ है जब नन्द सुन्दरी का शृंगार कर रहा होता है। उसी समय बौद्ध भिक्षुओं का स्वर 'बुद्धं शरणं गच्छामि...' सुनाई देता है और नन्द के हाथ से दर्पण गिर पड़ता है। दोनों में अत्यल्प संवाद होता है, पर नेपथ्य का स्वर और उसकी प्रतिक्रिया शब्द से भी पैनी होती है। फलतः सुन्दरी आहत हो उठती है और पूछ बैठती है : 'दर्पण का टूटना क्या अकारण था ?' और इस प्रकार नेपथ्य की वाणी नाटक के केन्द्रीय संघर्ष को जन्म देती है। **पैर तले की ज़मीन** में नेपथ्य का पर्याप्त प्रयोग है। टेलीफ़ोन के रूप में नेपथ्य में स्थित व्यक्ति की वाणी का उपयोग इस नाटक में कई बार हुआ है।

इसी दिशा में संवादों के बीच मौन का नाटकीय उपयोग शब्दों से भी अधिक

अर्थवान हो जाता है। स्वयं राकेश इस तथ्य से भली-भाँति परिचित थे[13] कि शब्दों के बीच की निस्तब्धता अपने में नाटकीय तनाव को वहन करने के कारण बहुत सार्थक हो सकती है। मौन या निस्तब्धता एक नकारात्मक स्थिति नहीं है। उससे पहले और बाद में प्रयुक्त शब्दों के बीच वह सेतु का काम ही नहीं करता वरन् उसके आधार पर एक अर्थ की सृष्टि भी करता है। राकेश मौन का भी भाषिक प्रयोग कर दिखाते हैं। वे जानते थे कि नाटक में 'वास्तविक अभिनय **शब्दों का नहीं,** शब्दों **के बीच में** होता है।' इसीलिए वे अपने संवादों में मौन, अनकहे शब्दों और रिक्त स्थानों के द्वारा भाषा की क्षमताओं को जगाते दिखाई देते हैं। **आधे अधूरे** में संवादहीनता की ऐसी ऐच्छिक स्थितियाँ आती हैं, जब संवाद चुक जाते हैं। भावना की अतिशयता निस्तब्धता में बदल जाती है और यही निस्तब्धता फिर संवाद की ओर ले जाती है। छोटी लड़की जब अपने पास स्कूल के प्रयोग के लिए बहुत-सी चीज़ें न होने की शिकायत करती है तो ख़ामोशी फैल जाती है। यह ख़ामोशी एक जीवन्त अहसास दे जाती है—यथार्थ का अहसास। बड़ी लड़की जब यह पूछती है कि वह कौन-सी चीज़ है जो वह संस्कार रूप से इस घर से ले गई है तो यह प्रश्न भी ख़ामोशी पैदा करता है। जिनसे प्रश्न पूछा गया है वे सब उत्तर एक-दूसरे की आँखों में खोजते हैं और टाल देते हैं। इसी तरह सावित्री जब यह कहती है कि उसने फ़ैसला कर लिया है तो जगमोहन की 'हूँ' को ख़ामोशी निगल जाती है। यह ख़ामोशी बेतुकेपन को उभारती है जिसको वह छिपा देना चाहता है, पर जो बरदाश्त नहीं होती।

खंडित और अधूरे संवादों की भी नाटक में अपनी व्यंजना होती है। मोहन राकेश ने अपने नाटकों में उनका भरपूर प्रयोग किया है। अपने कथ्य के कारण **आधे अधूरे** में इनका अपेक्षाकृत अधिक प्रयोग हुआ है। प्रायः पात्र कटु सत्य को मुँह पर न लाने की प्रवृत्ति के कारण वाक्य अधूरे छोड़ देते हैं, पर उनका अधूरापन अर्थ की विवृत्ति को और भी पुष्ट करता है। **आधे अधूरे** में स्त्री कहती है : 'कहना फ़र्ज़ होता है मेरा। आख़िर मेरा बॉस है। इसके उत्तर में अधूरी बात में ही महेन्द्र पूरा सत्य व्यंजित कर देता है : 'बॉस का मतलब यह नहीं कि...।' स्पष्टतः अनकही बात अश्लील संकेत उभारती है। वास्तविक बात को छिपाने के लिए अधूरे वाक्यों का राकेश के नाटकों में यत्र-तत्र प्रयोग हुआ है : 'ममा ऐसा भी होता है क्या कि ..।' 'मेरा पूछना इसलिए ग़लत है कि...।' अन्य नाटकों में भी अधिकांश खंडित वाक्यों में इसी प्रकार 'कि' के बाद बात अनकही छोड़ दी गई है। 'तुम यहाँ ऐसे सूत्र से बँधी नहीं हो कि'...(आषाढ़) 'सोचने का परिणाम यह हुआ कि...' (राजहंस) 'कि' के अतिरिक्त जैसे, जो, या, और कि बस, लेकिन, क्योंकि, तो आदि का प्रयोग भी इसी तरह हुआ है।

एक अच्छा संवाद अर्थ के कई स्तरों पर क्रियाशील होता है। उसकी एक विशेषता यह भी है कि वह बिम्बों की रचना करता है। राकेश के शब्दों में 'रंगमंच' में बिम्ब का उद्भव 'शब्दों के बीच से होता है।' एक स्थिति पर पहुँचकर ये बिम्ब ही संवादों के

साथ जुड़कर अनकहे अर्थों को व्यंजित करने लगते हैं। **आषाढ़ का एक दिन** में निर्मम राजसत्ता का बिम्ब उभरता है; किन्तु नाटक की रसवत्ता में मेघ का बिम्ब अपूर्व सार्थकता का निर्वाह करता है। मेघ के बरसने से उपत्यकाओं का भीगना और मल्लिका के अपने मन का रससिक्त होना, आकाश में बकुल-पंक्तियों को उड़ते देखने की कामना और स्वयं उसके अपने मन की उड़ान जैसे समानान्तर अनुभव दे जाते हैं। और फिर उभरती है उस क्षण को आत्मसात् कर लेने की कामना ('मैं चाहती थी उसे अपने में भर लूँ और आँख मूँद लूँ') किन्तु कामना भी जैसे अथाह हो चुकी है ('कितना पानी पिया है इन वस्त्रों ने')। वर्षा की नई-नई बूँदों से उद्भ्रान्त यौवन का उष्ण अनुभव सूक्ष्म सौन्दर्य को भी मांसल अनुभूतियों में ढाल देता है। ('मैं उसे छू सकती थी, देख सकती थी, पी सकती थी।') और यहीं वह कालिदास के काव्य का उत्स ढूँढ़ निकालती है। वह मेघ में सौंदर्य का ऐसा साक्षात्कार करती है जो उसके निजी अनुभव से जुड़कर काव्यमय हो उठता है। इस निजी अनुभव का विलक्षण उल्लास ही उसे कालिदास को, काव्य साधना के लिए उज्जयिनी भेजने के लिए तैयार करता है। जो मेघ मल्लिका को इतना रससिक्त और भाव-विभोर कर देते हैं, वे ही अम्बिका को रुला देते हैं। कालिदास के रहते जो मेघ घिरते थे, वे उसके चले जाने पर मल्लिका के जीवन को कितना अँधेरा कर जाते हैं। और तब मेघ और मल्लिका की आँखें एक साथ बरसने लगती हैं। एक साथ ही वह कालिदास की पीड़ा को भी समझती है : 'देखो माँ, चारों तरफ़ कितने मेघ घिरे हैं। कल ये मेघ उज्जयिनी की ओर उड़ जाएँगे।' मेघदूत के कवि कालिदास की आत्मीयता की वह जो कल्पना करती है उसे ये संवाद आधार देते हैं।

मल्लिका को ये मेघ पराए में बदलते से लगते हैं : 'सोचती थी तुम आओगे तो उसी तरह मेघ घिरे होंगे...परन्तु आज तुम आए हो तो सारा वातावरण ही और है। और...और...नहीं सोच पा रही कि तुम भी वही हो या...?' कालिदास के साथ ही मेघ भी बदले-बदले लगते हैं। राजसत्ता से जड़ीभूत कालिदास का संवेदनशून्य हृदय उसे भिगाता नहीं और फिर इसके साथ ही जीवन का यथार्थ उसे वहाँ पटक देता है जहाँ मेघ रस, यौवन, प्रेम के पुराने सारे सन्दर्भ से कट जाता है और वह विपत्ति का प्रतीक बन जाता है (मातुल के शब्दों में : 'यह आषाढ़ की वर्षा मेरे लिए काल हो रही है') और अन्ततः केवल मल्लिका के लिए ही नहीं, स्वयं कालिदास के लिए भी ये मेघ जीवन की विषय परिणति के प्रतीक बन जाते हैं और मल्लिका का द्वार धारासार वर्षा से बचने के लिए शरण-स्थल। विलोम के द्वार खटखटाने पर मल्लिका के पास बैठा कालिदास कहता है : 'देख तो लो, कौन आया है ?' मल्लिका उत्तर देती है : 'वर्षा का दिन है। **कोई भी** हो सकता है !' कालिदास की नियति को, मल्लिका के हृदय में रसी-बसी उदासीनता को शायद यही संवाद सही रूप में रखता है। अन्ततः कालिदास के लिए ये मेघ कितने कारुणिक, कितने त्रासद, कितने विडम्बनापूर्ण हो जाते हैं ! इस प्रकार **आषाढ़ का एक दिन** में मेघ एक पूर्ण सन्दर्भ बन जाता है और केन्द्रीय भाव के रूप में अतीत से लेकर वर्तमान तक पात्रों के जीवन पर छा जाता है। सारा नाटक मेघ की

बिम्ब-योजना से बुना गया है और यह अनुभूति को गहराई ही प्रदान नहीं करता वरन् कथ्य को भी समृद्ध करता है और रचना की रसवत्ता में भी वृद्धि करता है। वर्षा-सम्बन्धी परिवेश **पैर तले की ज़मीन** में भी है, किन्तु उसमें वर्षा की अर्थवत्ता स्थूल परिवेश से सम्बद्ध है और उसमें वह आन्तरिकता नहीं है जो **आषाढ़ का एक दिन** के मेघ में दिखाई देती है

लहरों के राजहंस में ऐसा अनुभूति-प्रवण बिम्ब-विधान कहीं नहीं है, क्योंकि उसमें अधिक बल प्रतीक तत्त्व पर है और प्रतीक भावना से कम और विचार से अधिक जुड़े हैं। ये प्रतीक पात्रों को समझने में मदद करते हैं। कहीं-कहीं उनमें बिम्ब की क्षमता भी दिखाई देती हैं। राजहंस का बिम्ब उसके प्रतीकत्व के साथ नाटक को प्रभावी बनाता है। इसके साथ ही 'बिना घाव अपनी ही क्लान्ति से मरा मृग' नन्द के साथ तादात्म्य स्थापित कर लेता है और उसके चिन्तन पर हावी हो जाता है। सारे कामोत्सव के आयोजन के बीच उसे लगता है जैसे 'घने वृक्षों की ओट में वह सदा से इसी तरह अटका है—इस आशा में कि 'कोई उसे वहाँ पड़ा देख लेगा और बाँहों में उठाकर ले जाएगा।' नन्द की बाद की मनोदशा का सही प्रतिनिधित्व यही बिम्ब करता है। राजहंस का प्रतीक नाटक पर बौद्धिक रूप से आरोपित होने के कारण आषाढ़ के मेघ के समान भावना से जुड़ नहीं पाता।

राकेश के नाटकों में बिम्ब-विधान संवादों से जुड़कर काव्यमय वातावरण की सृष्टि करता है। **आषाढ़ का एक दिन** में मेघ समुचित दृश्यबन्ध प्रस्तुत करता है और सारे नाटक में वह एक भावात्मक घटक बन जाता है। **लहरों के राजहंस** में भी राजहंस, मृग की मृत्यु, दर्पण के आगे सुन्दरी का प्रसाधन, 'बुद्धं शरणं गच्छामि' की नेपथ्य से आगत ध्वनि और दर्पण का टूटना—कई चीज़ें एक वातावरण बनाती हैं और विलक्षण असाधारणता धारण कर लेती हैं जिससे साधारण बातों में असाधारण अर्थ उभरते हैं। इसी प्रकार आरम्भ में श्यामांग के संवाद और अलका का स्वप्न अतिप्राकृतिक वातावरण बनाते हैं। नन्द जिस प्रकार पार्थिव और अपार्थिव द्वन्द्व में फँसता है, उसके लिए भी कामोत्सव के सहारे राकेश ने उपयुक्त वातावरण बनाया है। वस्तुतः एक विशिष्ट वातावरण की सृष्टि कर पात्रों को उसमें खड़ा कर देने की उनमें अद्भुत क्षमता है जिसमें पात्र ही नहीं, परिवेश भी जीवित हो जाता है। फलतः किसी वातावरण विशेष में पात्र विशिष्टता ग्रहण करते हैं और संवाद अतिरिक्त अर्थ वहन करने लग जाते हैं।

संवाद के रंगमंचीय तत्त्वों के प्रति भी राकेश पूरी तरह जागरूक थे। संवाद, दृश्य और शारीरिक क्रिया का जैसा समन्वय राकेश के नाटकों में मिलता है वैसा पूर्ववर्ती नाटकों में दुर्लभ है। डॉ. गिरीश रस्तोगी ने ठीक ही लिखा है कि 'हिन्दी के अधिकांश नाटकों में शब्द अलग मिलेंगे, शारीरिक क्रिया अलग। या इससे अधिक कुछ हुआ भी तो यह कि शब्द के अनुरूप ही क्रिया होगी। राकेश की नाट्य भाषा इस जड़ता को, अस्वाभाविकता को नकारती है। वहाँ शब्द स्वयं क्रिया का काम करते हैं और क्रिया

की भाषा को ढालते चलते हैं। अर्थात् भाषा और क्रिया का नियोजन, आन्तरिक गठन पहली बार मोहन राकेश में मिलता है।[14] वस्तुतः पात्र क्रिया के माध्यम से बोलता है और स्वयं क्रिया ही संवाद बन जाती है। **लहरों के राजहंस** और **आधे अधूरे** में इसका बड़ी कुशलता से प्रयोग हुआ है। श्यामांग के अपने मस्तिष्क की उलझन और पत्तियों को उलझाने की क्रिया दोनों एक स्तर पर घटित होती हैं। इसी प्रकार अग्निकाष्ठ को हाथ में लिए हुए भी दीपक को न जला पाना भी इसी प्रकार की क्रिया है। उपस्थित प्रश्न ने उसके सामने जो अन्धकार फैला दिया है उसमें वह किंकर्तव्य-विमूढ़ है। देवी यशोधरा के भिक्षुणी बनने की सूचना पर सुन्दरी का मदिरा-कोष्ठ के पास आकर चषक को भरने लगना भी अपने में सार्थक क्रिया है और फिर उद्यान में हंसों की जो ध्वनि सुनाई देती है और उनको लेकर जो संवाद उभरते हैं वे भी सुन्दरी के आन्तरिक संघर्ष और उसमें निहित केन्द्रीय अर्थ के साथ जुड़ जाते हैं। दर्पण का टूटना अपने में एक क्रिया है पर यह क्रिया अभिव्यंजना में संवाद की चरम परिणति भी है। ऐसी ही सुन्दर अभिव्यंजना का एक स्थल वहाँ पर आता है, जहाँ भिक्षु आनन्द के सामने कुछ घड़ी पहले 'नहीं, अभी मदिरा नहीं लूँगा भिक्षु' कहनेवाला नन्द सुन्दरी का सामना न कर सकने के कारण दूसरे क्षणों मदिरा ले लेता है। इस अवसर पर सुन्दरी का संवाद उसकी इस क्रिया पर सही टिप्पणी करता है—

सुन्दरी : शब्दों से अन्तर नहीं पड़ता। अर्थ आपका मेरे लिए स्पष्ट नहीं है।

नन्द : तुम न जाने किस अर्थ की बात कर रही हो।

सुन्दरी : वह जिसे छिपाने के लिए चषक में मदिरा डाल रहे हैं।...पर इस प्रयत्न से वह अर्थ छिप नहीं सकता।

आधे अधूरे में संवादों का क्रम, घर का परिवेश और क्रिया एक साथ चलते हैं। घर की गर्द, मेज़ पर बिखरे जूठे बर्तन, बेखरी कतरनें और इससे भी अधिक दीवाल पर झूलता गंदा पाजामा उस घर को साकार करते हैं। महेन्द्रनाथ का ध्यान बँटाने के लिए अख़बार पढ़ना, अशोक का कैंची से फ़िल्म अभिनेत्रियों की तस्वीरें काटना, सावित्री का माला को हाथ में लपेटना कुछ ऐसी ही क्रियाएँ हैं। सावित्री 'पाजामे को मरे हुए जानवर की तरह उठाकर देखती है और कोने में फेंकने को होकर फिर एक झटके से तहाने लगती है।' और इसके साथ ही जुड़ा हुआ संवाद एकदम महेन्द्रनाथ पर सीधी चोट करता है। बड़ी लड़की के यह पूछने पर के क्या चीज़ है वह जिसे मैं इस घर से ले गई हूँ, कुछ देर तक शब्द गुम हो जाते हैं और संवाद केवल क्रिया के द्वारा होता है। सिंघानिया के घर नें आने पर बड़े लड़के के संवाद जैसे क्रिया में ढल जाते हैं। लड़के का पैड निकालकर सिंघानिया का कार्टून बनाना ऐसी ही क्रिया है, जो इस संवाद के साथ जुड़कर अत्यन्त सार्थकता प्राप्त कर लेती है—

स्त्री : यह चेहरा कुछ-कुछ वैसा नहीं है ?

बड़ी लड़की : कैसा ?

स्त्री : तेरे डैडी जैसा।

बड़ी लड़की : डैडी जैसा ? नहीं तो...।

स्त्री : लगता तो है कुछ-कुछ।

बड़ी लड़की : वह तो इस आदमी का चेहरा बना रहा था...यह जो अभी गया है।

बड़े लड़के की हरकत इन संवादों से जुड़कर नाटक के एक केन्द्रीय सत्य को उजागर करती है, जिसके लिए नाटककार ने एक से ही चार पुरुषों की भूमिका करवाई है और जिसकी घोषणा अन्त में जाकर स्वयं सावित्री करती है : 'अलग-अलग मुखौटे, पर चेहरा ?—चेहरा सबका एक ही !' कभी ऐसा कोई वाक्य कथानक के अर्थकेन्द्र में आ जाता है। सावित्री का जगमोहन के पास जाने के लिए सजना-सँवरना, गले की माला को उँगली से लपेटना और झटका लगने से उसका टूटना, सफ़ेद बालों को कंघी से ढकना आदि ऐसे क्रिया-व्यापार हैं जो संवादों के समानान्तर नाटकीय अर्थ को ध्वनित करते हैं। यह दूसरी बात है कि ये क्रियाएँ रंगमंच पर घिसी-पिटी लगें, पर जैसी भी हैं, एक जागरूक लेखन का अंग ज़रूर हैं।

दृश्य, क्रिया आदि रंग तत्त्व से संवाद को जोड़ना रंग भाषा का पहला दायित्व होता है। मोहन राकेश ने **आषाढ़ का एक दिन** में मेघ के दृश्यत्व और **आधे अधूरे** में घर के बिखराव और पात्रों की हरकत का अद्भुत सामंजस्य किया है। **लहरों के राजहंस** में भी सुन्दरी की व्यस्तता/व्यग्रता तथा श्यामांग की अनन्यमनस्कता दृश्य के साथ संवादों में जुड़ती है। इस नाटक का दूसरा अंक इसी दृष्टि से रंगमंचीय उपलब्धि माना जाता है।

स्थिति और पात्र की अनुकूलता राकेश के संवादों की एक अन्य विशेषता है। मातुल, अनुस्वार, अनुनासिक, मल्लिका, अम्बिका, विलोम—सबकी अपनी भाषा है। अम्बिका जैसा ठोस चरित्र है, वैसे ही उसके संवाद भी हैं। विलोम अपनी मुँहफट प्रवृत्ति से कालिदास को उखाड़कर रख देता है। जहाँ वह एक ओर खल पात्र जैसा लगता है वहाँ दूसरी ओर उसके तीखे-पैने संवाद जीवन के अर्धसत्यों को भी झनकाते हैं और इसी से उसका चरित्र घृणित होने से बच जाता है। वह उन प्रश्नों को बड़ी प्रखरता से उभारता है जो अम्बिका या मल्लिका को उभारने चाहिए थे, किन्तु अम्बिका कालिदास का सीधा सामना नहीं करती और मल्लिका 'ऐसे महत्त्वपूर्ण क्षण' में अपने स्वार्थ की घोषणा नहीं करना चाहती। **लहरों के राजहंस** में श्यामांग के संवादों में उसका अपनापन है। श्यामांग की चिन्ताग्रस्त विक्षिप्तता का संकेत देने में उसके संवाद विशेष रूप से सहायक होते हैं। इसके साथ ही वे नाटक के कथ्य को भी असम्बद्ध बिम्बों के द्वारा उजागर करते हैं : 'कहाँ हूँ मैं ? क्यों हूँ यहाँ ?...मेरा स्वर पानी की लहरों का स्वर। सब-कुछ एक आवर्त्त में घूम रहा है।...एक चील...एक चील सब-कुछ झपटकर लिए जा रही है...।' इसी प्रकार सुन्दरी के संवाद उसके अहं के ही अनुरूप हैं। नन्द के संवाद उस उलझाव को साफ़ प्रकट करते हैं जो उसके व्यक्तित्व का अंग है, उसके संवाद गहरी बौद्धिकता से जुड़े हैं और तीव्र अनुभूति से भी। इसलिए वे मानवीय स्थितियों और अस्तित्व का जो बोध कराते हैं वह इकहरा नहीं है क्योंकि चरित्र स्वयं इकहरा नहीं है। इसीलिए

संवाद उसकी प्रवृत्ति के अनुरूप ही आत्मविश्लेषण भी करते हैं और स्थितियों पर टिप्पणी भी। **आधे अधूरे** के संवाद पात्रों की अपनी कुढ़न, झल्लाहट, तल्ख़ी और आक्रोश का प्रतिनिधित्व करते हैं। सावित्री हर वक़्त आमादा ही रहती है और महेन्द्रनाथ मिट्टी का लोंदा। छोटी लड़की किन्नी को हर वक़्त घरवालों से कोई-न-कोई शिकायत रहती है कि कोई उसकी बात नहीं सुनता। सबका चरित्र संवादों से उभरता है। सिंघानिया के सारे चरित्र का बिम्ब संवादों की ही देन है। बड़ा लड़का अशोक जैसा भी है, उसमें स्थिति का अहसास है और नई पीढ़ी का विद्रोही भाव भी। उसके संवादों में नई पीढ़ी का आक्रोश है जो स्थिति से समझौता नहीं करना चाहता, सीधा सामना करना चाहता है। उस 'नाकारा' लड़के की सीधी बातों से माँ तिलमिला उठती है। उसके संवादों में युवा पात्र के अनुरूप अद्भुत चमक और बिजली की-सी कौंध है। आक्रोश में पिसता आदमी जब विवशता का अनुभव करता है तो उसके पास केवल शब्द रह जाते हैं। **आधे अधूरे** के पात्र खार खाए बैठे हैं। इसलिए वे एक तीखी मार करते हैं।

नाटक में ध्वनि के रूप में भी शब्द की अपनी एक शक्ति होती है जो अर्थ के स्तर पर सक्रिय रहती है। सघोष, अघोष, तालव्य, मूर्द्धन्य आदि ध्वनियों और विविध ध्वन्यात्मक संयोजनों से अर्थ का प्रतिनिधित्व/प्रत्यंकन ही नहीं होता वरन् संवेदन की जाग्रति भी होती है। राकेश इस स्थिति से परिचित थे, उनके शब्दों में : 'कोई भी शब्द-योजना बिना अपनी एक आन्तरिक लय के प्राणवान नहीं होती, और यह लय या ध्वनि का ग्राफ़ ही उसकी वास्तविक अर्थवत्ता है।...शब्द मूलतः नादधर्मा हैं। इसीलिए नाद के आरोह-अवरोह में ही शब्द का आन्तरिक नाटक निहित है।'[15] राकेश के नाटकों में लय के प्रति जागरूकता के दर्शन होते हैं। वे शब्दों का ऐसा संयोजन करते हैं कि उनसे लयात्मक घटक का निर्माण हो सके। यह लय तत्त्व केवल शब्द के बाह्य रूप से ही निर्मित हो, ऐसी बात नहीं है। राकेश की नाट्य-भाषा की लयात्मकता आन्तरिक अनुभूति और संवेग की देन है। अनुभूति लय को जन्म देती है और लय अर्थ को। इसीलिए जहाँ भी भावना की तीव्रता है वहाँ लय है। जहाँ एक ही संवाद की शब्द योजना में एक ध्वनि-समूह और दूसरे में एक निश्चित लय है, वहाँ एक संवाद और दूसरे संवाद का परस्पर योग भी लय का सृजन करता है। उदाहरण के लिए, अनुस्वार और अनुनासिक के संवादों में एक ऐसी आन्तरिक लय आरोपित की गई है कि वे एक शैली में ढल गए हैं **आधे अधूरे** के संवाद लय की दृष्टि से सबसे अधिक सक्षम हैं। उनमें शब्दों और वाक्यों की लय, एक संवाद और दूसरे संवाद का परस्पर संघात, पात्रों का बोलने का लहजा, वाक्य-विन्यास, शब्द-क्रम सब मिलकर अद्भुत वाग्प्रवाह पैदा करते हैं। एक उदाहरण लीजिए—

बड़ी लड़की : तो तू सोचता है ममा जो कुछ भी करती है यहाँ...?

लड़का : मैं पूछता हूँ क्यों करती है ? किसके लिए करती है ?

बड़ी लड़की : मेरे लिए करती थी।

लड़का : तू घर छोड़कर चली गई।

बड़ी लड़की : किन्नी के लिए करती है।

लड़का : वह दिन-ब-दिन बदतमीज़ होती जा रही है।

बड़ी लड़की : डैडी के लिए करती है।

लड़का : उनकी हालत देखकर रहम नहीं आता ?

बड़ी लड़की : और सबसे ज़्यादा तेरे लिए करती है।

लड़का : और मैं ही शायद इस घर में सबसे ज़्यादा नकारा हूँ।

राकेश ने छोटे और लम्बे दोनों प्रकार के संवाद लिखे हैं। छोटे संवाद नपे-तुले हैं और लम्बे संवाद भावना से परिपूर्ण। **बीज नाटक** छोटे अर्थगर्भित संवादों का सुन्दर उदाहरण प्रस्तुत करते हैं खाली समय को भरनेवाले उनके संवाद अपनी निरर्थकता में भी सार्थक लगते हैं—

पुरुष : अच्छा बताओ...रात को खिड़की का क्या करना है ?

स्त्री : क्या करना है ?

पुरुष : इसे खुली रखना है या...?

स्त्री : जैसे कहो खुली रख लेंगे। बन्द कर देंगे।

पुरुष : बन्द करने से गर्मी लगेगी।

स्त्री : खुली रखने से ठंड लगेगी।

पुरुष : तो ?

स्त्री : तो कुछ नहीं...।

आषाढ़ का एक दिन में अन्तिम अंक में इसके विपरीत कालिदास कई पृष्ठों के एक लम्बे संवाद का प्रयोग करता है। **लहरों के राजहंस** में भिक्षु आनन्द के समक्ष नन्द के संवाद भी कम लम्बे नहीं हैं। इसी प्रकार, **आधे अधूरे** में पुरुष चार और सावित्री के संवादों में भी विस्तार है। इन लम्बे संवादों को आलोचकों ने अनाटकीय कहा है। राकेश स्वयं उनकी दुर्बलता से परिचित थे; पर उनकी अपरिहार्यता के प्रति वे विवश थे। वस्तुत : ये लम्बे-लम्बे संवाद नाटक के अन्त में कालिदास, नन्द, महेन्द्र और सावित्री के समस्त इतिहास को एक क्षण में समेटने का प्रयास करते हैं। इसी प्रयास में ये व्याख्यात्मक हो गए हैं और बहुत कुछ आत्मसुरक्षात्मक, क्योंकि पात्र अपनी वकालत में बहुत-कुछ कहना चाहते हैं। कालिदास जब थका-हारा लौट आता है, तो मल्लिका के सामने कठघरे में खड़ा होने को बाध्य है। उसमें जो अपराध-भाव अन्तःकरण की गहराइयों में दबा था वह एक लम्बे संवाद में उमड़ पाता है। और इसका सबसे बड़ा लाभ यह होता है कि पात्र के प्रति सहानुभूति होने लगती है। भिक्षु आनन्द के सामने नन्द अपना पक्ष प्रस्तुत करने के लिए बहुत कुछ कहने के लिए मुक्त है। बुद्ध के सामने बेहद चुपचाप बिना एक शब्द बोले केश कटवा देनेवाला नन्द आनन्द के सामने बेहद मुक्त हो जाता है। पर उसका परिणाम यह होता है कि उसमें चिन्तन की सम्भावना उभरती है; किन्तु उसके बावजूद अस्तित्व के सत्य को उद्घाटित करनेवाली बातें भिक्षु

आनन्द की ही है।

आधे अधूरे के अन्त में जो लम्बे संवाद आए हैं वे क्रिया-व्यापार की अपेक्षा परिस्थिति और पात्र के विश्लेषण पर आधारित हैं। ये संवाद महेन्द्रनाथ और सावित्री के पक्ष-विपक्ष को प्रस्तुत करते हैं और दुहरी दृष्टि से विरोधी तर्कों का जाल उपस्थित करते हैं। एक बार जुनेजा बड़ी लड़की के सामने यह बात रखता है : 'मैं समझता हूँ महेन्द्रनाथ ख़ुद ज़िम्मेदार है अपनी यह हालत करने के लिए,' दूसरी ओर वह सावित्री को ही सामने-सामने इसके लिए ज़िम्मेदार ठहराता है : 'तुमने इस तरह शिकंजे में कस रखा है उसे कि अपने दो पैरों पर चल सकने लायक़ भी नहीं रहा।' एक पक्ष जुनेजा प्रस्तुत करता है कि 'महेन्द्रनाथ इस औरत को इतना चाहता है, इतना चाहता है अन्दर से कि...' तो बड़ी लड़की दूसरा पक्ष रखती है : 'आप सोच भी नहीं सकते क्या-क्या होता रहा है यहाँ। डैडी का चीखते हुए मम्मी के कपड़े तार-तार कर देना...उसके मुँह पर पट्टी बाँधकर उसे बन्द कमरे में पीटना। खींचते हुए गुसलख़ाने में कमोड पर ले जाकर...।' सावित्री और जुनेजा के बीच जो संवाद चलते हैं, वे आरोप-प्रत्यारोप लिए हुए हैं—इस पर आरोप उन दोनों में बँट जाते हैं और सहानुभूति भी दोनों के प्रति होने लगती है।

तीनों नाटकों में प्रयुक्त ये लम्बे संवाद अंशतः सूच्य हैं, अंशतः तर्कपूर्ण, किन्तु इनकी सबसे बड़ी उपलब्धि भाव-निवृत्ति (रिलीफ) है या सम्बोधित पात्र में सही क़िस्म की प्रतिक्रिया पैदा करना। इनका प्रेक्षक पर एक समग्र प्रभाव पड़ता है। इसीलिए इन्हें अनाटकीय भी नहीं कहा जा सकता। रंगमंच पर इन्होंने अपने को पूरी तरह सार्थक सिद्ध किया है। वस्तुतः ये संवाद पाठक/प्रेक्षक को भावनात्मक स्तर पर इतना ऊँचा उठा लेते हैं और उसकी चेतना का इतना विस्तार हो जाता है कि उस समय वह स्वगत या लम्बे संवाद को स्वाभाविक रूप में स्वीकार कर लेता है। वस्तुतः उनमें चिन्तन, आत्मविश्लेषण और संवेदना की ऐसी समन्विति मिलती है कि वे मन पर पूरा प्रभाव छोड़ते हैं। उदाहरण के लिए कालिदास की सारी कैफ़ियत पिछले सन्दर्भों और संवादों की पुनरावृत्ति से इस प्रकार जुड़ी है कि वे नाटकीय बुनावट के अंग बन जाते हैं। और फिर बच्ची के रोने जैसे प्रसंग की अवतारणा करके राकेश ऐसी नाट्य-स्थिति का निर्माण करते हैं जहाँ स्थिति ही संवाद बन जाती है। **लहरों के राजहंस** में अन्त का एक लम्बा स्वगत थकाता अवश्य है। उसमें भी नन्द अपनी कैफ़ियत देता है, नेपथ्य में घटी घटनाओं का विवरण प्रस्तुत करता है, किन्तु विवरण लोगों की आकुलता को उजागर करता है। इसलिए अनाटकीय नहीं है—वह हृदय के अन्दर चलनेवाले नाटक को उभारकर रखता है। वह एकदम ऊब पैदा नहीं करता क्योंकि नाटककार ने उसके तुरन्त बाद एक अतिनाटकीय या अतिप्राकृत स्थिति को सुन्दरी के संवादों के बीच से उभारा है। **आधे अधूरे** में पुरुष चार महेन्द्र के पक्ष में लम्बी दलीलें देता है, बहुत-सा विश्लेषण करता है। किन्तु दोनों ओर के—विशेषतः सावित्री के—संवाद इतने तीखे, सहज और प्रवाहपूर्ण हैं कि नाटक की सारी कला भाषा और संवाद-शिल्प पर केन्द्रित हो जाती है।

राकेश की संवाद योजना की कला का दर्शन सच्चे अर्थों में उनके लम्बे संवादों में नहीं, छोटे नपे-तुले संवादों में होता है जहाँ वे विभिन्न शैल्पिक युक्तियों और मनोवैज्ञानिक गोपन और अभिव्यंजना की विधियों का प्रयोग करते हैं। **बीज नाटकों** के विषयांतर युक्त, असंबद्ध, मन के बिखराव को समेटने वाले संवाद ऐब्सर्ड भले ही लगें, पर राकेश की कला की चरम उपलब्धि कहे जा सकते हैं।

रंगानुभव

तथाकथित साहित्य नाटक साहित्य-कृति होते हुए भी नाटक नहीं होता। नाटककार के लिए वह आवश्यक है कि वह जो कुछ लिखता है आँख मूँदकर अपनी कल्पना के रंगमंच पर घटित हुए देखे। लिखा हुआ नाटक अपने में पूर्ण कृति नहीं होता।...लिखा हुआ नाटक हड्डियों के ढाँचे की तरह है जिसे रंगमंच का वातावरण मांसलता देता है।

***—मोहन राकेश :** साहित्यिक सांकृतिक दृष्टि, पृ. 98*

राकेश का सारा नाट्य लेखन गहरे रंगानुभव से जुड़ा है। इस कथन में अत्युक्ति न होगी कि वे अपने समय के अकेले नाटककार हैं जिन्होंने नाट्य लेखन में कलात्मक रंग-तत्त्वों की सही तलाश की है। इस लक्ष्य की उपलब्धि के लिए उन्होंने अपने नाटकों को बार-बार तराशा है। **लहरों के राजहंस** का पुनर्लेखन इसका एक प्रमाण है। इसका हवाला देते हुए उन्होंने स्वयं उन स्थितियों का उल्लेख किया है जो उन्हें इस नाटक के मंचन के समय सुप्रसिद्ध नाटक प्रयोक्ता श्यामानन्द जालान के साथ कई दिनों झेलनी पड़ीं। उनकी संपृक्ति की अभिव्यक्ति इन पंक्तियों में हुई है : 'पहली रात को मैं नाटक हॉल में बैठकर नहीं देख सका। गहरे तनाव की स्थिति में एक पार्श्व में खड़ा मंच की गतिविधियों और खेल में होनेवाली प्रतिक्रियाओं का जायजा लेता रहा। नाटक समाप्त होने तक माथे की नसें फड़कती रहीं। उनके चौथे दिन के अन्त तक मन सहज स्थिति में नहीं आ पाया।'[1]

यह बात दूसरी है कि वे 'अपने को उसका एक हिस्सा' मानने से कतराते थे क्योंकि रंगमंच की आज की गतिविधियों पर परिचानक बुरी तरह हावी है और उसमें नाटककार की स्थिति गौण हो चली है। परिचालक आज रंगमंच पर प्रयोग करने के लिए उतावला है, किन्तु उस प्रयोगशीलता में नाटक को उसकी चारदीवारी में छोड़ दिया जाता रहा है। फलतः नाटककार और रंगकर्मियों के सहयोग से जो सर्जनात्मक उपलब्धि सम्भव थी, उसके नकारे जाने पर केवल अनुकरणात्मक प्रयोगों को रंगमंच की सिद्धि माना जाने लगा है। राकेश इस स्थिति के विरोध में थे। वे रंगमंच को बाहर से 'नया' या 'आधुनिक'

रूप देने के बजाय भारतीय परिवेश और रंगधर्मिता के अन्दर से विकसित करने के पक्ष में थे। वे मानते थे कि पश्चिम के रंगमंच पर हमारी निर्भरता हमको 'बन्द गली की ओर' ले जा रही है और 'हम इस बन्द गली में इसलिए पहुँच गए हैं कि हमने किसी गली में मुड़ने की बात सोची ही नहीं। (पश्चिम का) तकनीकी रूप से समृद्ध और संश्लिष्ट रंगमंच भी अपने में विकास की एक दिशा है, परन्तु उससे हटकर एक दूसरी दिशा भी है और मुझे लगता है कि हमारे प्रयोगशील रंगमंच की वही दिशा हो सकती है। वह दिशा रंगमंच के शब्द और **मानव पक्ष** को समृद्ध बनाने की है—अर्थात **न्यूनतम उपकरणों** के साथ संश्लिष्ट से संश्लिष्ट प्रयोग कर सकने की। यहीं रंगमंच में **शब्दकार का स्थान महत्त्वपूर्ण** हो उठता है—उससे कहीं अधिक महत्त्वपूर्ण जितना कि हम समझ पाए हैं।'[2]

पश्चिम में दृश्य की क्षमता को ज़रूरत से ज़्यादा इस्तेमाल करनेवाला रंगमंच आज श्रीहीन होता जा रहा है। इसीलिए रंगकर्मी वहाँ भी यह अनुभव करने लगे हैं कि अब लौटने के अतिरिक्त कोई चारा नहीं है। राकेश नाटक में शब्द को स्थापित करने की जिस आकांक्षा से भरे थे, वह आज के रंगमंच की सही दिशा है। दृश्य का अधिकाधिक प्रयोग करने की दृष्टि से सिनेमा और टेलीविज़न उपयुक्त माध्यम हैं। नाटक उनकी तुलना में नहीं आ सकता। उसे शब्द की क्षमताओं पर निर्भर रहना पड़ेगा। इसी आधार पर राकेश रंगमंच को दृश्य की अपेक्षा एक श्रव्य माध्यम मानते हैं।[3] यह अवधारणा एकदम विरोधी नहीं है। दृश्यत्व के बावजूद नाटक में शब्द की आधारभूत भूमिका से इनकार नहीं किया जा सकता क्योंकि उसमें शब्द ही दृश्य को जन्म देते हैं और दृश्य शब्द की सहचारिता में ही अर्थ ग्रहण करता है। यह धारणा भारतीय नाट्य दृष्टि से सर्वथा प्रतिकूल नहीं है—हमारे यहाँ नाटक केवल दृश्य ही नहीं, **काव्य** भी रहा है और वह मूलतः शब्द पर निर्भर रहा है।

आषाढ़ का एक दिन की भूमिका में मोहन राकेश ने हिन्दी रंगमंच के स्वरूप के सम्बन्ध में एक महत्त्वपूर्ण प्रश्न उठाया था : 'हिन्दी रंगमंच को हिन्दी-भाषी प्रदेश की सांस्कृतिक पूर्तियों और आकांक्षाओं का प्रतिनिधित्व करना होगा, रंगों और राशियों के विवेक को व्यक्त करना होगा। हमारे दैनंदिन जीवन के रागरंग को प्रस्तुत करने के लिए, हमारे संवेदों और स्पन्दनों को अभिव्यक्त करने के लिए जिस रंगमंच की आवश्यकता है वह पाश्चात्य रंगमंच से कहीं भिन्न होगा।' इस भिन्नता का आधार शब्द और दृश्य की मान्यताएँ ही हैं। पाश्चात्य रंगमंच सुविधाओं, तकनीकी उपकरणों पर जिस रूप में आश्रित है, उस रूप में हिन्दी रंगमंच के लिए उसका अनुकरण न संभव है और न ग्राह्य। राकेश पाश्चात्य रंगमंच का अनुमोदन इसलिए नहीं करते, क्योंकि वह 'चकाचौंध का पर्याय' बन चुका है, जबकि वे रंगमंच का मूल तर्क सादगी और प्रतीकात्मकता को मानते हैं। पाश्चात्य रंगमंच परिचालक का रंगमंच है, भारतीय रंगमंच शब्द का रंगमंच है जिसमें नाटककार का दायित्व सबसे ऊपर है। राकेश 'नाटककार के रंगमंच' के सबसे बड़े समर्थक थे। वस्तुतः रंगमंच का सारा रूप-विधान रचना नहीं कर सकता। राकेश की अपनी एक रंग-दृष्टि थी, यह कम महत्त्वपूर्ण तथ्य नहीं है। उन्होंने बड़ी रूढ़ियाँ तोड़ी

हों या बिल्कुल नई तलाश की हो, ऐसा तो नहीं कहा जा सकता, किन्तु इतना सत्य है कि रंगमंच के बारे में एक नए सिरे से सोचने की शुरुआत उन्होंने ज़रूर की है।

राकेश के नाटक रंगानुभति से परिपूर्ण हैं। यों भी कहा जा सकता है कि रंग-तत्त्वों के समावेश और रंगमंचीय क्षमता की दृष्टि से वे अद्वितीय हैं। वे साहित्य और अभिनय पद्धति दोनों का अपूर्व समन्वय प्रस्तुत करते हैं। एक ओर उनके नाटक साहित्यिक तत्त्व से परिपूर्ण हैं, दूसरी ओर रंगीय क्रिया-व्यापार से। उनमें रक्त-मांस और क्रिया-व्यापार की सारी जीवन्त सुन्दरता सम्भावनाओं के साथ विद्यमान है। अपनी नाटकीयता में वे इतने पूर्ण हैं कि वे मंच पर अमिट सम्भावनाओं को सिद्ध करते हैं। उनका कथ्य प्रेक्षक की संभावना को एकदम पकड़ लेता है। वस्तुतः पकड़ की यह क्षमता घटनाओं में न होकर चरित्र और उसके द्वन्द्व तथा इससे भी अधिक मानव-नियति के अंकन में निहित है। राकेश के नाटकों की रंग-क्षमता उन स्थितियों पर भी निर्भर करती है जिनमें वे पात्र को खड़ा करते हैं। वस्तुतः उनकी नाट्य-स्थितियाँ और पात्र उनकी नाट्य-कला का परिचय देते हैं। वे अजीब नाटकीय विडम्बनाओं और व्यंजनाओं से परिपूर्ण दिखाई देते हैं। अनुभूति का तत्त्व एक और उपादान है जो राकेश के रंगमंच को काव्यपूर्ण संस्कार प्रदान करता है। **आषाढ़ का एक दिन** में मेघ के बदलते रंग रंगमंच पर एक पूरी कविता रच डालते हैं। उसमें राकेश सच्चे अर्थों में 'रंगमंच के कवि' दिखाई देते हैं। यह नाटक अनुभूति और उससे सिक्त शब्दाश्रयी बिम्बों का जैसा प्रयोग करता है वह रंगमंच पर एक विलक्षण अनुभव दे जाता है। पर इतना निश्चित है कि जो काव्यमय वातावरण वे अपने नाटकों में निर्मित करते हैं, वह काव्य-भाषा से नहीं, वरन् स्थितियों से निर्मित है। वे मंच पर जो कविता रचते हैं, वह शब्द की कविता नहीं, पूरी तरह रंगमंच का काव्य है जो शब्द पर निर्भर तो है पर उसके साथ-साथ पात्र की स्थितियों, मनःस्थितियों, मंचीय उपकरणों, प्रतीकों और बिम्बों पर भी आश्रित है। उदाहरण के लिए कालिदास और मल्लिका, मल्लिका और विलोम की नाट्य-स्थिति एक ओर प्रेम की रससिक्ति तो दूसरी ओर संघर्ष अथवा विवशता की पीड़ा में नाटकीय ही नहीं, प्रेक्षक के लिए समसामयिक आस्वाद देनेवाली भी है। उनके बीच जो घटित होता है, वह विडम्बना कथानक की क्षमता के कारण उसे बाँधे रखती है। कथावस्तु की बुनावट में, क्रिया-व्यापार, बिम्ब विधान, पात्रों की सम-विषम सृष्टि तथा तनावपूर्ण कारुणिक कथा-स्थितियाँ ऐसा बहुत कुछ है जो उसे एक गहन अनुभव दे जाता है। इसी प्रकार **लहरों के राजहंस** में कामोत्सव का आयोजन, श्यामांग का प्रलाप, मृग-प्रकरण, नेपथ्य में बौद्ध भिक्षुओं का समवेत स्वर, सुन्दरी का प्रसाधन, नन्द का मुंडित-शिर लौटना—सब अद्भुत भाव सृष्टि करते हैं। अनुभूति की इसी प्रधानता के कारण इन दोनों नाटकों की आत्मा काव्यमयी हो गई है। नाटक के पात्र का शब्दों में काव्यमय होना कोई गुण नहीं पर राकेश के इन नाटकों की गुणवत्ता तो इस बात में है कि उनमें काव्य शब्द के आधार पर नहीं, संवेदना के आधार पर है और इसीलिए काव्य स्वतंत्र न रहकर नाटक में ढल गया है। **आधे अधूरे** समस्त गद्यात्मक क्रिया-कलाप और घिसी-पिटी स्थितियों के बावजूद

नाटककार की अनुभूति और रंग विधाचिनी शक्ति का स्वर इतना जीवन्त है कि प्रेक्षकों ने आलोचकों के सारे आक्षेपों को झूठा साबित कर दिखाया है। इस नाटक में टूटे-फूटे फर्नीचर, गर्द-भरी फाइलों जूठे चाय के बर्तनों और गन्दे पाजामे के द्वारा राकेश ने घर का जीर्ण चित्र ही नहीं उभारा है वरन् एक वातावरण भी निर्मित किया है। यह स्थिति गद्यात्मक है, पर इसके भी अपनी भावपूर्ण व्यंजनाएँ हैं जो असाधारण नाट्य प्रस्तुतीकरण से उभरकर आती हैं।

सर्जक अनुभूति के साथ वातावरण की सृष्टि राकेश के रंग-शिल्प की सबसे बड़ी विशेषता है। वस्तुतः वे अपने नाटकों के लिए ऐसा दृश्य-विधान चुनते हैं जो इस घटना या पात्र को पृष्ठभूमि ही प्रदान नहीं करता वरन् मानवीय क्रिया-कलाप के लिए एक आधार भी प्रस्तुत करता है। मोहन राकेश दृश्य-विधान के द्वारा एक अन्तरंग वातावरण जुटाने में अद्भुत कौशल दिखाते हैं। **आषाढ़ का एक दिन** में मेघ द्वारा प्रस्तुत वातावरण नाटक के केन्द्र में स्थित हो जाता है। वह मल्लिका, कालिदास, मातुल, विलोम, अम्बिका—सबको अपने-अपने ढंग से छूता है। इसके अतिरिक्त इस नाटक में अर्थपूर्ण दृश्य-विधान का आयोजन हुआ है जिसका प्रमुख अंग मल्लिका का वह प्रकोष्ठ है जो उसी के साथ हर अंक में उजड़ता जाता है। पहले अंक में वह पुता है, उसमें बीच में गेरू के स्वस्तिक चिह्न बने हैं। हल्दी से कमल और शंख बनाए गए हैं। दूसरे अंक में वे सब बुझे-बुझे दिखाई देते हैं; पुताई उजड़ जाती है, बर्तन कम होते जाते हैं और वस्त्र फटते जाते हैं। तीसरे अंक में सब-कुछ जर्जर और अस्त-व्यस्त है—कुम्भ फूटा है; स्वस्तिक चिह्न लगभग मिट चुके हैं और बर्तनों पर स्याही चढ़ चुकी है। यह दृश्य-विधान स्वयं ही मल्लिका के विघटन की कहानी कहता है। इसी प्रकार का एक सार्थक प्रयोग **आधे अधूरे** में हुआ है जिसमें दृश्य का आधार एक कमरा है—'सब रूपों में इस्तेमाल होनेवाला एक कमरा जिसमें घर के व्यतीत स्तर के कई अवशेष—सोफासेट, डाइनिंग टेबल, कबर्ड आदि—किसी-न-किसी तरह अपने लिए जगह बनाए हैं। जो कुछ भी है, वह अपनी अपेक्षाओं के अनुसार न होकर कमरे की सीमाओं के अनुसार एक और ही अनुपात से है।...सामान में कहीं एक तिपाई, कहीं दो-एक मोढ़े, कहीं फटी-पुरानी किताबों का एक शेल्फ और कहीं पढ़ने की एक मेज़-कुर्सी भी है। गद्दे, परदे, मेज़पोश और पलंगपोश अगर हैं तो इस तरह घिसे-फटे या सिले हुए कि समझ में नहीं आता कि उनका न होना क्या होने से बेहतर नहीं था।' सारा दृश्य-विधान बिखराव, बेतरतीबी, जहालत और जलालत को मुखर करता है। दृश्य की इस पृष्ठभूमि पर कुछ पात्र विसंगति के साथ उभरते हैं : पुरुष एक, दो, तीन, चार, सावित्री, बिन्नी, किन्नी, अशोक—सब अपने-अपने विसंगत चेहरे और उन पर वैषम्य का भाव लिए हुए। और इन चेहरों के साथ सब रूपों में इस्तेमाल होनेवाला वह कमरा जैसे घुलमिल जाता है। यह कमरा पात्रों की अपनी सीमाओं का प्रतीक बन जाता है। राकेश की महानता इस बात में है कि वे सारे परिवेश की सृष्टि बहुत ही सामान्य उपकरणों को लेकर करते हैं और उनका प्रतीकात्मक प्रयोग कर दिखाते हैं। यहीं पात्र और वातावरण एक हो जाते

हैं। पर वास्तविक स्थिति में राकेश दोनों को एक नहीं होने देते—परिवेश पात्र पर सदा हावी रहता है।

राकेश की रंगमंचीय अवधारणा में अनुभूति और वातावरण का तो महत्त्व है। वातावरण की सृष्टि में पात्रों की अनुभूतियों को वे बड़ी बारीकी से चित्रित करते हैं। कुंवरजी अग्रवाल ने ठीक ही कहा है : 'मोहन राकेश की नाटककार के रूप में सफलता का एक रहस्य यह भी है कि वे पात्रों के तात्कालिक क्रमिक अनुभव परिणतियों को ब्यौरे के साथ उपस्थित करते हैं जिससे दर्शक अन्तर्मुख हो अपने ही अनुभव संसार को टटोलने की ओर उन्मुख होता है।'[4] आधुनिक नाटक यद्यपि रसानुभूति पर बल नहीं देता, किन्तु इन तत्त्वों को राकेश के नाटकों में साधारणीकरण और रसानुभूति में सहायक मानने में कोई आपत्ति नहीं होनी चाहिए। राकेश के नाटक एक पूरा भावात्मक वातावरण उपस्थित करते हैं और प्रेक्षकों में अद्भुत संवेदन-क्षमता जगाते हैं। उनके नाटकों में भावों का प्रवाह और नाटकीय लय-रचना मिलती है।

रंगीयता का भी इस दृष्टि से कम महत्त्व नहीं है। **लहरों के राजहंस** में इसी दृष्टि से प्रतीकों और अतिप्राकृतिक तत्त्व का उपयोग हुआ है। ये दीपाधारों पर स्थित पुरुष और स्त्री-मूर्ति के माध्यम से पूरे नाटक के काव्य को स्थापित करने का प्रयास भी एक महत्त्वपूर्ण युक्ति है। **आधे अधूरे** में फाइलें झाड़ने, कैंची से तस्वीर काटने, कार्टून बनाने, तुतला या हकलाकर बोलने, पैंट में कीड़ा घुसने का नाटक करने जैसी सामान्य घिसी-पिटी रंगचर्याओं में रंगीयतावादी दृष्टि का परिचय मिलता है।

पात्रों को भी राकेश कहीं-कहीं सामान्य रहने नहीं देते। मल्लिका, विलोम, अम्बिका, श्यामांग एक ऐसे वैचित्र्य के साथ उभरते हैं कि उनके रंग यथार्थ पर पुत जाते हैं। इस प्रकार के प्रयोग नाट्यानुभव को कहीं प्रतीकात्मक वातावरण प्रदान करते हैं तो कहीं नाटककार की आत्मपरकता को उभारते हैं। राकेश सामान्य को असामान्य और असामान्य को सामान्य रूप में प्रस्तुत करते हैं। कालिदास महान कवि होकर भी सामान्य हैं और विलोम सामान्य होकर भी असामान्य है। वे उन्हें मानवीय धरातल पर चित्रित करते हैं। अम्बिका भी ऐसी ही है सावित्री भी सामान्य पात्र होते हुए भी असामान्य दीखती है। वस्तुतः राकेश के कई पात्रों के खंडित व्यक्तित्व, विक्षुब्धता और विक्षिप्तता के पीछे कोई युक्ति कार्य करती दिखाई देती है जो प्रेक्षक के मन पर अपना प्रभाव छोड़ती है।

राकेश की नाट्य संरचना में रंगमंच की कई अन्य विधियों का भी सफल प्रयोग मिलता है। उदाहरण के लिए वे अपने नाटकों में घटना-क्रम, नाट्य-स्थिति आदि का इस प्रकार नियोजन करते हैं कि दृश्य-बिम्ब प्रभावशाली रूप से उभर सकें। पात्रों की भावमुद्राएँ उभारने में भी वे कुशल हैं विशेषतः उनके नाटकों के अन्त अर्थपूर्ण भावमुद्राओं की सृष्टि करते हैं। **लहरों के राजहंस** में नन्द की स्तब्धता, फिर आहत भाव से चला जाना और सुन्दरी का सिसकते हुए हथेलियों पर औंधी हो जाना, एक पूरा प्रभाव पैदा करता है **आधे अधूरे** में लाठी टेकते महेन्द्रनाथ जिस तरह घर लौटता है,

वह एक पूरा दृश्य-बिम्ब प्रस्तुत करता है। **आषाढ़ का एक दिन** में मूसलाधार वर्षा में कालिदास का मल्लिका के द्वार से चला जाना और मेघ का गर्जन और वर्षण का दृश्य-बिम्ब नाटक को अर्थवत्ता देता है। राकेश ऐसा दृश्यबन्ध प्रदान करते हैं जिसमें पात्रों का भाव जगत् प्रतिबिम्बित होता है।

राकेश रंगमंच के समर्थ सर्जक के रूप में सामने आते हैं। वे अपने नाटकों के रंग-तत्त्व को नाट्य-स्थितियों, पात्र के द्वन्द्व, उसकी भंगिमाओं और बदलती आकृतियों और संवादों में ढालने में कुशल हैं। उनके नाटकों का शब्द-रूप दृश्य-रूप के अनुरूप है और वे सभी नाटकीय तत्त्वों का इस तरह प्रयोग कर दिखाते हैं कि रंगमंच एक सही बिम्ब उभारने में सहायक हो। उनके रंग-संकेत इस बात के उदाहरण हैं कि वे कितने जागरूक कलाकार थे। राकेश ने रंग-संकेत उसी सर्जक कलाकार की कलात्मक भावना से लिखे हैं, जिस कुशलता से उन्होंने अपने नाटकों के संवाद रचे हैं। ये रंग-संकेत नाटक की भावात्मक अपेक्षाओं को रेखांकित करते हैं और साथ ही परिवेश का संकेत भी देते हैं। वे रंगमंच पर बिम्ब-विधान करते हैं। **आषाढ़ का एक दिन** का प्राकृतिक परिवेश बहुत विस्तृत है। यद्यपि सभी नाटक कमरे में घटित होते हैं, पर कमरे के बाहर के परिवेश को भी वे नहीं भूलते। **लहरों के राजहंस** में भी बाहर उद्यान का संकेत किया गया है। **आधे अधूरे** ही एक ऐसा नाटक है जिसमें कथ्य के अनुरूप अँधेरे बन्द कमरे का प्रभाववादी बिम्ब बिल्कुल प्राकृतिक परिवेश से कटकर उभरता है क्योंकि सारा नाटक घरेलू है। ये नाटक अन्त में पात्र और परिवेश को लेकर जो रंगीय बाह्यता उभारते हैं, वह अपूर्व प्रभाव छोड़ जाता है—लौटता हुआ महेन्द्रनाथ, वर्षा में मल्लिका के घर से निकलता कालिदास और बाँह छुड़ाती पत्नी की उपेक्षा से आहत नन्द रंगमंच पर अपनी छाप छोड़ जाते हैं।

राकेश ने अपने नाटकों के लिए जिस मंच की कल्पना की थी, वह सम्भवतः 'बॉक्स सेट' पर आधारित थी। इस परिकल्पना का सबसे बड़ा लाभ यह है कि उनके नाटकों में दृश्यों की बहुलता नहीं है। वे एक ही दृश्यबन्ध पर खेले जा सकते हैं। एक अंक और दूसरे अंक के बीच उनमें काल का व्यवधान भी अधिक नहीं। **लहरों के राजहंस** में संध्या, रात्रि के अन्तिम प्रहर और अगली रात तक सारी कथावस्तु घटित हो जाती है। **आधे अधूरे** में भी समय का व्यवधान नहीं के बराबर है। **आषाढ़ का एक दिन** में काल की अवधि का विस्तार अवश्य है; किन्तु मेघ के माध्यम से प्रवहमान चेतना और स्मृति सारे दृश्यों को एक सूत्र में पिरो देती है, जिससे काल के व्यवधान का कहीं आभास ही नहीं होता। वस्तुतः अखंड काल का प्रवाह मनोवैज्ञानिक युक्तियों से राकेश के नाटकों में सर्वत्र बना रहता है। वे चेतना-प्रवाह, बिम्ब-विधान की सूत्रता और रंगक्रियाओं का मनोवैज्ञानिक प्रयोग करते हैं। प्रेक्षक की अनुभूतियों की घनीभूत अन्विति भी उसका एक पहलू है। दिक् और काल की एक अन्विति भी राकेश के नाटकों को मंचीय सम्भावनाओं से परिपूर्ण करती है, किन्तु एक दृश्यबन्ध के कारण उनको पर्याप्त क्षति भी पहुँची है। एक ही स्थान की सीमा के कारण जो कुछ उस स्थान पर घटता है, उसे

मंच पर दिखाना सरल है, पर जो दूसरे स्थान पर दूसरी परिस्थिति में घटता है, उसे समस्त नाटकीय क्रिया-कलाप के साथ प्रस्तुत करना सम्भव नहीं होता। इसीलिए बहुत-सी बातों को, जो रंगमंच पर बहुत नाटकीय सिद्ध होतीं, नाटककार को सूच्य रूप में प्रस्तुत करने की विवशता का सामना करना पड़ा। इससे नाटक के अन्तिम अंक पर बहुत बोझ पड़ा है। उदाहरण के लिए कालिदास के उज्जयिनी प्रवास और काश्मीर के शासक होने के सारे अनुभव को एक लम्बे संवाद के रूप में प्रस्तुत किया गया है। वह काश्मीर जाते हुए मल्लिका से क्यों नहीं मिला या उनकी सर्जनात्मक क्षमताओं पर राजसत्ता का क्या प्रभाव पड़ा—इन्हीं प्रश्नों का उत्तर देने के लिए वह एक लम्बी व्याख्या देने को बाध्य होता है। **लहरों के राजहंस** में भी नन्द का सारा द्वन्द्व मंच के बाहर घटता है और मंच पर केवल उसकी रिपोर्टिंग—विवरण की प्रस्तुति मात्र होती है। पूरा घटनाक्रम एक ही स्थल पर घटित होता है। मनःस्थिति को व्यक्त करने के लिए सिर्फ़ उपादान और वातावरण बदलता है। बहुत कुछ मंच से बाहर घटित होता है। **आधे अधूरे** में जगमोहन और सावित्री के बीच जो घटता है, वह प्रत्यक्ष रंगमंच पर न आकर जुनेजा के शब्दों में वर्णित होता है : 'मैं बिना पूछे बता सकता हूँ कि क्या बात होगी। तुमने कहा : तुम बहुत-बहुत दुखी हो आज। उसने कहा : उसे बहुत-बहुत हमदर्दी है तुमसे। तुमने कहा : तुम जैसे भी हो अब इस घर से छुटकारा पा लेना चाहती हो। उसने कहा : कितना अच्छा होता अगर इस नतीजे पर तुम कुछ साल पहले पहुँच सकी होतीं। तुमने कहा...'आदि। यही नहीं, महेन्द्रनाथ के पक्ष में जुनेजा का सारा वक्तव्य नाटकीय न होकर कुछ-कुछ विवरणात्मक हो जाता है। यह दोष राकेश के तीनों नाटकों में है। इससे सम्भवतः वे परिचित भी थे, पर दृश्यबन्ध की सीमाओं के कारण वे इसके लिए बाध्य थे।

वस्तुतः सब-कुछ रंगमंच पर दिखाना सम्भव नहीं है। नेपथ्य भी नाटक और रंगमंच का अनिवार्य अंग है और उसकी योजना एकदम अरंगमंचीय हो, ऐसा नहीं कहा जा सकता। मोहन राकेश ने अपने नाटकों के अन्त में नेपथ्य में घटित कथा-स्थितियों का जैसा कथन किया है, उससे धारा-प्रवाह में भले ही बाधा आई है, परन्तु नाटक बिखराव से बचे हैं। दृश्यबन्ध की एकता के लिए एक ही दृश्यबन्ध पर दो या तीन अंकों में सारी कथावस्तु को समेटने के कारण सूच्य सामग्री का समावेश स्वाभाविक है। प्रश्न इतना ही है कि वह सब प्रेक्षक अपने में समाने में समर्थ है ?

दृश्यबन्ध की इस सीमा और घर के बन्द कमरों में तनाव से जूझते पात्रों को देखते हुए राकेश के नाटक 'बॉक्स सेट' के लिए ज़्यादा उपयुक्त ठहरते हैं। किन्तु रंग-दृष्टि से सीमित रंगमंच से बाहर निकलना भी ज़रूरी होता है। दिक् का व्यापक अनुभव देने के लिए मुक्ताकाशी रंगमंच ज़्यादा उपयुक्त रहता है। **आषाढ़ का एक दिन** के लिए उन्मुक्त वातावरण अपेक्षित है। इसलिए नाटककार की अपनी कल्पना से भी भिन्न रूप में मुक्ताकाशी रंगमंच पर भी बड़ी सफलता से खेला गया। **लहरों के राजहंस** का कथ्य जिन दो विराट् आयामों को छूता है, उसको देखते हुए, मुक्ताकाशी रंगमंच सर्वथा अनुपयुक्त नहीं है। भय इतना ही है कि मुक्ताकाशी रंगमंच पर नाटक के बिखर जाने

का ख़तरा रहता है, किन्तु यदि इस ख़तरे से सावधान रहा जाए तो 'बॉक्स सेट' से भिन्न दृश्यबन्ध भी उपयोगी हो सकते हैं। **आधे अधूरे** जैसा घरेलू नाटक भी ओम शिवपुरी ने मुक्ताकाशी रंगमंच पर बड़ी सफलता से खेलकर दिखा दिया। इस सम्बन्ध में उन्होंने लिखा है : 'मुझे सन्तोष है कि मेरा विश्वास सही साबित हुआ। बन्द और खुले प्रेक्षागृहों के अन्तर से नाटक की प्रभावान्विति पर कोई असर नहीं पड़ा। दर्शक के साथ तादात्म्य उतना ही तीव्र और गहन रहा।'[5] नाटक का दृश्य-विधान कमरे तक भले ही सीमित हो; पर अपने में वह बहुत व्यापक है। बेकेट का एक पात्र कहता है—'कमरा ही दुनिया है; उसके पास दूसरा नरक है।' **आधे अधूरे** का कमरा भी मुक्ताकाशी मंच पर दुनिया की विशालता ग्रहण कर लेता है। राजेन्द्रनाथ का विचार है कि राकेश के नाटकों के लिए प्रोसेनियम स्टेज अधिक उपयुक्त है। उनमें शब्द का तत्त्व प्रधान है; उसे उजागर करने के लिए प्रेक्षक और रंगमंच के बीच निकट का सम्बन्ध अनिवार्य है, क्योंकि दूरी शाब्दिक क्षमताओं को नष्ट कर देती है।

शब्द की क्षमता भी रंगमंच की बड़ी शक्ति होती है। वह भाषा पर भी निर्भर करती है और इसके साथ ही दृश्य पर भी। राकेश मानते थे कि दृश्य नाटक का अनिवार्य तत्त्व है पर वह भी भाषा पर निर्भर है। वे दृश्य को शब्द की ही एक परिणति मानते थे। संवाद के द्वारा नाट्य-स्थिति को दृश्य के अनुरूप गढ़ने में भी वे अपूर्व क्षमता दिखाते हैं। उसे दृश्य, स्थिति और अर्थ से जोड़कर वे विलक्षण प्रभाव उत्पन्न करते हैं। वे संवादों को एक विशिष्ट शिल्प में ढालकर, उन्हें व्यंजना, वक्रता, ध्वन्यात्मकता तथा पाठ से भिन्न अर्थ-गौरव प्रदान करते हैं। इस प्रकार की व्यंजना कहीं-कहीं शब्द, वाक्यों, पदलोप तथा अधूरे वाक्यों से भी साध्य हुई है। नाट्य-स्थिति की भाँति कहीं शब्दों पर आधारित नाटकीय व्यंग्य संवादों को अपूर्व चमत्कार प्रदान करता है। **आधे अधूरे** के संवादों में विशेषतः एक प्रकार की विदग्धता के भी दर्शन होते हैं और यह अनुभूति की तीव्रता और बौद्धिक तेजस्विता दोनों से जुड़ी है। यही राकेश की भाषा और संवादों को रंगीयता प्रदान करती है। वे शब्दों के आन्तरिक सम्बन्ध, आरोह, अवरोह और लय का प्रयोग संवाद की अर्थवत्ता बढ़ाने के लिए करते हैं। शब्द के अर्थ और ध्वनि के दोनों आयामों को स्वीकार करते हुए उन्होंने कहा है कि शब्द-योजना बिना अपनी आन्तरिक लय के प्राणवान नहीं होती। लय या ध्वनि का ग्राफ अर्थवत्ता प्रदान करता है और नाद के आरोह-अवरोह में ही शब्द का आन्तरिक नाटक निहित है।[6]

इसमें कोई सन्देह नहीं कि राकेश ने लय और संसर्ग से अर्थ के आयाम उभारे हैं। **छतरियाँ** के मंचीकरण में परोक्ष शब्दों से विभिन्न लयों के मूर्त बिम्ब उभरते हैं। **आषाढ़ का एक दिन** में अनुस्वार और अनुनासिक के संवादों की आन्तरिक लय स्थिति की विसंगति को उभारती है। अम्बिका के संवाद उसकी पीड़ा के मन्द लय में हैं। मल्लिका के संवादों में प्रारम्भ में क्षिप्रता है। श्यामांग की विक्षिप्ति ठहराव पैदा करती है। महेन्द्र और सावित्री के संवादों की कटुता की अभिव्यक्ति बहुत-कुछ संवादों में निहित लय पर ही निर्भर करती है। राकेश वाक्यों का गठन भावात्मक लय के अनुरूप

करते हैं। सुन्दरी की यह उक्ति लीजिए : 'तुम...! कितने-कितने बिन्दु खोजे हैं आज तक तुमने ! जाओ, एक और बिन्दु खोजो। कितने-कितने शब्दों में ढाँपा है उन बिन्दुओं को ? जाओ, कुछ और शब्द ढूँढो। परन्तु अन्त में कहाँ चले जाते हैं तुम्हारे वे शब्द ? फिर भी क्यों वही-के-वही बने रहते हो तुम ! वही...वही...।' स्पष्टतः इन वाक्यों में ऐसा विन्यास साधा गया है कि बोलने का लहजा, बलाघात, संवाद की अदायगी, उनकी भंगिमा, पुनरावृत्ति और अदायगी में निहित श्वास-प्रक्रिया मंच पर ऐसे संवादों को पूरा प्रभाव प्रदान करती है।

राकेश के नाटकों में सम्पूर्ण शब्द-तत्त्व रंग के साथ जुड़ा है। वे संवाद को क्रिया-व्यापार, दृश्य, मौन, नेपथ्य, बिम्ब आदि के साथ जोड़ने में सफल हुए हैं। संवाद को क्रिया से और क्रिया को नाटकीय अर्थ से जोड़ना राकेश के रंग-भाषणों की बहुत बड़ी उपलब्धि है। कई अवसरों पर उनमें शब्द क्रिया में बदल जाते हैं। भावना अस्तित्व का रूप ले लेती है और शब्द बहुत पीछे छूट जाते हैं। मल्लिका स्वेच्छा से कालिदास को उज्जयिनि भेज देती है, पर साथ ही अपने को सँभाल भी नहीं पाती। तब शब्द जैसे भाव, अनुभाव और संचारियों में बदल जाते हैं—

अम्बिका : मल्लिका !

मल्लिका रुक जाती है। पर कुछ भी उत्तर न देकर मुँह हाथों में छिपा लेती है। अम्बिका उठकर धीरे-धीरे उसके पास आती है और उसे बाँहों में ले लेती है। सारा शरीर रुलाई से काँपता रहता है, पर गले से स्वर नहीं निकलता। अम्बिका की आँखें भर आती हैं और वह उसके काँपते शरीर को अपने से सटाए उसकी पीठ पर हाथ फेरती रहती है। फिर ओंठों और गालों से उसके सिर को दुलारने लगती है।

अम्बिका : अब भी रोती है ? उसके लिए ? उस व्यक्ति के लिए जिसने...।

भाषा और क्रिया के सम्बन्धों के सबसे सुन्दर उदाहरण **आधे अधूरे** में देखने को मिलते हैं। इस नाटक की भाषा सर्वत्र सम्बन्धों और क्रियाओं की भाषा है। यदि उसके दो पृष्ठों (66-68) को लें तो स्त्री कभी गले में माला को लपेटती, झटका देती, गले से उतारती, डब्बे में रखती, कबर्ड को बन्द न होते देख झुँझलाती, कबर्ड के नीचे जूते-चप्पलों को खोजती, ठोकर मारती, फिर आईने को देखकर माला पहनती, गर्दन को मलती, झुर्रियाँ निकालने की कोशिश करती, ड्रेसिंग टेबल से चीज़ें उठाती-रखती, क्रीम की शीशी देखती, चेहरे पर लगाती, फिर बालों को देखती, कंघी से सफ़ेद बाल ढकती आदि न जाने कितनी क्रियाएँ करती दिखाई देती है। इस प्रकार राकेश ने कई स्थलों पर अति संक्षिप्त संवादों को क्रिया-व्यापार या भावमुद्रा का अनुषंगी बनाकर अपूर्व रंग-दृष्टि का परिचय दिया है। **आधे अधूरे** में इस निर्णायक क्षण का अंकन इसी रूप में हुआ है। सावित्री जगमोहन के पास चले जाने का निर्णय लेती है। बड़ी लड़की कहती है : 'और सोच लो थोड़ा...।' सावित्री आवेग में गले की माला को उँगली से लपेटती है और वह झटका लगने से टूट जाती है। यहाँ शब्द और क्रिया दोनों साथ चलते हैं।

फलतः सारी स्थिति प्रतीकार्थ को तो व्यक्त करती ही है, पर साथ ही दो-चार शब्दों के बीच मूकाभिनय भी अर्थ को नया विस्तार दे जाता है—

स्त्री : साल पर साल।...इसका यह हो जाए, उसका वह हो जाए। **मालाओं को डिब्बा रखकर कबर्ड को बन्द करना चाहती है। पर बीच की चीज़ों के अव्यवस्थित हो जाने से कबर्ड ठीक से बन्द नहीं होता।**

एक दिन...दूसरा दिन।

नहीं बन्द होता, तो उसे पूरा खोलकर झटके से बन्द करती है।

एक साल...दूसरा साल।

कबर्ड के नीचे रखे जूते-चप्पलों को पैर से टटोलकर एक चप्पल निकालने की कोशिश करती है। पर दूसरी चप्पल नहीं मिलती, तो सबको ठोकरें लगाकर पीछे हटा देती है।

अब भी और सोचूँ थोड़ा।

ड्रेसिंग टेबल के सामने चली जाती है। कुछ पल असमंजस में रहती है कि वहाँ क्योंई आई है। फिर आईने में देखकर माला पहनने लगती है—खाल को मलकर चेहरे की झुर्रियाँ निकालने की कोशिश करती है।

कब तक ?...क्यों ?

फिर समझ में नहीं आता कि क्या करना है।...क्रीम की शीशी हाथ में आ जाने पर पल भर उसे देखती रहती है। फिर खोल देती है।

घर दफ़्तर...घर दफ़्तर।

क्रीम चेहरे पर लगाते हुए ध्यान आता है कि वह इस वक़्त नहीं लगानी थी। उसे तौलिये से पोंछकर एक और शीशी उठा लेती है...।

सोचो...सोचो !

ध्यान सिर के बालों पर अटक जाता है। ..उँगलियों से टटोलकर देखती है कि कहाँ सफ़ेद बाल ज़्यादा हैं। कंघी ढूँढ़ती है, पर वह मिलती नहीं। ..कंघी वहीं तौलिये के नीचे मिल जाती है।

चख् चख्...किट् किट्...चख् चख्...किट् किट्...!

क्या सोचो ?

कंघी से सफ़ेद बालों को ढकने लगती है। ध्यान आँखों की झाइयों पर चला जाता है तो कंघी रखकर उन्हें सहलाने लगती है।

शब्दों के बीच के मूकाभिनय को लेकर राकेश प्रयोगशील थे। रंगमंच का जहाँ वे शब्दों को माध्यम मानते थे, वहाँ शब्दों के अतिरिक्त मोह, उनकी बाढ़ को रंगानुभव के लिए बाधक भी मानते थे। इसीलिए उन्होंने अपने नाटकों में शब्द और मूक अभिनय का सुन्दर समन्वय किया है। उन्हीं के शब्दों में, 'नाटकीय रंगमंच के अन्तर्गत मूक अभिनय भी लम्बी निःशब्दता की तरह शब्दों के बीच की एक कड़ी है। शब्दों से उद्‌भूत बिम्ब में से एक बिम्ब यह भी हो सकता है, होता है। हमारा भाषा संस्कार इस बात

का प्रमाण है कि शब्दों की यात्रा में बहुत बार बहुत-कुछ अनकहे शब्दों द्वारा कहा जाता है। ये अनकहे शब्द बिम्ब के साथ-साथ यात्रा करते हुए बिना ध्वनियों के भी अपना अर्थ ध्वनित कर देते हैं।'[7] यह भी ध्यान योग्य है कि राकेश के सभी नाटकों का अन्त मुख्य पात्र के मूकाभिनय से होता है जहाँ मुद्रा और बिम्ब संवाद से भी अधिक व्यंजक हो जाते हैं। उदाहरण के लिए, नन्द मूक और स्तब्ध दिखाई देता है और फिर आहत भाव से चला जाता है। सुन्दरी भी कुछ नहीं कहती, केवल सिसकते हुए हथेलियों पर औंधी हो जाती है। मल्लिका झरोखे पर खड़ी कालिदास को जाते देखती है। उसके पैर 'बाहर को बढ़ने लगते हैं परन्तु बच्ची को बाँहों में देखकर जैसे वहीं जकड़ जाती है।' और **आधे अधूरे** में लड़के की बाँहें थामे महेन्द्रनाथ की धुँधली आकृति घर के अन्दर प्रवेश करती दिखाई देती है। ये सारे दृश्य बिम्ब निश्चित रूप से राकेश की रंग-चित्र प्रस्तुत करने की क्षमता को व्यक्त करते हैं।

शब्दों के बीच मौन का भी राकेश ने रंगमंचीय उपयोग किया है। मौन भी उनके नाटकों में एक नाटकीय क्षण होता है, जहाँ शब्द चुक जाते हैं, निरर्थक हो जाते हैं या जानबूझकर अनुपस्थित रहते हैं। राकेश ऐसा लम्बा शब्दहीन प्रयोग भी करते हैं जहाँ शब्द क्रिया में ढल जाता है और ऐसी निःशब्दता का भी जो क्षणिक होती है और प्रायः शब्दों के बीच होती है। ऐसी निःशब्दता के बारे में उनका कहना है, कि वह नाटकीय तनाव को वहन करने में सहायक होती है और वह प्रायः संवादों के बीच के अन्तराल पर निर्भर करती है। इसीलिए उन्होंने उसे शब्दों की यात्रा का एक पड़ाव माना है अथवा दोनों ओर के शब्दों को जोड़नेवाला अन्तराल।[8]

शब्दों के बीच की निःशब्दता के कलात्मक प्रयोग के राकेश बड़े हामी थे। **आषाढ़ का एक दिन** में मल्लिका के द्वारा धारासार वर्षा में अपने उल्लास का बखान और उसकी प्रतिक्रिया में सब-कुछ अनसुना कर अपने काम में लगा रहना सार्थक मौन का सुन्दर उदाहरण है। **लहरों के राजहंस** में नन्द और सुन्दरी के बीच औघड़ ख़ामोशी के कई अवसर आते हैं। उसमें कभी उत्तर न देना, कुछ न कहना सिर्फ़ देखते रहना जैसी कई रंग युक्तियाँ प्रयुक्त हुई हैं। **आधे अधूरे** में तो मौन और मुखरता का या जान-बूझकर अनसुना करने के कई क्षण आते हैं; किन्तु कालिदास के लौट आने पर मल्लिका के आगे खड़े होने का क्षण मौन की सबसे बड़ी मर्मान्तक अभिव्यक्ति लगता है।

आषाढ़ का एक दिन और **लहरों के राजहंस** में एकाध स्थल पर शब्द के सन्दर्भ में नेपथ्य बड़ा अर्थवान हो जाता है। नेपथ्य प्रेक्षक की कल्पना को ही नहीं जगाता, रंगमंच का विस्तार भी करता है। **लहरों के राजहंस** में श्यामांग के प्रलाप और कराह को नेपथ्य से दिखाया गया है। कुछ आलोचकों ने इसकी कड़ी आलोचना की है। किन्तु सच्चाई यह है कि श्यामांग जैसा चरित्र अपने नेपथ्य-संवादों में जीवन्त लगता है। मंच से दूरी उसको एक अमूर्त सत्ता जैसा अस्तित्व देती है और वह एक अहसास, एक अनचाहे भाव के रूप में अपना प्रभाव पैदा करता है। राकेश ने **छतरियाँ** में (अंग्रेज़ी में **मैड डिलाइट**) इस तरह का जान-बूझकर एक प्रयोग किया है। इस सम्बन्ध में उन्होंने

कहा है : 'मैंने दृश्य से शब्द की सरूपता को अलग करने का प्रयास किया है। शब्द ध्वनि को अलग करके नेपथ्य में रख दिया गया है क्योंकि मैं इस बात को जानना चाहता था कि क्या नेपथ्य में प्रयुक्त शब्द/ध्वनि एक मुख्य तत्त्व की भूमिका निभा सकते हैं।'[9] उनका यह प्रयोग **छतरियाँ** में बेहद सफल रहा। श्यामाग के नेपथ्य संवाद भी रंगमंच पर अपूर्व प्रभाव पैदा करते हैं। इसी प्रकार **आषाढ़ का एक दिन** में (नाटक के अन्त में) जहाँ नेपथ्य में मेघ का गर्जन सुनाई देता है, वहाँ अन्दर से बच्ची के कुनमुनाने और रोने का शब्द भी कहीं नेपथ्य जैसी सार्थकता प्राप्त कर लेता है और इसी रूप में वह प्रभाव भी छोड़ता है। इसमें कोई सन्देह नहीं कि राकेश के नाटकों में रंगमंच के पीछे जो घटता है, वह रंगमंच पर घटित क्रिया-व्यापार को सघन बनाता है।

'राकेश की मंच-चेतना परदों से प्रकाश की ओर बढ़ी है'—इस स्थापना में पर्याप्त सत्य है।[10] इसमें कोई सन्देह नहीं कि उनके पहले दो नाटक परदा उठने के निर्देश से प्रारम्भ होते हैं। **आषाढ़ का एक दिन** में अंक-विभाजन के लिए परदे को स्वीकार किया गया है। **लहरों के राजहंस** में परदे का उल्लेख आरम्भ में तो है, पर अंक विभाजन के लिए प्रकाश-योजना का सहारा लिया गया है। **आधे अधूरे** प्रकाश का पूरा-पूरा उपयोग करता है। ये तीनों नाटक राकेश के विकसित होते रंगानुभव को प्रकट करते हैं। अपने पहले नाटक के लिखते वक़्त संभवतः प्रकाश के रंगीय उपयोग के बारे में वे बहुत स्पष्ट नहीं थे। वे चाहते तो प्रकाश ही नहीं, अन्धकार (शेड) के उपयोग के सन्दर्भ में कई निर्देश दे सकते थे। बादलों के घिरने से हुआ अन्धकार मल्लिका, अम्बिका के जीवन के अन्धकार को व्यक्त करने में प्रायः सहायक ही होता ! वर्षा के अँधेरे दिन का उल्लेख प्रायः मल्लिका करती सुनाई देती है। कालिदास के उज्जयिनी जाने के निर्णायक क्षण का चित्रण करते हुए राकेश ने प्रकाश के क्षीण होने की बात कही है जो विलोम के इस व्यंजनापूर्ण संवाद कि 'घिरे हुए मेघों ने आज असमय अन्धकार कर दिया है' के सन्दर्भ में महत्त्वपूर्ण युक्ति बन जाती है। अपने दूसरे नाटक में राकेश श्यामांग के नेपथ्य-संवादों के साथ प्रकाश-योजना का कलात्मक प्रयोग करते हैं। नाटक के महत्त्वपूर्ण प्रसंग रात में घटते हैं, इसलिए प्रभात और दीपाधारों का प्रकाश अर्थवान हो उठता है। यही नहीं, प्रकाश की विभिन्न स्थितियाँ भावमुद्रा को व्यक्त करने में कितनी सहायक हो सकती हैं, इसका एक सुन्दर प्रयोग दूसरे अंक के अन्त में मिलता है जहाँ नन्द बुद्ध के पीछे चला जाता है और सुन्दरी कहीं गहरे आहत होकर अन्तर्मुख हो जाती है। इसे व्यक्त करने के लिए नाटककार ने प्रकाश-योजना का कलात्मक सहारा लिया है : 'बाहर का प्रकाश बढ़ने और अन्दर का प्रकाश धीमा पड़ने से उनकी आकृति एक छायाकृति में बदलने लगती है।'

प्रकाश-योजना के साथ-साथ संगीत की संभावनाओं को लेकर भी राकेश अपने नाटकों में प्रयोग कर रहे थे। **पैर तले की ज़मीन** में संगीत 'बिखराव और ख़ालीपन को रेखांकित' करता है। **लहरों के राजहंस** में प्रकाश-योजना के साथ संगीत भी जुड़ जाता है। **आधे अधूरे** में जाकर प्रकाश का निर्देश अधिक जागरूक दृष्टि से किया गया है; और प्रकाश का अन्धकार और संगीत का सम्मिलित बिम्ब नाटक के अन्त को

प्रभावशाली बना देता है : 'प्रकाश, खंडित होकर स्त्री और बड़ी लड़की तक सीमित रह जाता है। स्त्री स्थिर आँखों से बाहर की तरफ आहिस्ता से कुरसी पर बैठ जाती है। हलका-सा मातमी संगीत उभरता है जिसके साथ उन दोनों पर भी प्रकाश मद्धिम पड़ने लगता है। तभी लगभग अँधेरे में लड़के की बाँह थामे पुरुष एक की धुँधली आकृति अन्दर आती दिखाई देती है।...उन दोनों के आगे बढ़ने के साथ संगीत अधिक स्पष्ट और अँधेरा अधिक गहरा होता जाता है।'

तात्पर्य यह है कि राकेश गहरे रंगानुभव से जुड़े थे जिसका प्रमाण उनके रंग-शिल्प में विद्यमान है। कहा जाता है कि 'अभिनेता की हर गति और रंगचर्या को नाटककार ने बाँध देना चाहा है। इसको परिचालक और अभिनेता के विषय-क्षेत्र में दख़ल भले ही कुछ लोग मान लें, किन्तु इतना निश्चित है कि नाट्यानुभूति की भाँति रंगानुभूति भी नाटककार की अपनी चीज़ है। इसलिए नाटक की सृष्टि करते हुए एक विशेष रंग-शिल्प में ढालने और उसका स्वरूप निर्धारित करने का उसे पूरा अधिकार है। राकेश 'नाटककार के रंगमंच के प्रबल समर्थक' थे। और इस दृष्टि से उनका रंग-विधान एक जागरूक नाटककार की पहचान करवाता है।

कुछ लोगों ने राकेश की रंग-अवधारणा में कुछ दोष ढूँढ़ निकाले हैं। वस्तुतः उनके नाटक सामान्य अभिनेता या परिचालक के लिए नहीं हैं। उन्हें मंच पर प्रस्तुत करने के लिए रंगानुभव और सर्जन क्षमता चाहिए। **लहरों के राजहंस** का तीसरा अंक निश्चयतः साधारण रंगकर्मियों के लिए समस्या पैदा करेगा। तीसरा अंक जमता नहीं है; और यह दोष नाटकीय अवधारणा के साथ जुड़ा है; इसलिए उससे सर्वथा मुक्त होना तो सम्भव नहीं; पर उसे सूझ-बूझ के साथ सँभाला तो जा सकता है। किसी ने यह भी कहा है कि नन्द का मुंडित शिर उपस्थित होना मंच पर हास्य पैदा करता है। ऐसा केवल परिचालक और अभिनेता की लापरवाही से हो सकता है। इसी प्रकार **आषाढ़ का एक दिन** के मंचन में विलोम, कालिदास, मातुल, अनुस्वार, अनुनासिक आदि की प्रस्तुति में थोड़ी-बहुत सतर्कता चाहिए। विलोम को सामान्य होने पर भी एक असाधारण कवि के रूप में उभारना भी ज़रूरी है। विवेकदत्त झा के अनुसार मातुल का अतिनाटकीय होना प्रभाव पैदा कर सकता है।[11] अनुस्वार और अनुनासिक को भी प्रभाव और विषयान्तर तथा कुछ-कुछ हास्य उपजाने की दृष्टि से एक विशिष्ट गति और लय के साथ प्रस्तुत किया जाना उचित रहेगा।

आधे अधूरे में एक पुरुष द्वारा चार पुरुषों की भूमिका निभाने का एक प्रयोग प्रेक्षक और अभिनेता को निकट लाता है। ब्रेख़्त के निरपेक्षता सिद्धान्त (ऐलियेनेशन) के आधार पर यह प्रयोग प्रेक्षक की तन्मयता को भंग कर उसे इस बात का संकेत देता है कि वह जीवन की प्रतिरूपता को नहीं, नाटक को देख रहा है। यह स्थिति उसे वस्तुस्थिति को पहचानने और उसका विश्लेषण करने का भी अवसर प्रदान करती है। कुछ लोगों ने इस प्रयोग को रंगमंच की दृष्टि से अनुपयोगी और अव्यावहारिक बताया है; पर ओम शिवपुरी ने एक ही साथ चार भूमिकाएँ अदा करके वैविध्य को

सफलतापूर्वक निभाया है। दूसरी ओर चार अलग-अलग व्यक्तियों का प्रयोग कर राजेन्द्रनाथ ने भी यह अनुभव किया है कि एक ही पुरुष द्वारा चार भूमिकाएँ अदा करना अधिक कलात्मक अनुभव देता है। पर यह अनुभव जागरूक रंग-कल्पना की माँग करता है। कभी सारी बात प्रेक्षकों के अनुभव तक नहीं पहुँच पाती तो वह प्रभाव से वंचित हो जाती है ! फिर यह यह प्रस्तावना चाहे वैचारिक और नाटकीय रूप-विधान की दृष्टि से महत्त्वपूर्ण कही जा सकती है, पर फिर भी वह सम्पूर्ण नाटक से अलग-थलग पड़ जाती है। प्रस्तावना का आरोप नाटक पर आगे नहीं होता और इसीलिए उसकी सारी अर्थवत्ता विस्मृत होती जाती है। वस्तुतः यह सारा प्रयोग आधा-अधूरा रह जाता है।[12]

इस रंगयुक्ति की उपेक्षा करते हुए कुछ निर्देशकों ने एक ही व्यक्ति से विभिन्न भूमिकाएँ कराने के बजाय पाँच अलग अभिनेताओं को लेकर प्रयोग किए। बादल सरकार के **पगला घोड़ा** में भी एक स्त्री चार स्त्रियों की भूमिका में आती है। शम्भु मित्र ने चार स्त्री पात्रों के लिए अलग-अलग अभिनेत्रियों का प्रयोग किया। **आधे अधूरे** की रंगयुक्ति प्रेक्षक को चमत्कृत अवश्य करती है, पर उसके लिए बड़े कौशल की अपेक्षा होती है। केवल एक ही व्यक्ति को चार-पाँच तरह के कपड़े पहनाकर या व्यवहार की विचित्रता प्रदान कर कोई प्रभाव पैदा करना सम्भव नही लगता। विभिन्न पात्रों की भूमिका में समानता का अहसास कराना बहुत आवश्यक है। फिर भी नाटक के दोनों प्रकार के प्रदर्शन हुए और दोनों प्रकार की रायें अभी तक बनी हुई हैं। एक अभिनेता द्वारा कई भूमिकाएँ निभाने में भी कुछ कमियाँ नहीं और जिन निर्देशकों ने अलग-अलग अभिनेताओं से विभिन्न भूमिकाएँ करवाईं वे भी कोई ख़ास प्रभाव नहीं पैदा कर पाए। अमाल़ अल्लाना ने अपनी प्रस्तुति में प्रस्तावना को टुकड़ों में बाँटकर अंकों के बीच प्रस्तुत किया। श्यामानन्द जालान ने जो प्रदर्शन किया उसमें भी प्रस्तावना थोपी हुई लगी। बहुधा रंगमंच नाटक के साथ न्याय नहीं कर पाता नाटक में भी दोष होते हैं। किन्तु इतना निश्चित है कि राकेश के नाटकों ने रंगमंच के क्षेत्र में एक नया अभियान खड़ा कर दिखाया। वे नाटक और रंगमंच के क्षेत्र में एक अग्रदूत बनकर आए। हिन्दी के उपेक्षित नाटक को रंगमंच दिलाने में आज भी चार दशक तक वे महत्त्वपूर्ण भूमिका निभा रहे हैं।

पाद-टिप्पणियाँ

एक यात्रा : पद-चिह्नों के आर-पार

1. 'मिस जूली' की भूमिका।
2. मोहन राकेश : 'बक़लम ख़ुद', पृ. 97
3. हैजेल ई. बार्नीस : 'ऐन एक्जिस्टेंशियलेस्ट इथिक्स'; के. गुरुदत्त : 'एक्जिस्टेंशियलिज़्म ऐंड इंडियन थॉट', तथा मार्गरिट एल. विले : 'क्रियेटिव स्केप्टिक्सि'।
4. 'बक़लम ख़ुद', पृ. 34
5. वही, पृ. 35

आत्महंता यायावर

1. कथूरिया द्वारा सम्पादित 'नाटककार मोहन राकेश' में राजेन्द्रपाल का लेख 'अन्तर्विरोधी व्यक्तित्व', पृ. 22
2. अनीता राकेश : 'चंद सतरें और', पृ. 86
3. वही, पृ. 88
4. वही, पृ. 98
5. वही, पृ. 86
6. वही, पृ. 87
7. 'आइने के सामने', पृ. 194 अथवा 'परिवेश', पृ. 14
8. वही।
9. वही।
10. 'परिवेश', पृ. 15
11. चंद सतरें और, पृ. 87
12. वही, पृ. 88
13. 'आइने के सामने', पृ. 194
14. वही, पृ. 99
15. सुन्दरलाल कथूरिया द्वारा सम्पादित 'नाटककार मोहन राकेश', पृ. 31
16. राकेश की डायरी 'व्यक्तिगत', सारिका, अगस्त 1968
17. 'चंद सतरें और' पृ. 94
18. 'नाटककार मोहन राकेश', पृ. 31
19. 'चंद सतरें और', पृ. 58

20. वही, पृ. 78
21. 'सारिका', मार्च 1973
22. 'इनैक्ट', अंक 73-74, राजेन्द्रपाल द्वारा राकेश का इंटरव्यू।
23. चंद सतरें और, भूमिका।
24. डायरी।
25. 'बक़लम ख़ुद', पृ. 112
26. इन्द्रनाथ मदान : 'आधुनिकता और हिन्दी साहित्य'
27. 'चंद सतरें और', पृ. 99
28. इब्राहिम अल्काजी, पु. ल. देशपांडे तथा सुरेश अवस्थी द्वारा सम्पादित 'आज के रंग नाटक' में उद्धृत, पृ. 13
29. वही।

खोलो द्वार

1. नीत्शे : 'जॉय-फुल विजडम', पृ. 100
2. 'द मिथ ऑफ सिसिफस', पृ. 77
3. बर्दयेव : 'द बिगनिंग ऐंड द ऐंड,' पृ. 173
4. 'द मिथ ऑफ सिसिफस', पृ. 79
5. देखिए 'लहरों के राजहंस' की भूमिका, पृ. 15
6. ऐजाज गुल की एक कविता से।
7. 'सारिका', मोहन राकेश अंक, मार्च 1973, पृ. 80 अथवा मोहन राकेश : 'साहित्यिक और सांस्कृतिक दृष्टि', पृ. 22
8. भूमिका, 'लहरों के राजहंस', पृ. 8
9. 'अनसाइंटिफिक पोस्ट-स्क्रिप्ट' पृ. 119-20
10. मोहन राकेश : 'साहित्यिक और सांस्कृतिक दृष्टि', पृ. 24
11. 'बक़लम ख़ुद', पृ. 68
12. डॉ. जगदीश शर्मा : 'मोहन राकेश की रंग-सृष्टि', पृ. 26
13. मनोहर अभय का लेख : 'समसामयिक युग बोध', संकलित सुन्दरलाल कथूरिया द्वारा सम्पादित 'नाटककार मोहन राकेश', पृ. 136
14. 'आषाढ़ का एक दिन', विशिष्ट संस्करण, भूमिका, पृ. 13
15. भूमिका, 'लहरों के राजहंस', पृ. 8
16. वही, पृ. 8
17. वही, पृ. 9
18. 'लिरिकल ऐंड द क्रिटिकल', पृ. 145

दो दीपाधारों की जलती लौ

1. 'गौतमनन्द', पृ. 20
2. भूमिका, 'लहरों के राजहंस', पृ. 23
3. 'लहरों के राजहंस', पृ. 24
4. वही, पृ. 10
5. मोहन राकेश : 'साहित्यिक और सांस्कृतिक दृष्टि', पृ. 65-66

6. वही, पृ. 165
7. वही, पृ. 166-67
8. भूमिका, 'लहरों के राजहंस', छात्र संस्करण, पृ. 8-9
9. वही, पृ. 23
10. 'मोहन राकेश की रंग-सृष्टि', पृ. 35
11. 'मिथ ऑफ सिसिफस', पृ. 98
12. मिगनेल डी उनामुनो : 'द ट्रैजिक सेंस ऑफ लाइफ', पृ. 226
13. वही, पृ. 47-48
14. दाउ कांस्ट नॉव प्रूव दैट दाउ आर्ट स्पिरिट अलोन
    नौर कांस्ट दाउ प्रूव दैट दाउ आर्ट बोथ इन वन।
    × × ×
    फॉर नथिंग वर्दी प्रूविंग कैन वी प्रूवेन,
    नौर येट डिस्प्रूवेन : व्हेरफोर दाउ बी वाइज।
    क्लीव ऐवर टु द सनियर साइड ऑफ डाउट,
    क्लिंग टु फेथ बियोंड द फौर्म्स ऑफ फेथ।
15. भूमिका, 'लहरों के राजहंस', पृ. 17
16. मोहन राकेश : 'साहित्यिक और सांस्कृतिक दृष्टि', पृ. 195
17. भूमिका, 'लहरों के राजहंस', पृ. 27

घर की तलाश

1. राकेश का 'किसी के नाम पत्र', मोहन राकेश स्मृति अंक, 'सारिका', पृ. 69
2. भूमिका, 'मिस जूली' : द हाफ वोमन इज ए टाइप कनिंग मोर ऐंड मोर इंटु प्रोमिनेंस सेलिंग हरसेल्फ नाउ-ए-डेज फॉर पॉवर, डेकोरेशंस, डिस्टिंक्शंस, डिप्लोमाज, ऐज फौर्मली फॉर मनी ऐंड द टाइप इंडिकेट्स डीजेनेरेशन। इट इज नॉट ए गुड टाइप बट अनफॉर्चुनेटली इट हैज द पॉवर ऑफ रिप्रोड्यूसिंग इट सेल्फ ऐंड इट्स मिजरी टू वन मोर जेनेरेशन। ऐंड डीजेनेरेट मेन सीम इंसटिंक्टिवली टु मेक देअर सेलेक्शन फ्रॉम दिस काइंड ऑफ वीमेन—।
3. 'अन्तराल', पृ. 127
4. ऐंथॉनी स्टोर : ह्यूमन एग्रेशन', पृ. 66
5. वही, पृ. 66
6. वही, पृ. 67
7. कार्ल यास्पर्स : 'मैन इन द मॉडर्न एज', पृ. 50
   व्हाट हैपंस क्विकली ऐंड इज सून फारगॉटन। पीपुल देअरफोर टेंड टु बिहेव ऐज इफ दे वर ऑल ऑफ द सेम एज। चिल्ड्रन बिकम लाइक ग्रो-अप्स ऐज सून ऐज दे पॉसिब्ली कैन...। व्हेन द ओल्ड प्रीटेंड टु बी यंग, ऑफ कोर्स द यंग हैव नो रिवरेंस फॉर देअर ऐल्डर्स।...यूथ ऐज द पीरियड ऑफ हाइयेस्ट वाइटल ऐफिसियंसी ऐंड ऑफ इरोटिक एक्जौलटेशन बिकम्स द डिजायर्ड टाइप ऑफ लाइफ इन जेनरल ऐंड इफ यूथ इज ओवर ही विल स्टिल स्ट्राइव टु शो इट्स सेंबलेंस।
8. राजेन्द्र यादव : 'एक दुनिया समानान्तर', पृ. 40
9. काल यास्पर्स : 'मैन इन द मॉडर्न एज', पृ. 71
10. वही, पृ. 74
11. कमलेश्वर : 'मेरा हमदम मेरा दोस्त', लेख।

12. कार्ल यास्पर्स : 'एक्जिस्टेंशियलिज़्म ऐंड ह्यूमैनिज़्म', पृ. 63
13. मोहन राकेश : 'साहित्यिक और सांस्कृतिक दृष्टि', पृ. 173
14. सार्त्र : 'बींग ऐंड नथिंगनेस', पृ. 239-40
15. देखिए सार्त्र : 'व्हाट इज लिटरेचर', पृ. 22

रचना दृष्टि : बिन्दु से बिन्दु तक

1. 'परिवेश', पृ. 157 तथा पृ. 198
2. 'द मिथ ऑफ सिसिफस', पृ. 198
3. 'आधुनिकता और हिन्दी साहित्य', पृ. 29
4. 'बक़लम ख़ुद', पृ. 141
5. वही, पृ. 144
6. वही।
7. मोहन राकेश : 'साहित्यिक और सांस्कृतिक दृष्टि', पृ. 115
8. वही, पृ. 158-59
9. 'परिवेश', पृ. 198
10. 'आज के रंग नाटक', पृ. 24

रूपबन्ध : रोयें-रेशे

1. 'बक़लम ख़ुद' पृ. 71
2. 'परिवेश', पृ. 203
3. मोहन राकेश : 'साहित्यिक और सांस्कृतिक दृष्टि', पृ. 64-65
4. वही, पृ. 163
5. डेविड थॉमसन : 'एम्स ऑफ हिस्ट्री', पृ. 10
6. 'इनैक्ट', अंक 73-74, जी.आर तनेजा : 'ऐन आवर विद मोहन राकेश'।
7. पुष्पा बंसल : 'मोहन राकेश का नाट्य साहित्य', पृ. 3
8. राकेश : डायरी, सारिका, अगस्त 68
   इस सम्बन्ध में कुंवरजी अग्रवाल का मन्तव्य है : 'उस रचना की दृष्टि से **आधे अधूरे** का भाव-प्रवाह दोषपूर्ण लगता है। लगातार दो-तीन घंटों के लिए रंगशाला में दर्शकों की समवेत उपस्थिति कहानी की तुलना में नाटक में एक भिन्न प्रकार की और अधिक सावधान लय के निर्माण की अपेक्षा रखती है। इस नाटक में संघर्ष सतह के ऊपर उभरा और काफ़ी तीखा है। उससे निरन्तर तनावों की सृष्टि होती है...पर तनाव की क़िस्म और मात्रा एक जैसी है।' 'रंगमंच : एक माध्यम', पृ. 78
9. उद्धृत, मॉडर्न अमेरिकन थियेटर (सं. आल्विन बी. कर्नेल), पृ. 165
10. मोहन राकेश : 'साहित्यिक और सांस्कृतिक दृष्टि', पृ. 166-67
11. 'द मिथ ऑफ सिसिफस', पृ. 30
12. मोहन राकेश : 'साहित्यिक और सांस्कृतिक दृष्टि', पृ. 173

भाषा : शब्द की बुनावट

1. 'बक़लम ख़ुद', पृ. 94
2. वही, पृ. 95

3. 'नई कहानी की भूमिका', पृ. 204
4. 'इनैक्ट' 73-74 में मोहन महर्षि के द्वारा लिया गया राकेश का इंटरव्यू, चेंजिंग रोल ऑफ वर्ड'।
5. सुरेश अवस्थी : 'लास्ट डेज विद राकेश', लेख 'इनैक्ट', अंक 73-74
6. 'रंगमंच और शब्द', लेख। 'मोहन राकेश : साहित्यिक और सांस्कृतिक दृष्टि', पृ. 92
7. वही।
8. मोहन राकेश : 'परिवेश', पृ. 193
9. गिरीश रस्तोगी : 'मोहन राकेश और उनके नाटक', पृ. 152
10. भूमिका, 'आषाढ़ का एक दिन', विशिष्ट संस्करण।
11. 'बक़लम ख़ुद' पृ. 67
12. 'इनैक्ट' 73-74 में प्रकाशित नाटकीय शब्द के सम्बन्ध में अध्यायों की रूपरेखा तथा 'रंगमंच और शब्द' लेख जो अब, 'मोहन राकेश : साहित्यिक और सांस्कृतिक दृष्टि' में संकलित है।
13. वही।
14. गिरीश रस्तोगी : 'मोहन राकेश और उनके नाटक', पृ. 157
15. राकेश का 'शब्द और ध्वनि', लेख देखिए : मोहन राकेश : 'साहित्यिक और सांस्कृतिक दृष्टि', पृ. 91-93

रंगानुभव

1. 'लहरों के राजहंस', भूमिका।
2. मोहन राकेश : 'साहित्यिक और सांस्कृतिक दृष्टि', पृ. 86
3. वही, पृ. 91
4. कुंवरजी अग्रवाल : 'रंगमंच : एक माध्यम', पृ. 79
5. 'आज के रंग नाटक', पृ. 348
6. राकेश का लेख 'शब्द और ध्वनि', पृ. 11
7. मोहन राकेश : 'साहित्यिक और सांस्कृतिक दृष्टि', पृ. 92
8. वही, पृ. 90
9. मोहन महर्षि द्वारा लिया गया इंटरव्यू, 'चेंजिंग रोल ऑफ वर्ड, 'इनैक्ट', अंक 73-74
10. डॉ. जगदीश शर्मा : 'मोहन राकेश की रंग-सृष्टि' पृ. 52
11. वही, पृ. 27
12. यह कोई राकेश का अपना मौलिक प्रयोग नहीं है। सदामोब 'पैरोडी' में यह प्रयोग कर चुका था। हैंडकेने 'कैस्पर' भी यह युक्ति प्रयुक्त कर चुका था। बादल सरकार के 'पगला घोड़ा' में एक स्त्री चार भूमिकाएँ निभाती हैं।

●●●